Contaminated Soils, Sediments and Water: Issues and Challenges

Contaminated Soils, Sediments and Water: Issues and Challenges

Edited by Brett Green

SYRAWOOD
PUBLISHING HOUSE
New York

Published by Syrawood Publishing House,
750 Third Avenue, 9th Floor,
New York, NY 10017, USA
www.syrawoodpublishinghouse.com

Contaminated Soils, Sediments and Water: Issues and Challenges
Edited by Brett Green

International Standard Book Number: 978-1-64740-428-4 (Hardback)

Cataloging-in-publication Data

Contaminated soils, sediments and water : issues and challenges / edited by Brett Green.
p. cm.
Includes bibliographical references and index.
ISBN 978-1-64740-428-4
1. Soil remediation. 2. Water--Purification. 3. Soil conservation. 4. Water--Pollution. I. Green, Brett.
TD878 .C66 2023
628.55--dc23

TABLE OF CONTENTS

Preface......VII

Chapter 1 **Use of Larval Morphological Deformities in *Chironomus plumosus* (Chironomidae: Diptera) as an Indicator of Freshwater Environmental Contamination**......1
Enzo Goretti, Matteo Pallottini, Sarah Pagliarini, Marianna Catasti, Gianandrea La Porta, Roberta Selvaggi, Elda Gaino, Alessandro Maria Di Giulio and Arshad Ali

Chapter 2 **Using Mixed Active Capping to Remediate Multiple Potential Toxic Metal Contaminated Sediment for Reducing Environmental Risk**......18
Meng-Yuan Ou, Yu Ting, Boon-Lek Ch'ng, Chi Chen, Yung-Hua Cheng, Tien-Chin Chang and Hsing-Cheng Hsi

Chapter 3 **AQUALIFE Software: A New Tool for a Standardized Ecological Assessment of Groundwater Dependent Ecosystems**......32
Giovanni Strona, Simone Fattorini, Barbara Fiasca, Tiziana Di Lorenzo, Mattia Di Cicco, Walter Lorenzetti, Francesco Boccacci and Diana M. P. Galassi

Chapter 4 **Reduction of Nutrient Leaching Potential in Coarse-Textured Soil by using Biochar**......49
Yu-Lin Kuo, Chia-Hisng Lee and Shih-Hao Jien

Chapter 5 **The Application of Artificial Mussels in Conjunction with Transplanted Bivalves to Assess Elemental Exposure in a Platinum Mining Area**......64
Marelize Labuschagne, Victor Wepener, Milen Nachev, Sonja Zimmermann, Bernd Sures and Nico J. Smit

Chapter 6 **Biodegradation of Amoxicillin, Tetracyclines and Sulfonamides in Wastewater Sludge**......84
Chu-Wen Yang, Chien Liu and Bea-Ven-Chang

Chapter 7 **Physiological Profiling and Functional Diversity of Groundwater Microbial Communities in a Municipal Solid Waste Landfill Area**......102
Marco Melita, Stefano Amalfitano, Elisabetta Preziosi, Stefano Ghergo, Eleonora Frollini, Daniele Parrone and Annamaria Zoppini

Chapter 8 **Aqueous Mercury Removal with Carbonaceous and Iron Sulfide Sorbents and their Applicability as Thin-Layer Caps in Mercury-Contaminated Estuary Sediment**......117
Boon-Lek Ch'ng, Che-Jung Hsu, Yu Ting, Ying-LinWang, Chi Chen, Tien-Chin Chang and Hsing-Cheng Hsi

Chapter 9 **Preliminary Analysis of the Diet of *Triturus carnifex* and Pollution in Mountain Karst Ponds in Central Apennines**.....134
Mattia Iannella, Giulia Console, Paola D'Alessandro, Francesco Cerasoli, Cristina Mantoni, Fabrizio Ruggieri, Francesca Di Donato and Maurizio Biondi

Chapter 10 **Amendment of Husk Biochar on Accumulation and Chemical Form of Cadmium in Lettuce and Pak-Choi Grown in Contaminated Soil**.....149
Kuei-San Chen, Chun-Yu Pai and Hung-Yu Lai

Chapter 11 **Marble Slurry's Impact on Groundwater**.....161
Leonardo Piccini, Tiziana Di Lorenzo, Pilario Costagliola and Diana Maria Paola Galassi

Chapter 12 **Evaluation of the Effect of Gold Mining on the Water Quality**.....177
Alison Martín, Juliana Arias, Jennifer López, Lorena Santos, Camilo Venegas, Marcela Duarte, Andrés Ortíz-Ardila, Nubia de Parra, Claudia Campos and Crispín Celis Zambrano

Chapter 13 **The Toxicity and Uptake of As, Cr and Zn in a Stygobitic Syncarid (Syncarida: Bathynellidae)**.....194
Grant C. Hose, Katelyn Symington, Maria J. Lategan and Rainer Siegele

Chapter 14 **Metallic Pollution and the use of Antioxidant Enzymes as Biomarkers in *Bellamya unicolor* (Olivier, 1804) (Gastropoda: Bellamyinae)**.....207
Mohammed Othman Aljahdali and Abdullahi Bala Alhassan

Chapter 15 **Potential of a Trait-Based Approach in the Characterization of an N-Contaminated Alluvial Aquifer**.....221
Tiziana Di Lorenzo, Alessandro Murolo, Barbara Fiasca, Agostina Tabilio Di Camillo, Mattia Di Cicco and Diana Maria Paola Galassi

Permissions

List of Contributors

Index

PREFACE

Soil contamination, also called soil pollution and land pollution, is a part of land degradation caused by the spillage, burying or migration of hazardous contaminants from local waste, untreated industrial effluents, agrochemicals, manufacturing wastes and other human-made activities or substances into the soil ecosystem. Industrial activity, agricultural chemicals and improper waste disposal are the leading causes of soil contamination. Sediments refer to silt, loose sand, clay and other soil particles that are present at the bottom of a water body. Sediments are produced by decomposition of animals and plants, as well as soil erosion. When sediments are mixed with water, it leads to water contamination. The risk of flooding increases when sediments clog catch basins and storm drains, which are used to divert water away from residential areas and roads. Soil and water contamination, both have a negative impact on the health of plants, animals and humans. This book aims to shed light on some of the unexplored aspects of soil and water contamination. Students, researchers and experts associated with this field will benefit alike from it.

This book is a comprehensive compilation of works of different researchers from varied parts of the world. It includes valuable experiences of the researchers with the sole objective of providing the readers (learners) with a proper knowledge of the concerned field. This book will be beneficial in evoking inspiration and enhancing the knowledge of the interested readers.

In the end, I would like to extend my heartiest thanks to the authors who worked with great determination on their chapters. I also appreciate the publisher's support in the course of the book. I would also like to deeply acknowledge my family who stood by me as a source of inspiration during the project.

Editor

1

Use of Larval Morphological Deformities in *Chironomus plumosus* (Chironomidae: Diptera) as an Indicator of Freshwater Environmental Contamination (Lake Trasimeno, Italy)

Enzo Goretti [1,*], Matteo Pallottini [1], Sarah Pagliarini [1], Marianna Catasti [1], Gianandrea La Porta [1], Roberta Selvaggi [1], Elda Gaino [1], Alessandro Maria Di Giulio [2] and Arshad Ali [3]

1 Dipartimento di Chimica, Biologia e Biotecnologie, Università degli Studi di Perugia, 06123 Perugia, Italy; matteo.pallottini@unipg.it (M.P.); sarah.pagliarini87@gmail.com (S.P.); marianna.catasti@gmail.com (M.C.); gianandrea.laporta@unipg.it (G.L.P.); roberta.selvaggi@unipg.it (R.S.); elda.gaino@unipg.it (E.G.)

2 Servizio Disinfestazione, USLUmbria1, 06127 Perugia, Italy; alessandro.digiulio@uslumbria1.it

3 Entomology and Nematology Department, University of Florida, Gainesville, FL 32611, USA; umar@ufl.edu

* Correspondence: enzo.goretti@unipg.it

Abstract: The mentum deformity incidence in *Chironomus plumosus* larvae to assess the environmental contamination level in Lake Trasimeno, Central Italy, was investigated. The survey lasted from May 2018 to August 2019. Fifty-one samplings were carried out: 34 in the littoral zone and 17 in the central zone. The deformity assessment was based on 737 and 2767 larval specimens of *C. plumosus* collected from the littoral and central zones, respectively. Comparison of the larval morphometric variables between normal and deformed specimens highlighted that the deformities did not cause alterations of the larval growth. The deformity incidence amounted to 7.22% in the whole Trasimeno's ecosystem, reaching 8.28% in the littoral zone and 6.94% in the central zone. Among the different seasonal cohorts, the spring cohort had overall the highest deformity value (11.41%). The deformity type assessment protocol highlighted that the most common deformity type was "round/filed teeth" (64%). The results of this 2018–2019 survey revealed a low deformity incidence, within the background range of relatively low-impacted freshwaters. Comparison with previous investigations (2000–2010) of the same habitat showed a clear decrease of the deformity incidence. This study further contributes to the evaluation of the mentum deformity in chironomids that represent an indicator endpoint of the anthropogenic contamination level in freshwaters.

Keywords: Chironomidae; *Chironomus plumosus* larvae; mentum deformities; freshwater contamination; Lake Trasimeno

---

1. Introduction

Anthropic activities continuously add urban, industrial, and agricultural wastes into the environment. Common and widespread groups of pollutants, such as heavy metals, pesticides, polycyclic aromatic hydrocarbons, and polychlorinated biphenyls are dispersed into the environment forming a complex mixture whose toxic effects are difficult to quantify [1–3].

Normally, the set of contaminants contributing to the total toxic burden is not well known and, even when the exact composition is known, it is difficult to interpret the estimate of potential effects of simple additivity or synergism or antagonism [4]. In fact, the measurement of concentrations of toxic elements or compounds in the environment does not imply knowledge of the resulting toxicity of their mixture on the biota [5]. Furthermore, the peculiar trophic network of living organisms [6] determines bioaccumulation and biomagnification processes of contaminants [7–9].

For an evaluation of the anthropogenic contamination in a given territory, the biological monitoring of freshwaters seems to be the most appropriate strategy. The rains draining the atmosphere and the soil carry pollutants into the water receptor of the underlying hydrographic basin [10,11].

Among aquatic organisms, benthic macroinvertebrates inhabiting the surficial layer of the bottom sediments in inland waters are valid biological indicators because they are subject to the resulting action of pollutants and thus respond to many ecological stressors [12–16]. Aquatic Diptera typically represent the predominant component of the benthic biocoenosis, and among them chironomids (Chironomidae: Diptera) in particular are characterized by diverse species compositions and have a key role in the trophic network of freshwater ecosystems especially those impaired by substantial organic loads [17]. Therefore, chironomid communities represent an important food source for a wide range of aquatic and terrestrial animals [18], also considering in some cases, the role of pestiferous chironomid species [19].

Some chironomid species are frequently used as endpoints (as test organisms) in ecotoxicological bioassays for the assessment of ecological risks; for example, larval growth rate, adult emergence, and survival [20]. In particular, widely used in bioassays are the larvae of aquatic midge species belonging to the genus *Chironomus*, such as *Chironomus plumosus* (Linnaeus, 1758), *Chironomus riparius* (Meigen, 1804), *Chironomus tentans* (Fabricius, 1805), *Chironomus sancticaroli* (Strixino and Strixino, 1981), and *Chironomus tepperi* (Skuse, 1889) [21–25].

In freshwater ecosystems, these bioindicator chironomid species are common, abundant, and tolerant to polluted sediments. The exposure to potentially contaminated sediments is due to their benthic habits and feeding on fine particulate organic matter (POM: 0.5 μm–1 mm) collected from sediments [18]. Thus, they exhibit suitability in monitoring the pollutant effects in inland waters.

In several laboratory and field studies, an association between environmental contamination level (metals, such as As, Cd, Cr, Cu, Hg, Mn, Ni, Pb, Zn, and organic compounds, such as polychlorinated biphenyls (PCBs) and pesticides) and the onset of sublethal effects as deformities of mouthparts (i.e., mentum, mandibles, and pectin epipharyngis) in chironomid larvae has been observed. In this regard, studies on chironomid mouthpart deformities have been developed since the 1970s [26–28] and a vast amount of literature based on laboratory bioassays [22,23,29–41] and field surveys [21,42–47] endorses that the incidence of these deformities is well associated with the degree of sediment toxicity, whereas no relationship has been detected between deformed phenotypes and organic enrichment; hence, the water quality assessment indices are not always consistent with the use of chironomid deformities [48]. However, some studies have expressed concerns about the association between deformity incidence and toxicants [24,49].

The mouthpart deformities probably are a consequence of the interaction of contaminants, playing as endocrine-disrupting chemicals, with hormones structurally related to estrogen, such as ecdysone, that leads and regulates insect growth and metamorphosis [50]. Therefore, the "altered" ecdysone might interfere during the molting processes in larval development, causing phenotypic anomalies, as the mouthpart deformities in chironomids [51].

Thus, morphological deformities in chironomids are an aspecific response to different contaminants, such as heavy metals and pesticides, and an ecotoxicological endpoint to evaluate the total effects of the exposure to a toxic mixture of peculiar toxicants present in the freshwater environments [4,22]. For this reason, chironomid deformities are a reliable biomarker for assessing environmental toxicity.

The purpose of the present research was: (i) to use the incidence of mentum deformities in *C. plumosus* larvae as an effective tool for biological assessment of the environmental contamination entity in Lake Trasimeno, Central Italy; (ii) to compare the results with previous investigation in Lake Trasimeno (nearly ten years ago [43]) and evaluate the enhancement of environmental contamination over time.

2. Materials and Methods

2.1. *Study Area*

Lake Trasimeno, the fourth largest lake in Italy, is located within the Tiber River basin in Umbria, Central Italy. It is the most extended lake of peninsular Italy (124.3 km^2 surface). It is a shallow lake of tectonic origin, at hydrometric zero (257.33 meters above sea level (m a.s.l.)) and has an average depth of 4.7 m with a maximum depth of 6.3 m [52,53]. It is characterized by a Mediterranean climate, and its hydrological regime depends on local precipitations, which shows considerable fluctuations in water level over the years [54,55]. These peculiar features and the diffuse anthropogenic sources of pollution (agriculture, livestock farms, industry, and urban settlements) in the lake surrounding land areas make this biotope susceptible to contamination [56]. For its naturalistic and ecological importance, Lake Trasimeno is considered as a Special Area of Conservation (IT5210018; Habitat Directive, 1992/43/EEC) and Special Protection Area (IT5210070; Bird Directive, 2009/147/EC), and it is also a regional park.

2.2. *Field Sampling Campaign*

The field sampling campaign was carried out in the littoral and central zones of the lake. In the littoral zone, six sampling sites were selected along a 350 m area, representative of the main bottom types (sandy, silty, and macrophyte-covered) present in Lake Trasimeno, situated along the coastline (1.5 m average depth) close to the town of Castiglione del Lago. This area among the littoral zone is most subject to the anthropic impact on the lake. In the central zone of the lake, three sampling sites were selected in the same bathymetric zone (4–5 m depth) (Figure 1). The sampling design in the present study was based on a previous investigation [43] carried out from 2000 to 2010 at Lake Trasimeno.

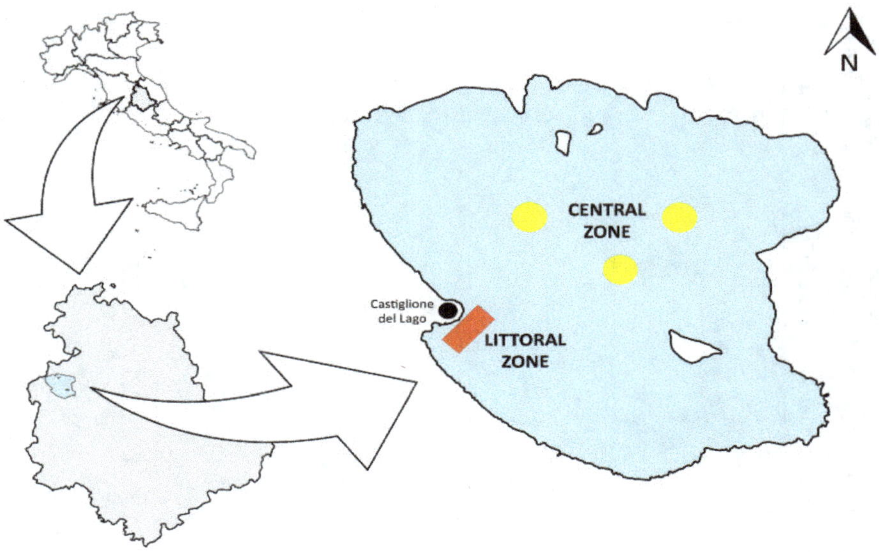

Figure 1. Lake Trasimeno, Umbria, Italy, and location of the field sampling sites.

The survey lasted for 16 months: from May 2018 to August 2019. A total of 51 samplings were carried out: 34 in the littoral zone and 17 in the central zone (Table 1). For each of the two zones, at least one monthly survey was conducted (with the exception of June 2018 for the central zone), with a higher sampling frequency during the spring–summer period.

During each sampling occasion, in the littoral zone 6, quantitative sub-samples of the bottom sediments were collected at each sampling site (a total of 36 sub-samples) with a hand dredge (112 cm^2 sampling surface); in the central zone the sampling technique consisted of a quantitative sampling of the bottom sediments with a compressed air dredge (400 cm^2 sampling surface, 21 mesh per cm nylon net), repeated 5 times per site, with a duration of 1 min each (a total of 15 sub-samples).

Table 1. *Chironomus plumosus* individuals (N), *C. plumosus* relative percentages within the chironomid community (%), Chironomidae densities (individuals m^{-2}), Chironomidae relative percentages within the total macroinvertebrates (%), Naididae densities (individuals m^{-2}), Naididae relative percentages within the total macroinvertebrates (%), other taxa densities (individuals m^{-2}), other taxa relative percentages within the total macroinvertebrates (%), macroinvertebrate densities (individuals m^{-2}); minimum (Min), median, mean, standard deviation (SD), and maximum (Max) of these parameters; (**a**) littoral zone, (**b**) central zone.

(a) Littoral Zone									
Date	*C. plumosus*	*C. plumosus*	Chironomidae	Chironomidae	Naididae	Naididae	Other Taxa	Other Taxa	Macroinvertebrates
	(N)	(% of Chironomidae)	(ind. m^{-2})	(%)	(ind. m^{-2})	(%)	(ind. m^{-2})	(%)	(ind. m^{-2})
11/05/2018	2	2.94	183	9.24	1793	90.40	7	0.36	1983
18/05/2018	6	4.44	469	28.08	1188	71.17	12	0.74	1669
25/05/2018	4	2.26	646	35.67	1154	63.68	12	0.65	1812
01/06/2018	16	11.85	518	35.13	935	63.36	22	1.51	1476
14/06/2018	7	3.87	599	42.62	578	41.12	228	16.26	1405
22/06/2018	2	1.60	426	39.29	536	49.41	122	11.29	1084
28/06/2018	6	8.33	239	23.76	706	70.23	60	6.01	1006
11/07/2018	16	23.53	242	23.29	670	64.56	126	12.15	1037
19/07/2018	12	29.27	122	9.51	1099	86.02	57	4.47	1277
25/07/2018	8	19.05	152	14.32	814	76.98	92	8.70	1058
01/08/2018	35	51.47	242	18.11	993	74.41	100	7.48	1334
09/08/2018	28	20.90	452	26.55	1182	69.48	68	3.97	1702
22/08/2018	21	24.42	376	20.10	1451	77.51	45	2.39	1872
03/09/2018	72	46.15	490	27.51	1250	70.20	41	2.29	1781
18/09/2018	102	76.69	439	35.76	743	60.60	45	3.64	1226
27/09/2018	12	37.50	179	13.77	1103	85.02	16	1.21	1297
10/10/2018	15	12.93	567	25.65	1641	74.23	3	0.12	2211
17/10/2018	33	13.25	1079	34.09	2066	65.27	20	0.64	3166
24/10/2018	75	35.55	843	47.95	903	51.34	12	0.71	1758
06/11/2018	8	6.11	457	48.29	490	51.71	0	0	947
04/12/2018	9	6.47	472	58.18	339	41.82	0	0	811
16/01/2019	3	1.75	533	53.00	462	45.95	11	1.04	1006
12/02/2019	16	10.53	505	51.70	472	48.30	0	0.00	977
15/03/2019	0	0	258	34.90	469	63.42	12	1.68	739
04/04/2019	6	4.17	463	32.93	944	67.07	0	0	1407
02/05/2019	13	3.75	1008	41.73	1375	56.91	33	1.36	2415
21/05/2019	7	7.95	304	25.70	829	70.19	48	4.10	1181
04/06/2019	16	7.69	722	32.59	1409	63.61	84	3.81	2215
18/06/2019	33	17.19	623	38.20	952	58.45	55	3.35	1629
27/06/2019	11	14.47	252	42.67	328	55.56	11	1.78	591
09/07/2019	68	83.95	262	48.66	262	48.66	14	2.67	539
23/07/2019	24	55.81	130	13.97	760	81.64	41	4.38	931

Table 1. *Cont.*

(a) Littoral Zone									
Date	*C. plumosus* (N)	*C. plumosus* (% of Chironomidae)	Chironomidae (ind. m^{-2})	Chironomidae (%)	Naididae (ind. m^{-2})	Naididae (%)	Other Taxa (ind. m^{-2})	Other Taxa (%)	Macroinvertebrates (ind. m^{-2})
06/08/2019	72	40.22	656	24.74	1885	71.13	110	4.14	2651
22/08/2019	38	17.33	880	15.13	4655	80.01	283	4.86	5818
Min	0	0.00	122	9.24	262	41.12	0	0.00	539
Median	14	13.09	460	32.76	939	64.91	37	2.34	1369
Mean	23	20.69	464	31.55	1072	64.98	53	3.46	1589
SD	25	21.41	246	13.17	782	12.90	64	3.83	955
Max	102	83.95	1079	58.18	4655	90.40	283	16.26	5818

(b) Central Zone									
Date	*C. plumosus* (N)	*C. plumosus* (% of Chironomidae)	Chironomidae (ind. m^{-2})	Chironomidae (%)	Naididae (ind. m^{-2})	Naididae (%)	Other Taxa (ind. m^{-2})	Other Taxa (%)	Macroinvertebrates (ind. m^{-2})
29/05/2018	107	81.68	252	16.76	1250	83.24	0	0	1502
18/07/2018	396	98.02	685	75.69	220	24.31	0	0	905
08/08/2018	274	90.43	577	50.96	548	48.45	7	0.59	1132
22/08/2018	274	89.54	542	55.75	430	44.25	0	0	972
21/09/2018	117	100	202	16.40	1028	83.60	0	0	1230
16/10/2018	68	100	120	21.24	445	78.76	0	0	565
28/11/2018	66	100	113	12.88	767	87.12	0	0	880
20/12/2018	107	98.17	208	22.12	732	77.70	2	0.18	942
31/01/2019	82	100	143	22.34	498	77.66	0	0	642
21/02/2019	61	96.83	112	27.57	293	72.43	0	0	405
08/03/2019	86	100	145	27.44	383	72.56	0	0	528
12/04/2019	38	97.44	65	25.49	190	74.51	0	0	255
17/05/2019	69	100	123	28.57	308	71.43	0	0	432
13/06/2019	78	100	145	21.22	538	78.78	0	0	683
11/07/2019	74	98.67	137	37.61	225	61.93	2	0.46	363
31/07/2019	598	99.50	1115	37.54	1853	62.40	2	0	2970
08/08/2019	567	100	1008	43.94	1287	56.06	0	0	2295
Min	38	81.68	65	12.88	190	24.31	0	0.00	255
Median	86	99.50	145	27.44	498	72.56	0	0.00	880
Mean	180	97.07	335	31.97	647	67.95	1	0.08	982
SD	180	5.11	330	16.57	461	16.62	2	0.18	715
Max	598	100.00	1115	75.69	1853	87.12	7	0.59	2970

Benthic macroinvertebrates were separated from the sediment, sorted, and then preserved in 70% ethanol for later examination in the laboratory. The main taxa composing the benthic community of each sample were firstly identified to the family level using suitable taxonomic keys [57]. Chironomid specimens were subsequently identified to genus or species (*Chironomus plumosus*) level through the observation of peculiar features of their head capsules [58–61].

2.3. Mouthpart Deformities of *Chironomus plumosus* Larvae

Body length and head capsule dimensions (width and length) of chironomid larvae were measured to identify the larval stage (instar), and head capsules were then mounted on slides and fixed with Faure liquid (modified composition [22,43]).

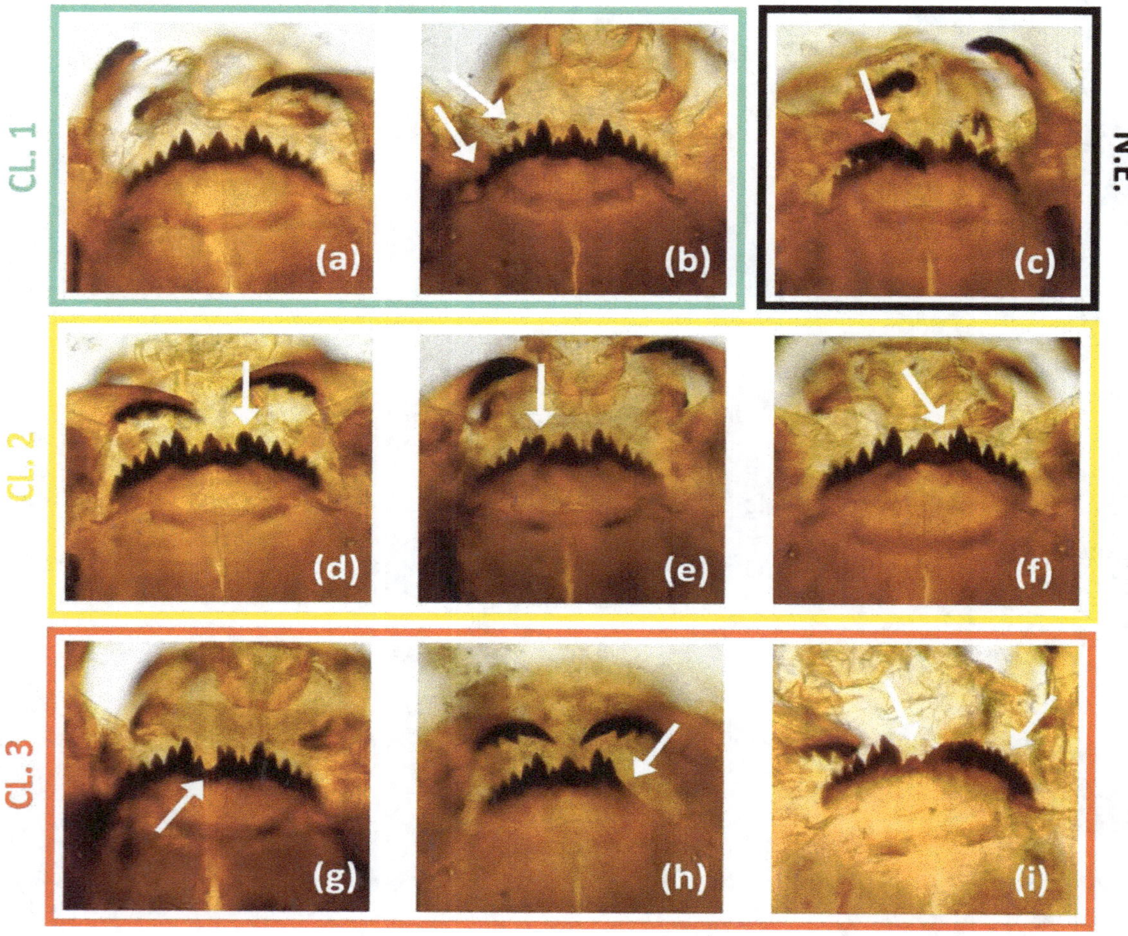

Figure 2. *Chironomus plumosus* images showing Class 1 (CL. 1): (**a**) the normal-shaped mentum, (**b**) the normal-shaped mentum with mechanical damage (arrows), (**c**) the mentum not evaluable (N.E., not completely visible, arrow), and the various mentum deformity types of Class 2 (CL. 2) and Class 3 (CL. 3). (**d**) CL. 2: mentum with a rounded tooth (arrow), (**e**) mentum with a filed tooth (arrow), and (**f**) mentum with an additional tooth (arrow). (**g**) CL. 3: mentum with a Köhn gap (arrow), (**h**) mentum with missing teeth (arrow), and (**i**) mentum with a combination of different deformities (arrows).

Only *Chironomus plumosus* head capsules were considered for mouthpart deformity assessment and examined under a microscope. Specimens that did not have completely visible mentum or those with a cephalic capsule highly damaged during fixation on the microscope slides were excluded from further analyses. *Chironomus plumosus* specimens were classified according to the protocol for morphological response of the mentum, consisting of three classes [22,42–44,62]: Class 1 (CL. 1) specimens without morphological deformity or with mechanical damages (broken or worn menta or teeth) [4,5,63–65]; Class 2 (CL. 2) specimens with weak deformity (one or two round/filed teeth; one missing or additional tooth; one bifid tooth; one serrate tooth; two joined teeth; weak asymmetry); Class 3 (CL. 3) specimens with severe deformity (very round/filed teeth; two or more missing or additional teeth; two or more bifid teeth; serrate teeth; three or more joined teeth; severe asymmetry; Köhn gap; combination of different deformities). Examples of mentum deformity classes of *C. plumosus* larvae are shown in Figure 2.

2.4. Statistical Analysis

Ljung–Box test was used to analyze temporal autocorrelation. Generalized linear models (GLM) were used to analyze differences in head width and length among the different deformity classes, setting larval instar as a covariate and Gaussian family as the link function. The test of equal proportions was carried out to examine differences in the deformity incidences among the seasons [66]. Each deformity class was tested independently and the pairwise comparisons for proportions with the Bonferroni's correction were used as a post-hoc procedure. The results were regarded as significant at a two-tailed p-value of 0.05. All calculations were performed using the glm function and the prop.test of the R software [67].

3. Results

3.1. Macroinvertebrates and Chironomids

During this survey at Lake Trasimeno, a total of 31,240 benthic macroinvertebrate specimens were collected: 21,220 in the littoral zone and 10,020 in the central zone. The benthic communities were mainly composed of Oligochaeta (Naididae) and Diptera: (Chironomidae). The littoral zone community was composed of the following means (±SD): Naididae: 64.98% (±12.90%), Chironomidae: 31.55% (±13.17%), and other taxa: 3.46% (±3.83%). A very similar benthic community composition was found in the central zone: Naididae: 67.95% (±16.62%), Chironomidae: 31.97% (±16.57%), and other taxa: 0.08% (±0.18%). The other macroinvertebrate taxa collected were essentially alien invasive species (i.e., *Dreissena polymorpha* (Pallas, 1771) and *Dikerogammarus villosus* (Sowinsky, 1894)) (Table 1).

During the whole sampling period, the mean (±SD) density of Chironomidae was 464 (±246) and 335 (±330) individuals m^{-2}, in the littoral and central zones, respectively; the mean (±SD) density of Naididae was 1072 (±782) and 647 (±461) individuals m^{-2}, in the littoral and central zones, respectively; and the mean (±SD) density of other taxa was 53 (±64) and 1 (±2) individuals m^{-2}, in the littoral and central zones, respectively (Table 1).

A total of 3857 *C. plumosus* larvae was collected, 796 from the littoral zone and 3061 from the central zone. In the littoral zone, *C. plumosus* represented a mean (±SD) of 20.69% (±21.41%) of the total chironomid community (ranging from 0% to 83.95%), while it was always the dominant taxon of the central zone chironomid community (mean 97.07% ± 5.11%), ranging from 81.68% to 100% (Table 1).

The analysis of biometric variables identified the respective instar of each specimen. In the littoral zone most (60.93%) specimens belonged to instar IV, 29.27% to instar III, and 9.80% to instar II. Almost all specimens (99.25%) from the central zone belonged to instar IV, while only 0.72% and 0.03% belonged to instar III and II, respectively.

3.2. Mentum Deformity in Chironomus plumosus

For mentum deformity evaluation of *C. plumosus* larvae, 59 specimens collected from the littoral zone and 294 collected from the central zone of Lake Trasimeno were excluded due to damages to the cephalic capsule or due to incomplete visibility of mentum. Thus, deformity assessment was based on 737 specimens from the littoral zone and on 2767 specimens from the central zone. Ljung–Box test revealed no temporal autocorrelation within deformity classes among samplings. The deformity incidence amounted to 7.22% (1.51% in CL. 3) in the whole Lake Trasimeno's ecosystem, reaching 8.28% (3.12% in CL. 3) in the littoral zone and 6.94% (1.08% in CL. 3) in the central zone of the lake (Table 2).

Table 2. *Chironomus plumosus* incidence of the deformity classes: CL. 1, no deformity; CL. 2, weak deformity; CL. 3, strong deformity and CL. (2 + 3); individuals (N) and percentages (%); (**a**) littoral zone; (**b**) central zone; (**c**) Lake Trasimeno.

(a) Littoral Zone			**(b) Central Zone**			**(c) Lake Trasimeno**		
CLASS	N	%	CLASS	N	%	CLASS	N	%
CL. 1	676	91.72	CL. 1	2575	93.1	CL. 1	3251	92.78
CL. 2	38	5.16	CL. 2	162	5.85	CL. 2	200	5.71
CL. 3	23	3.12	CL. 3	30	1.08	CL. 3	53	1.51
CL. (2 + 3)	61	8.28	CL. (2 + 3)	192	6.94	CL. (2 + 3)	253	7.22

The polyvoltine biological cycle of *C. plumosus* at Lake Trasimeno exhibited one winter cohort, one spring cohort, and different summer cohorts during the year. Thus, the mentum deformity classes were analyzed in the different seasonal cohorts as shown in a box-plot of the relative statistical variables in Figure 3. In particular, in the littoral zone (34 samplings), the deformed specimens (CL. 2 + 3) had mean values of 1.0 ± 0, 5.0 ± 3.6, and 10 ± 3.0 in the winter, spring, and summer cohorts, respectively; while in the central zone (17 samplings) the deformed specimens (CL. 2 + 3) had mean values of 4.5 ± 1.5, 9.3 ± 7.5, and 45.7 ± 33.1 in the winter, spring, and summer cohorts, respectively.

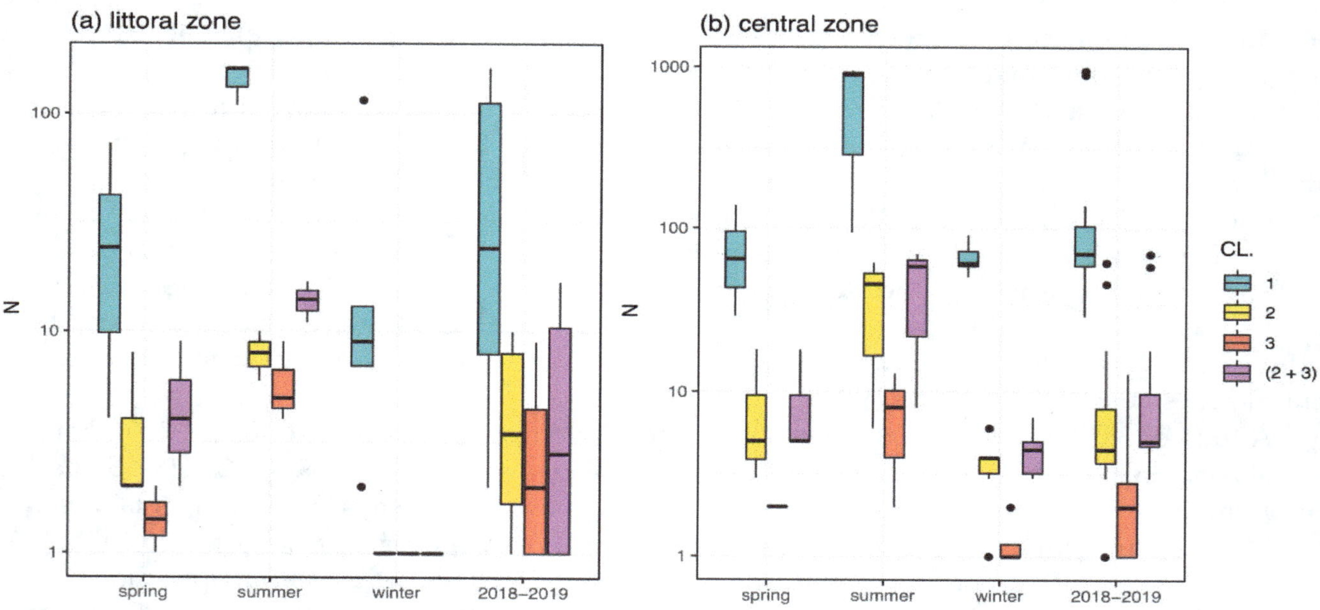

Figure 3. Boxplot of *C. plumosus* specimens (N, log scale) in seasonal cohorts and overall 2018–2019 survey, grouped by deformity classes: CL. 1 (green, no deformity), CL. 2 (yellow, weak deformity), CL. 3 (red, strong deformity), and CL. (2 + 3) (purple). (**a**) Littoral zone; (**b**) central zone. Box represents the interquartile range, thick lines represent the median, whiskers represent the minimum and maximum values, and dots represent the outliers.

In the littoral zone the deformity incidence of the winter cohort was 2.67% (1.33% in CL. 3), the spring cohort was 12.93% (2.59% in CL. 3), and the summer cohort was 8.92% (3.82% in CL. 3); while in the central zone the deformities in the winter cohort amounted to 6.32% (1.17% in CL. 3), 10.73% in the spring cohort (0.77% in CL. 3), and 6.59% in the summer cohort (1.11% in CL. 3) (Figure 4).

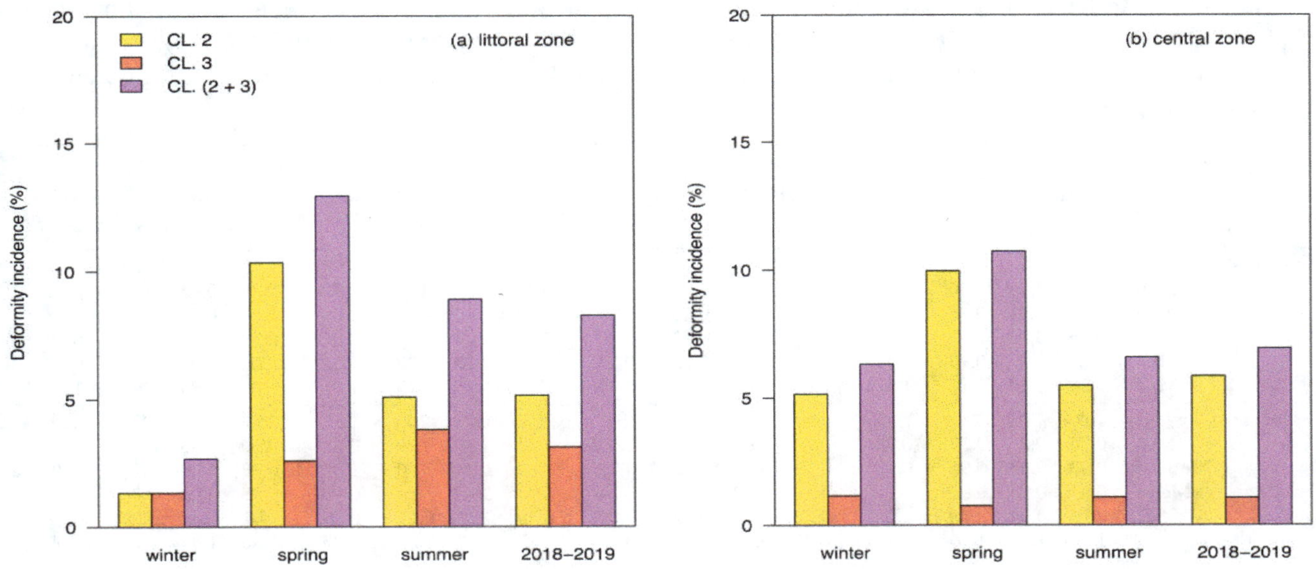

Figure 4. Deformity incidence (%): CL. 2 (yellow, weak deformity); CL. 3 (red, strong deformity); and CL. (2 + 3) (purple) in *C. plumosus* seasonal cohorts and overall 2018–2019 survey; (**a**) littoral zone; (**b**) central zone.

In the littoral zone, the test of equal proportions showed significant differences in the rate of deformity incidence among the seasons for CL. 2 ($p < 0.01$). In contrast, no differences were observed for CL. 3 ($p > 0.6$). Combining CL. 2 and CL. 3, significant differences were detected with an increase in the proportion of specimens with deformities in spring ($p < 0.01$). In the central zone, significant differences were observed for CL. 2, with a deformity incidence in spring significantly higher than that recorded in summer ($p < 0.01$), while no differences were detected for CL. 3 ($p > 0.9$). Combining CL. 2 and CL. 3, no statistical differences emerged.

The deformity assessment in relation to the different larval stages (instars) highlighted that in the littoral zone 7.33% of the deformed specimens (2.85% in CL. 3) belonged to instar IV and only 0.95% (0.27% in CL. 3) to instar III, while in the central zone the instar IV deformed specimens were 6.90% (1.08% in CL. 3) and only 0.04% (none in CL. 3) belonged to instar III. No deformed specimens in both zones belonged to instar II (Table 3).

Figure 5 shows the distribution and relative head size (head length and width) of normal and deformed larvae among instars, in the littoral and central zones of Lake Trasimeno. In the littoral zone, the head length and width means of instar IV specimens (about 61% of total *C. plumosus* larvae) belonging to CL. 1 (normal), CL. 2 (weak deformity), and CL. 3 (strong deformity) were 0.93 and 0.73 mm, 0.93 and 0.73 mm, and 0.95 and 0.75 mm, respectively. Similar trends of head length and width of instar IV specimens (about 99% of total *C. plumosus* larvae) were recorded in the central zone, the mean sizes were 0.98 and 0.78 mm (CL. 1), 0.98 and 0.78 mm (CL. 2), and 1.00 and 0.80 mm (CL. 3) (Table 3). No significant differences were observed in the biometric variables of the head among the three deformity classes ($p > 0.05$).

Table 3. Head length and width (mm) of *C. plumosus* larvae (all instars, instar IV, instar III, and instar II); CL. 1, no deformity; CL. 2, weak deformity; CL. 3, strong deformity; and CL. (2 + 3); number (N), minimum (Min), median, mean, standard deviation (SD), and maximum (Max); (**a**) littoral zone; (**b**) central zone.

		(a) Littoral Zone								(b) Central Zone							
		Head Lenght (mm)				Head Width (mm)				Head Lenght (mm)				Head Width (mm)			
CL		1	2	3	2 + 3	1	2	3	2 + 3	1	2	3	2 + 3	1	2	3	2 + 3
All Instars	N	676	38	23	61	676	38	23	61	2575	162	30	192	2575	162	30	192
	Min	0.18	0.48	0.40	0.40	0.13	0.38	0.38	0.38	0.30	0.53	0.85	0.53	0.25	0.43	0.70	0.43
	Median	0.83	0.91	0.95	0.93	0.65	0.75	0.75	0.75	1.00	1.00	1.00	1.00	0.78	0.78	0.79	0.78
	Mean	0.75	0.88	0.90	0.89	0.58	0.69	0.72	0.70	0.98	0.98	1.00	0.98	0.78	0.78	0.80	0.78
	SD	0.24	0.17	0.16	0.17	0.20	0.13	0.12	0.13	0.09	0.10	0.09	0.10	0.08	0.08	0.08	0.08
	Max	1.20	1.10	1.10	1.10	0.93	0.95	0.90	0.95	1.25	1.25	1.25	1.25	1.00	1.05	1.00	1.05
Instar IV	N	402	33	21	54	402	33	21	54	2555	161	30	191	2555	161	30	191
	Min	0.70	0.75	0.75	0.75	0.50	0.60	0.63	0.60	0.70	0.68	0.85	0.68	0.50	0.55	0.70	0.55
	Median	0.93	0.93	0.95	0.95	0.75	0.75	0.75	0.75	1.00	1.00	1.00	1.00	0.78	0.78	0.79	0.78
	Mean	0.93	0.93	0.95	0.94	0.73	0.73	0.75	0.74	0.98	0.98	1.00	0.98	0.78	0.78	0.80	0.79
	SD	0.10	0.09	0.08	0.08	0.08	0.07	0.07	0.07	0.08	0.09	0.09	0.09	0.07	0.07	0.08	0.07
	Max	1.20	1.10	1.10	1.10	0.93	0.95	0.90	0.95	1.25	1.25	1.25	1.25	1.00	1.05	1.00	1.05
Instar III	N	214	5	2	7	214	5	2	7	19	1	-	1	19	1	-	1
	Min	0.40	0.48	0.40	0.40	0.20	0.38	0.38	0.38	0.50	0.53	-	0.53	0.30	0.43	-	0.43
	Median	0.53	0.50	0.45	0.50	0.40	0.40	0.39	0.40	0.55	0.53	-	0.53	0.40	0.43	-	0.43
	Mean	0.53	0.51	0.45	0.49	0.40	0.40	0.39	0.40	0.55	0.53	-	0.53	0.41	0.43	-	0.43
	SD	0.05	0.03	0.07	0.05	0.05	0.03	0.02	0.03	0.04	-	-	-	0.04	-	-	-
	Max	0.70	0.55	0.50	0.55	0.50	0.45	0.40	0.45	0.63	0.53	-	0.53	0.45	0.43	-	0.43
Instar II	N	60	-	-	-	60	-	-	-	1	-	-	-	1	-	-	-
	Min	0.18	-	-	-	0.13	-	-	-	0.30	-	-	-	0.25	-	-	-
	Median	0.33	-	-	-	0.23	-	-	-	0.30	-	-	-	0.25	-	-	-
	Mean	0.31	-	-	-	0.22	-	-	-	0.30	-	-	-	0.25	-	-	-
	SD	0.05	-	-	-	0.04	-	-	-	-	-	-	-	-	-	-	-
	Max	0.40	-	-	-	0.33	-	-	-	0.30	-	-	-	0.25	-	-	-

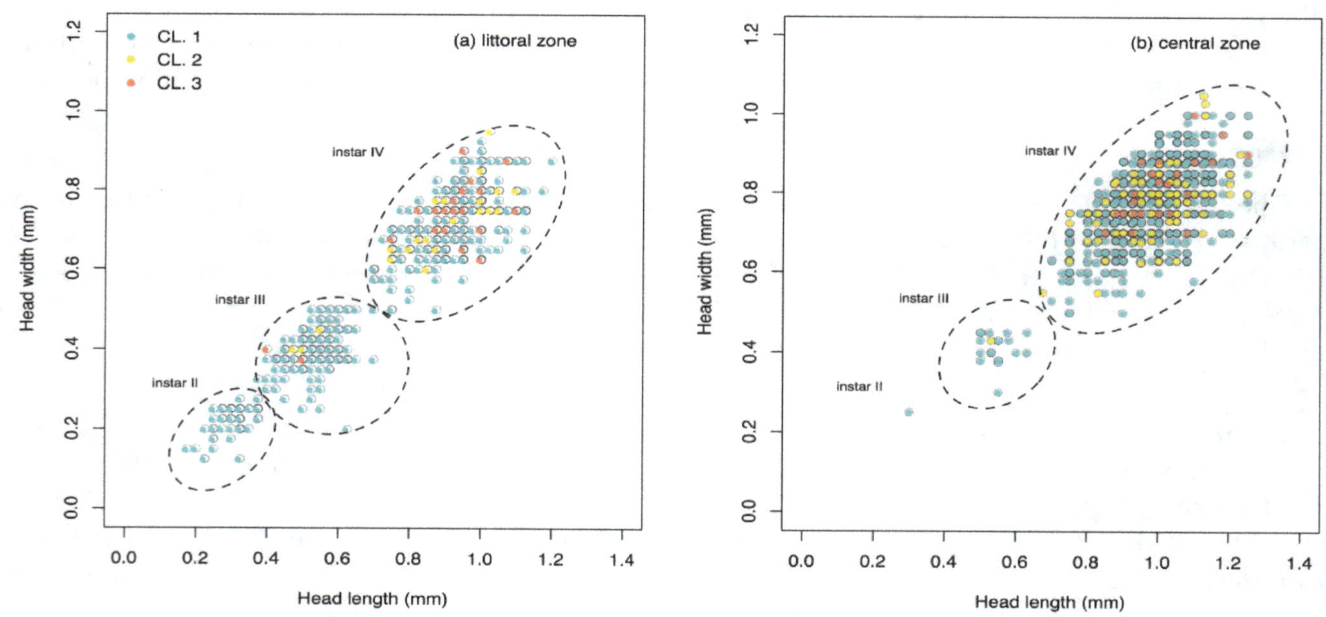

Figure 5. Size (length and width; mm) of *C. plumosus* larvae, different instars (II, III, and IV) are surrounded by dotted circles, mentum in CL. 1 (green, no deformity), CL. 2 (yellow, weak deformity), and CL. 3 (red, strong deformity). (**a**) littoral zone; (**b**) central zone.

Lastly, in Table 4 the different types of *C. plumosus* mentum deformities in Lake Trasimeno are reported. The main CL. 2 deformity types in the littoral zone, central zone, and the whole lake, respectively, were: "one or two round/filed teeth" with 81.58%, 69.75%, and 72.00%; "one missing tooth" with 2.63%, 17.90%, and 15.00%; "one additional tooth" with 5.26%, 3.09%, and 3.50%; "one serrate tooth" with 10.53%, 3.70%, and 5.00%; "one bifid tooth" with 0%, 4.94%, and 4.00%. The main CL. 3 deformity types, listed in the same order, were: "very round/filed teeth" with 39.13%, 33.33%, and 35.85%; "two or more missing teeth" with 21.74%, 16.67%, and 18.87%; "two or more additional teeth" with 8.70%, 0%, and 3.77%; "Köhn gap" with 0%, 20.00%, and 11.32%.

Table 4. *Chironomus plumosus* mentum deformity types: CL. 2 (weak deformity), CL. 3 (strong deformity), and CL. (2 + 3); numbers (N) and percentages (%); (**a**) littoral zone; (**b**) central zone; (**c**) Lake Trasimeno, Central Italy.

Class and Deformity Type	(a) Littoral Zone		(b) Central Zone		(c) Lake Trasimeno	
	(N)	(%)	(N)	(%)	(N)	(%)
CL. 2	**38**		**162**		**200**	
One or Two Round/Filed Teeth	31	81.58	113	69.75	144	72.00
One Missing Tooth	1	2.63	29	17.90	30	15.00
One Additional Tooth	2	5.26	5	3.09	7	3.50
One Bifid Tooth	0	0	8	4.94	8	4.00
One Serrate Tooth	4	10.53	6	3.70	10	5.00
Two Joined Teeth	0	0	1	0.62	1	0.50
Weak Asymmetry	0	0	0	0	0	0
CL. 3	**23**		**30**		**53**	
Very Round/Filed Teeth	9	39.13	10	33.33	19	35.85
Two or More Missing Teeth	5	21.74	5	16.67	10	18.87
Two or More Additional Teeth	2	8.70	0	0	2	3.77
Two or More Bifid Teeth	0	0	0	0	0	0
Serrate Teeth	0	0	0	0	0	0
Three or More Joined Teeth	0	0	0	0	0	0
Severe Asymmetry	0	0	0	0	0	0
Köhn Gap	0	0	6	20.00	6	11.32
Combination of Different Deformities	7	30.43	9	30.00	16	30.19
CL. (2 + 3)	**61**		**192**		**253**	
Round/Filed Teeth	40	65.57	123	64.06	163	64.43
Missing Teeth	6	9.84	34	17.71	40	15.81
Additional Teeth	4	6.56	5	2.60	9	3.56
Bifid Teeth	0	0	8	4.17	8	3.16
Serrate Teeth	4	6.56	6	3.13	10	3.95
Joined Teeth	0	0	1	0.52	1	0.40
Asymmetry	0	0	0	0	0	0
Köhn Gap	0	0	6	3.13	6	2.37
Combination of Different Deformities	7	11.48	9	4.69	16	6.32

4. Discussion

This study analyzed mentum deformities in *C. plumosus* larvae as an endpoint for evaluating the effects of anthropogenic stressors in the shallow Lake Trasimeno. The results of the survey showed a low deformity incidence, amounting to about 7% (1.5% in CL. 3) in the whole Lake Trasimeno's ecosystem, with similar values in the littoral and central zones of the lake, about 8% (3% in CL. 3) and 7% (1% in CL. 3), respectively.

Paleolimnological studies, representing natural background levels during the pre-industrial age, showed deformity incidences between 0% and 0.8% [68,69]. In the contemporary age [70,71] the incidence of background deformities in low-impacted areas is lower than 8% suggesting this threshold as the reference background for deformity incidence in natural inland waters. Therefore, the results of the present research affirm that this biotope has a deformity incidence value comparable to that of inland waters subject to low anthropic impact. Perhaps, this can be associated to a low degree of contamination of the Trasimeno's ecosystem, in particular of sediments that represent a preferential sink for toxic compounds and, at the same time, serve as a habitat for chironomid larvae.

The deformity incidence analysis among the different cohorts of the polyvoltine biological cycle of *C. plumosus* showed that the spring cohort had overall the highest deformity values in Lake Trasimeno, corresponding to 11.41% (1.33% in CL. 3). Perhaps, this peak of deformities in the spring cohort is due to the compromise between metabolic activity and residence time in sediments, compared to the winter cohort (low metabolic activity and long residence time) and to summer cohorts (high metabolic activity and short residence time).

Head size (length and width) of normal and deformed larvae grouped according to their instar showed similar trends, regardless of whether they were found in the littoral or in the central zone of the lake. These results highlight that mentum deformity did not cause alterations of the head size of deformed specimens, meaning a similar larval growth of the latter compared to normal specimens [62].

A comparison of current results with previous investigations (2000–2010) conducted in Lake Trasimeno using the same sampling methods showed a clear decrease in the incidence of mentum deformities in *C. plumosus* larvae, both in the littoral and in the central zone of the lake [43]. In fact, the deformity incidences that were about 20% (9.5% in CL. 3) in the littoral zone of Castiglione del Lago during 2006–2009 and about 13% (3.8% in CL. 3) in the central zone of the lake during 2000–2010, decreased to 8% and 7% in the respective lake areas during the present survey (2018–2019). This 10-year comparison demonstrated a clear decrease in the current deformity index values. Over the years, in effect, there has been a confirmation of this decreasing trend of deformity incidence that, in the littoral zone, decreased from 31% in 2006 to 10% in 2009, until reaching the current level of 8%. A similar phenomenon tended to occur in the central zone of the lake as well, which showed a deformity index of 11% in 2000, 17% in 2001, 10% in 2002, and 13% in October 2010, until reaching the current level of 7%.

In summary, the surveys carried out at Lake Trasimeno allow a space–time biological assessment of lacustrine sediments' contamination showing a qualitative recovery over time of the lake ecosystem both in the central and littoral zone; the latter zone, more easily subject to the introduction of pollutants, showed an incidence of severe deformities (CL. 3) of 3% compared to 1% in the central zone.

However, the relationship between the incidence of strong (CL. 3) and weak (CL. 2) deformities is not easy to interpret, in fact, even if severe deformed phenotypes (CL. 3) were found in specimens subject to sediments contaminated both with organic and inorganic pollutants, their increase is probably influenced not only by the quantity but also by the type of compounds that form the toxic mixture [4,22]. Consequently, as a reliable biomarker, it is more appropriate to consider the two deformity classes together (CL. 2 + 3) for an overall assessment of the toxic effects of sediment contamination [22].

In this regard, some studies based on ecotoxicological laboratory experiments queried the relationship between chironomid deformities and toxicity assessment in freshwater ecosystems [24,49,72–76]. However, the high variability of those experimental results can be influenced by different variables, such as: (i) the endpoint used (mentum, mandibles, pectin epipharyngis); (ii) the deformity types considered; (iii) the larval instars used in the experiment; (iv) the type of cultures used which often cause a high incidence of deformities in the control test; (v) the exposure time of the larvae to the contaminants; and (vi) the type of contaminant mixture used in the test.

On the other hand, in natural environments the association between incidence of chironomid deformities and levels of chemical contamination is very consistent, despite the lack of a standardized protocol for the deformity assignment which can generate cases of inconsistent interpretation. In this respect, in the ring test (based on digital images) [74] involving 25 international experts to assess the status (normal or deformed) of 211 chironomid larvae menta (IV instar) from Lake Saimaa (Finland), the majority of them (at least 2/3) highlighted clearly that the incidence of mentum deformities was high in the polluted site (about 30%) and low in the reference site (<10%), with no significant differences between blind or open assessment.

In addition, the present study contributes to the elaboration of a detailed protocol for the definition of which types of mentum deformities to be considered for an assessment of the deformity incidence in *Chironomus* populations of freshwater biotopes.

The present study data showed that, among deformed menta of *C. plumosus* found in Lake Trasimeno (littoral and central zones), the most common deformity types were "round/filed teeth" with 64% followed by "missing teeth" (16%), a "combination of different deformities" (about 6%), "serrate teeth" (4%), "additional teeth" (4%), "bifid teeth" (3%), "Köhn gap" (2%), and "joined teeth" (0.4%) (Table 4).

In conclusion, the present study on Lake Trasimeno provides a further contribution to the evaluation of mentum deformities in chironomid populations of inland waters. In the biomonitoring programs, the deformity incidence can provide an easy biological tool to assess the effects of sediment-associated toxicants. The mentum deformities assessment is complementary and non-substitutive of the chemical instrumental analyses; in fact, a high incidence of deformities in a freshwater ecosystem is a warning (indicator) for the start of a chemical analytical monitoring program aimed at qualitative and quantitative detection of the toxic compounds causing environmental contamination.

Author Contributions: Conceptualization, E.G. (Enzo Goretti), M.P., S.P., M.C., G.L.P., E.G. (Elda Gaino), and A.A.; Data curation, E.G. (Enzo Goretti), M.P., S.P., M.C., and G.L.P.; Methodology, E.G. (Enzo Goretti), M.P., S.P., M.C., G.L.P., R.S., and A.M.D.G.; Investigation, E.G. (Enzo Goretti), M.P., S.P., M.C., G.L.P., R.S., and A.M.D.G.; Writing—original draft, E.G. (Enzo Goretti), M.P., G.L.P., E.G. (Elda Gaino), and A.A.; Writing—review and editing, E.G. (Enzo Goretti), M.P., G.L.P., E.G. (Elda Gaino), and A.A. All authors have read and agreed to the published version of the manuscript.

Acknowledgments: We thank Andrea Pagnotta (amateur fisherman), Marco Chiappini (Provincia di Perugia), Alberto Fais, and Francesco Giglietti (USL Umbria1), Rosalba Padula and Luca Nicoletti (ARPAUmbria) for their great support in the sampling campaign.

References

1. Rosenberg, D.M.; Resh, V.H. Introduction to freshwater biomonitoring and benthic macroinvertebrates. In *Freshwater Biomonitoring Benthic Macroinvertebrates*; Chapman and Hall: New York, NY, USA, 1993.
2. Maldonado, A.C.D.; Wendling, B. Manejo de ecossistemas aquáticos contaminados por Metais pesados. *Revista Agropecuária Técnica* **2009**, *30*, 21–32. [CrossRef]
3. Goretti, E.; Pallottini, M.; Ricciarini, M.I.; Selvaggi, R.; Cappelletti, D. Heavy metals bioaccumulation in selected tissues of red swamp crayfish: An easy tool for monitoring environmental contamination levels. *Sci. Total Environ.* **2016**, *559*, 339–346. [CrossRef] [PubMed]
4. Vermeulen, A.C. Elaboration of chironomid deformities as bioindicators of toxic sediment stress: The potential application of mixture toxicity concepts. *Ann. Zool. Fenn.* **1995**, *32*, 265–285.
5. Warwick, W.F. Morphological deformities in Chironomidae (Diptera) larvae as biological indicators of toxic stress. In *Toxic Contaminants and Ecosystem Health; A Great Lakes Focus*; Evans, M.S., Ed.; John Wiley and Sons: New York, NY, USA, 1988; pp. 281–320.
6. Apostolico, F.; Vercillo, F.; La Porta, G.; Ragni, B. Long-term changes in diet and trophic niche of the European wildcat (*Felis silvestris silvestris*) in Italy. *Mamm. Res.* **2016**, *61*, 109–119. [CrossRef]
7. Davutluoglu, O.I.; Seckin, G.; Ersu, C.B.; Ylmaz, T.; Sari, B. Heavy metal content and distribution in surface sediments of the Seyhan River, Turkey. *J. Environ. Manag.* **2011**, *92*, 2250–2259. [CrossRef]
8. Goretti, E.; Pallottini, M.; Cenci Goga, B.T.; Selvaggi, R.; Petroselli, C.; Vercillo, F.; Cappelletti, D. Mustelids as bioindicators of the environmental contamination by heavy metals. *Ecol. Indicat.* **2018**, *94*, 320–327. [CrossRef]
9. Goretti, E.; Pallottini, M.; Rossi, R.; La Porta, G.; Gardi, T.; Cenci Goga, B.T.; Elia, A.C.; Galletti, M.; Moroni, B.; Petroselli, C.; et al. Heavy metal bioaccumulation in honey bee matrix, an indicator to assess the contamination level in terrestrial environments. *Environ. Pollut.* **2019**, *256*, 113388. [CrossRef]
10. Holt, M.S. Sources of chemical contaminants and routes into the freshwater environment. *Food Chem. Toxicol.* **2000**, *38*, 21–27. [CrossRef]

11. Chapman, P.M. The sediment quality triad approach to determining pollution-induced degradation. *Sci. Total Environ.* **1990**, *97*, 815–825. [CrossRef]
12. Cummins, K.W.; Wilzbach, M.A. *Field Procedures for the Analysis of Functional Feeding Groups in Stream Ecosystems*; Contribution 1611; Appalachian Environmental Laboratory, University of Maryland: Frostburg, MD, USA, 1985.
13. Céréghino, R.; Park, Y.S.; Compin, A.; Lek, S. Predicting the species richness of aquatic insects in streams using a limited number of environmental variables. *J. N. Am. Benthol. Soc.* **2003**, *22*, 442–456. [CrossRef]
14. Dolédec, S.; Statzner, B. Reponses of freshwater biota to human disturbances: Contribution of J-NABS to developments in ecological integrity assessments. *J. N. Am. Benthol. Soc.* **2010**, *29*, 286–311. [CrossRef]
15. Pallottini, M.; Goretti, E.; Gaino, E.; Selvaggi, R.; Cappelletti, D.; Cereghino, R. Invertebrate diversity in relation to chemical pollution in an Umbrian stream system (Italy). *C. R. Biol.* **2015**, *338*, 511–520. [CrossRef] [PubMed]
16. Pallottini, M.; Cappelletti, D.; Fabrizi, E.; Gaino, E.; Goretti, E.; Selvaggi, R.; Cereghino, R. Macroinvertebrate functional trait responses to chemical pollution in agricultural-industrial landscapes. *River Res. Appl.* **2017**, *33*, 505–513. [CrossRef]
17. Adler, P.H.; Courtney, G.W. Ecological and Societal Services of Aquatic Diptera. *Insects* **2019**, *10*, 70. [CrossRef]
18. Armitage, P.D.; Pinder, L.C.; Cranston, P. *The Chironomidae: Biology and Ecology of Non-Biting Midges*; Springer Netherlands: Heidelberg, Germany, 1995.
19. Goretti, E.; Coletti, A.; Di Veroli, A.; Di Giulio, A.M.; Gaino, E. Artificial light device for attracting pestiferous chironomids (Diptera): A case study at Lake Trasimeno (Central Italy). *Ital. J. Zool.* **2011**, *78*, 336–342. [CrossRef]
20. OECD. *Test No. 235: Chironomus sp., Acute Immobilisation Test*; OECD Publishing: Paris, France, 2011. [CrossRef]
21. Božanić, M.; Marković, Z.; Živić, M.; Dojčinovć, B.; Perić, A.; Stanković, M.; Živić, I. Mouthpart Deformities of *Chironomus plumosus* Larvae Caused by Increased Concentrations of Copper in Sediment from Carp Fish Pond. *Turk. J. Fish. Aquat. Sci.* **2018**, *19*, 251–259. [CrossRef]
22. Di Veroli, A.; Goretti, E.; León Paumen, M.; Kraak, M.H.S.; Admiraal, W. Mouthpart deformities in *Chironomus riparius* larvae exposed to toxicants. *Environ. Pollut.* **2012**, *166*, 212–217. [CrossRef]
23. Martinez, E.A.; Moore, B.C.; Schaumloffel, J.; Dasgupta, N. Morphologic and Growth Responses in *Chironomus tentans* to Arsenic Exposure. *Arch. Environ. Contam. Toxicol.* **2006**, *51*, 529–536. [CrossRef]
24. Gagliardi, B.S.; Long, S.M.; Pettigrove, V.J.; Griffin, P.C.; Hoffmann, A.A. A re-evaluation of chironomid deformities as an environmental stress response: Avoiding survivorship bias and testing noncontaminant biological factors. *Environ. Toxicol. Chem.* **2019**, *38*, 1658–1667. [CrossRef]
25. Corbi, J.J.; Bernegossi, A.C.; Moura, L.; Felipe, M.C.; Issa, C.G.; Silva, M.R.L.; Gorni, G.R. *Chironomus sancticaroli* (Diptera, Chironomidae) as a Sensitive Test Species: Can We Rely on Its Use After Repeated Generations, Under Laboratory Conditions? *Bull. Environ. Contam. Toxicol.* **2019**, *103*, 213–217. [CrossRef] [PubMed]
26. Saether, O.A. A survey of the bottom fauna of the Okanagan valley, British Columbia. *Can. Fish. Mar. Serv. Techn. Rep.* **1970**, *196*, 1–29.
27. Hamilton, A.L.; Saether, O.A. The occurrence of characteristic deformities in the chironomid larvae of several Canadian lakes. *Can. Entomol.* **1971**, *103*, 363–368. [CrossRef]
28. Hare, L.; Carter, J.C.H. The distribution of *Chironomus* (s.s.)? Cucini (*salinarius* group) larvae (Diptera: Chironomidae) in Parry Sound, Georgian Bay, with particular reference to structural deformities. *Can. J. Zool.* **1976**, *54*, 2129–2134. [CrossRef]
29. Janssens de Bisthoven, L.; Vermeulen, A.; Ollevier, F. Experimental Induction of Morphological Deformities in *Chironomus riparius* Larvae by Chronic Exposure to Copper and Lead. *Arch. Environ. Contam. Toxicol.* **1998**, *35*, 249–256. [CrossRef]
30. Reynolds, S.K.; Ferrington, L.C. Differential Morphological Responses of Chironomid Larvae to Severe Heavy Metal Exposure (Diptera: Chironomidae). *J. Kans. Entomol. Soc.* **2002**, *75*, 172–184.
31. Martinez, E.A.; Moore, B.C.; Schaumloffel, J.; Dasgupta, N. Morphological abnormalities in *Chironomus tentans* exposed to cadmium and copper-spiked sediments. *Ecotoxicol. Environ. Saf.* **2003**, *55*, 204–212. [CrossRef]

32. Péry, A.R.R.; Mons, R.; Garric, J. Energy-based modeling to study population growth rate and production for the midge *Chironomus riparius* in ecotoxicological risk assessment. *Ecotoxicology* **2004**, *13*, 647–656. [CrossRef]
33. Ebau, W.; Rawi, C.S.M.; Din, Z.; Al-Shami, S.A. Toxicity of cadmium and lead on tropical midge larvae, *Chironomus kiiensis* Tokunaga and *Chironomus javanus* Kieffer (Diptera: Chironomidae). *Asian Pac. J. Trop. Biomed.* **2012**, *2*, 631–634. [CrossRef]
34. Gagliardi, B.; Pettigrove, V. Removal of intensive agriculture from the landscape improves aquatic ecosystem health. *Agric. Ecosyst. Environ.* **2013**, *176*, 1–8. [CrossRef]
35. Grebenjuk, L.P.; Tomilina, I.I. Morphological Deformations of Hard Chitinized Mouthpart Structures in Larvae of the Genus *Chironomus* (Diptera, Chironomidae) as the Index of Organic Pollution in Freshwater Ecosystems. *Inland Water Biol.* **2014**, *7*, 273–285. [CrossRef]
36. Beneberu, G.; Mengistou, S. Head capsule deformities in *Chironomus* spp. (Diptera: Chironomidae) as indicator of environmental stress in Sebeta River, Ethiopia. *Afr. J. Ecol.* **2014**, *53*, 268–277. [CrossRef]
37. Arimoro, F.O.; Odume, O.N.; Meme, F.K. Environmental drivers of head capsule deformities in *Chironomus* spp. (Diptera: Chironomidae) in a stream in north central Nigeria. *Zool. Ecol.* **2015**, *25*, 70–76. [CrossRef]
38. Odume, O.N.; Palmer, C.G.; Arimoro, F.O.; Mensaha, P.K. Chironomid assemblage structure and morphological response to pollution in an effluent-impacted river, Eastern Cape, South Africa. *Ecol. Indic.* **2016**, *67*, 391–402. [CrossRef]
39. Arimoro, F.O.; Auta, Y.I.; Odume, O.N.; Keke, U.N.; Mohammed, A.Z. Mouthpart deformities in Chironomidae (Diptera) as bioindicators of heavy metals pollution in Shiroro Lake, Niger State, Nigeria. *Ecotoxicol. Environ. Saf.* **2018**, *149*, 96–100. [CrossRef]
40. Akyildiz, G.K.; Bakir, R.; Polat, S.; Duran, M. Mentum Deformities of Chironomid Larvae as an Indicator of Environmental Stress in Büyük Menderes River, Turkey. *Inland Water Biol.* **2018**, *11*, 515–522. [CrossRef]
41. Deliberalli, W.; Cansian, R.L.; Pereira, A.A.M.; Loureiro, R.C.; Hepp, L.U.; Restello, R.M. The effects of heavy metals on the incidence of morphological deformities in Chironomidae (Diptera). *Zoologia* **2018**, *35*. [CrossRef]
42. Di Veroli, A.; Selvaggi, R.; Pellegrino, R.M.; Goretti, E. Sediment toxicity and deformities of chironomid larvae in Lake Piediluco (Central Italy). *Chemosphere* **2010**, *79*, 33–39. [CrossRef]
43. Di Veroli, A.; Selvaggi, R.; Goretti, E. Chironomid mouthpart deformities as indicator of environmental quality: A case study in Lake Trasimeno (Italy). *J. Environ. Monit.* **2012**, *14*, 1473–1478. [CrossRef]
44. Di Veroli, A.; Santoro, F.; Pallottini, M.; Selvaggi, R.; Scardazza, F.; Cappelletti, D.; Goretti, E. Deformities of chironomid larvae and heavy metal pollution: From laboratory to field studies. *Chemosphere* **2014**, *112*, 9–17. [CrossRef]
45. Savić-Zdravković, D.; Jovanović, B.; Đurđević, A.; Stojković-Piperac, M.; Savić, A.; Vidmar, J.; Milošević, D. An environmentally relevant concentration of titanium dioxide (TiO_2) nanoparticles induces morphological changes in the mouthparts of *Chironomus tentans*. *Chemosphere* **2018**, *211*, 489–499. [CrossRef] [PubMed]
46. Tomilina, I.I.; Grebenjuk, L.P. Induction of Deformities of the Hard-Chitinized Mouthpart Structures of Larvae *Chironomus riparius* Meigen under Various Contents of Persistent Organic Substances in Bottom Sediments. *Inland Water Biol.* **2019**, *12*, 346–355. [CrossRef]
47. Park, K.; Kwak, I.S. Disrupting effects of antibiotic sulfathiazole on developmental process during sensitive life-cycle stage of *Chironomus riparius*. *Chemosphere* **2018**, *190*, 25–34. [CrossRef] [PubMed]
48. Pallottini, M.; Goretti, E.; Selvaggi, R.; Cappelletti, D.; Dedieu, N.; Cereghino, R. An efficient semi-quantitative macroinvertebrate multimetric index forthe assessment of water and sediment contamination in streams. *Inland Waters* **2017**, *7*, 314–322. [CrossRef]
49. Arambourou, H.; Planelló, R.; Llorente, L.; Fuertes, I.; Barata, C.; Delorme, N.; Noury, P.; Herrero, Ó.; Villeneuve, A.; Bonnineau, C. *Chironomus riparius* exposure to field-collected contaminated sediments: From subcellular effect to whole-organism response. *Sci. Total Environ.* **2019**, *671*, 874–882. [CrossRef]
50. He, H.; Xi, G.; Lu, X. Molecular cloning, characterization, and expression analysis of an ecdysone receptor homolog in *Teleogryllus emma* (Orthoptera: Gryllidae). *J. Insect Sci.* **2015**, *15*. [CrossRef]
51. Vermeulen, A.C.; Liberloo, G.; Dumont, P.; Ollevier, F.; Goddeeris, B.R. Exposure of *Chironomus riparius* larvae (Diptera) to lead, mercury, and β-sitosterol: Effects on mouthpart deformation and moulting. *Chemosphere* **2000**, *41*, 1581–1591. [CrossRef]
52. Marchegiano, M.; Francke, A.; Gliozzi, E.; Wagner, B.; Ariztegui, D. High-resolution palaeohydrological reconstruction of central Italy during the Holocene. *Holocene* **2018**, *29*, 481–492. [CrossRef]

53. Frondini, F.; Dragoni, W.; Morgantini, N.; Donnini, M.; Cardellini, C.; Caliro, S.; Melillo, M.; Chiodini, G. An Endorheic Lake in a Changing Climate: Geochemical Investigations at Lake Trasimeno (Italy). *Water* **2019**, *11*, 1319. [CrossRef]
54. Ludovisi, A.; Gaino, E. Meteorological and water quality changes in Lake Trasimeno (Umbria, Italy) during the last fifty years. *J. Limnol.* **2010**, *69*, 174–188. [CrossRef]
55. Marchegiano, M.; Francke, A.; Gliozzi, E.; Ariztegui, D. Arid and humid phases in central Italy during the Late Pleistocene revealed by the Lake Trasimeno ostracod record. *Palaeogeogr. Palaeoclimatol. Palaeoecol.* **2018**, *490*, 55–69. [CrossRef]
56. Mancinelli, G.; Papadia, P.; Ludovisi, A.; Migoni, D.; Bardelli, R.; Fanizzi, F.P.; Vizzini, S. Beyond the mean: A comparison of trace- and macroelement correlation profiles of two lacustrine populations of the crayfish *Procambarus clarkii*. *Sci. Total Environ.* **2018**, *624*, 1455–1466. [CrossRef] [PubMed]
57. Tachet, H.; Richoux, P.; Bournaud, M.; Usseglio-Polatera, P. *Invertébrés d'eau douce. Systématique, biologie, écologie*; CNRS Éditions: Paris, France, 2010.
58. Ferrarese, U.; Rossaro, B. Chironomidi, 1. In *Guida Per Il Riconoscimento Delle Specie Animali Delle Acque Interne Italiane*; CNR AQ/1/129; CNR: Roma, Italy, 1981.
59. Rossaro, B. Chironomidi, 2. In *Guida Per Il Riconoscimento Delle Specie Animali Delle Acque Interne Italiane*; CNR AQ/1/171; CNR: Roma, Italy, 1982.
60. Ferrarese, U. Chironomidi, 3. In *Guida Per Il Riconoscimento Delle Specie Animali Delle Acque Interne Italiane*; CNR AQ/1/204; CNR: Roma, Italy, 1983.
61. Nocentini, A.M. Chironomidi, 4. In *Guida Per Il Riconoscimento Delle Specie Animali Delle Acque Interne Italiane*; CNR AQ/1/233; CNR: Roma, Italy, 1985.
62. Janssens De Bisthoven, L.; Yts, P.; Goddeeris, B.; Ollevier, F. Sublethal parameters in morphologically deformed *Chironomus* larvae: Clues to understand their bioindicator value. *Freshw. Biol.* **1998**, *39*, 179–191. [CrossRef]
63. Madden, C.P.; Austin, A.D.; Sutter, P.J. Pollution monitoring using chironomid larvae: What is a deformity? In *Chironomids from Genes to Ecosystems*; Craston, P., Ed.; CSRIO Publication: East Melbourne, Australia, 1995; pp. 89–94.
64. Nazarova, L.B.; Riss, H.W.; Kahlheber, A.; Werding, B. Some observations of buccal deformities in chironomid larvae (Diptera: Chironomidae) from the Ciénaga Grande de Santa Marta, Colombia. *Caldasia* **2004**, *26*, 275–290.
65. Cao, Y.; Zhang, E.; Tang, H.; Langdon, P.; Ning, D.; Zheng, W. Combined effects of nutrients and trace metals on chironomid composition and morphology in a heavily polluted lake in central China since the early 20th century. *Hydrobiologia* **2016**, *779*, 147–159. [CrossRef]
66. Newcombe, R.G. Interval estimation for the difference between independent proportions: Comparison of eleven methods. *Stat. Med.* **1998**, *17*, 873–890. [CrossRef]
67. R Core Team. *R: A Language and Environment for Statistical Computing*; R Foundation for Statistical Computing: Vienna, Austria, 2018; Available online: https://www.R-project.org/ (accessed on 13 December 2019).
68. Warwick, W.F. Paleolimnology of the Bay of Quinte, Lake Ontario: 2800 years of cultural influence. *Can. Bull. Fish. Aquat. Sci.* **1980**, *206*, 1–118.
69. Wiederholm, T. Incidence of deformed chironomid larvae (Diptera Chironomidae) in Swedish Lakes. *Hydrobiologia* **1984**, *109*, 243–249. [CrossRef]
70. Warwick, W.F. Morphological abnormalities in Chironomidae (Diptera) larvae as measures of toxic stress in freshwater ecosystems: Indexing antennal deformities in *Chironomus* Meigen. *Can. J. Fish. Aquat. Sci.* **1985**, *42*, 1881–1914. [CrossRef]
71. Bird, G.A. Use of chironomid deformities to assess environmental degradation in the Yamaska River, Quebec. *Environ. Monit. Assess.* **1994**, *30*, 163–175. [CrossRef]
72. Vogt, C.; Langer-Jaesrich, M.; Elsässer, O.; Schmitt, C.; Van Dongen, S.; Köhler, H.R.; Oehlmann, J.; Nowak. Effects of inbreeding on mouthpart deformities of *Chironomus riparius* under sublethal pesticide exposure. *Environ. Toxicol. Chem.* **2013**, *32*, 423–425. [CrossRef] [PubMed]
73. Arambourou, H.; Beisel, J.N.; Branchu, P.; Debat, V. Exposure to sediments from polluted rivers has limited phenotypic effects on larvae and adults of *Chironomus riparius*. *Sci. Total Environ.* **2014**, *484*, 92–101. [CrossRef] [PubMed]

74. Salmelin, J.; Vuori, K.M.; Hämäläinen, H. Inconsistency in the analysis of morphological deformities in Chironomidae (Insecta: Diptera) larvae. *Environ. Toxicol. Chem.* **2015**, *34*, 1891–1898. [CrossRef]
75. Salmelin, J.; Karjalainen, A.K.; Hämäläinen, H.; Leppänen, M.T.; Kiviranta, H.; Kukkonen, J.V.K.; Vuori, K.M. Biological responses of midge (*Chironomus riparius*) and lamprey (*Lampetra fluviatilis*) larvae in ecotoxicity assessment of PCDD/F-, PCB- and Hg-contaminated river sediments. *Environ. Sci. Pollut. Res.* **2016**, *23*, 18379–18393. [CrossRef] [PubMed]
76. Gagliardi, B.S.; Pettigrove, V.J.; Long, S.M.; Hoffmann, A.A. A Meta-Analysis Evaluating the Relationship between Aquatic Contaminants and Chironomid Larval Deformities in Laboratory Studies. *Environ. Sci. Technol.* **2016**, *50*, 12903–12911. [CrossRef] [PubMed]

2

Using Mixed Active Capping to Remediate Multiple Potential Toxic Metal Contaminated Sediment for Reducing Environmental Risk

Meng-Yuan Ou [1], Yu Ting [1], Boon-Lek Ch'ng [1], Chi Chen [1], Yung-Hua Cheng [1], Tien-Chin Chang [2] and Hsing-Cheng Hsi [1,*]

1 Graduate Institute of Environmental Engineering, National Taiwan University, No. 1, Sec. 4, Roosevelt Rd., Taipei 10671, Taiwan; auanlie@outlook.com (M.-Y.O.); yuting821216@gmail.com (Y.T.); r06541135@ntu.edu.tw (B.-L.C.); q2461015@gmail.com (C.C.); r08541121@ntu.edu.tw (Y.-H.C.)

2 Institute of Environmental Engineering and Management, National Taipei University of Technology, No. 1, Sec. 3, Zhongxiao E. Rd., Taipei 10608, Taiwan; tcchang@ntut.edu.tw

* Correspondence: hchsi@ntu.edu.tw

Abstract: In this study, kaolinite, carbon black (CB), iron sulfide (FeS), hydroxyapatite (HAP), and oyster shell powder (OSP) were selected as potentially ideal amendments to immobilize metals in sediment, including Ni, Cr, Cu, Zn, and Hg. In aqueous batch experiments, the five adsorbents were tested for capturing the five potential toxic metals individually at various concentrations. HAP and OSP showed the largest removal efficiencies towards Ni (OSP: 76.47%), Cr (OSP: 100.00%), Cu (HAP: 98.39%), and Zn (HAP: 64.56%), with CB taking the third place. In contrast, FeS and CB played a more significant role in Hg removal (FeS: 100.00%; CB: 86.40%). In the modified six-column microcosm experiments, five mixing ratios based on various considerations using the five adsorbent materials were tested; the water samples were collected and analyzed every week for 135 days. Results showed that caps including CB could immobilize the release of Hg and methylmercury (MeHg) better than those with FeS. More economical caps, namely, with a higher portion of OSP in the mixed capping, could not reach comparable effects to those with more HAP for immobilizing Ni, but performed almost the same for the other four metals. All columns with active caps showed greater metal immobilization as compared to the controlled column without caps.

Keywords: active capping; toxic metal; sediment; remediation; multiple materials

1. Introduction

Wastewater containing potential toxic metals originating from anthropogenic activities discharged to river streams is a widespread environmental issue nowadays, processing significant toxicity to aquatic organisms and accumulating by food chain, finally causing various diseases and disorders [1]. Natural processes are frequently inadequate to deal with the elevated metal loading, therefore there is an urgent need for remediation measures [2]. Thin-layer capping has been applied as an economically-feasible in-situ method for sediment remediation, reducing contaminants release from sediment to overlying water, subsequently reducing ecological and human health risk [3]. Sediment can be seen as an important sink of various organic and inorganic compounds, resulting in the simultaneous existence of several different contaminants [4–6]. However, most of the previous studies focused on only one or two capping materials. To cope with highly complex conditions in sediment, mixed capping with multiple materials was proposed. Notably, using cheap and effective alternatives for the removal of potential toxic metals could reduce operating costs, reduce the prices of products, improve competitiveness, and benefit the environment [7].

Several relatively cheap, environment-friendly materials have been proposed and examined. Based on previous research [7–22], kaolinite, carbon black (CB), iron sulfide (FeS), oyster shell powder (OSP), and hydroxyapatite (HAP; $Ca_{10}(PO_4)_6(OH)_2$,) were considered as potentially ideal amendments to be part of the mixed caps. Previous reports provided excellent results showing that kaolinite can be used as a cheap and naturally occurring adsorbent to remove Pb^{2+}, Cd^{2+}, Ni^{2+}, Co^{2+}, Cr^{6+}, Zn^{2+}, and Cu^{2+} from aqueous solution in both single and multi-metal ions [8–11]. Sulfide minerals are shown to be ideal materials to scavenge Hg^{2+} by immobilizing it through adsorption or co-precipitation as a discrete sulfide phase (HgS) and solid solution formation with iron sulfides [12,13]. HAP jumped out recently because of its high adsorption capacity of various metals, which brings the perspective for removal of Cd^{2+}, Zn^{2+}, Pb^{2+}, U^{6+}, Co^{2+} ions, and so on [14–17]. HAP is a nanostructured material, which was recently synthesized from some high-calcium biological wastes as raw materials, such as seashells and eggshells [18]. In the hunt of such low-cost and efficient raw materials for the production of HAP, OSP has emerged as a suitable one. As for OSP itself, owing to its low price and basicity, it has been employed widely for stabilization/solidification of many kinds of potential toxic metals (As, Pb, Ni, and Cu, etc.) [7,19–22]. The solution pH of OSP greatly affected the adsorption process towards Cu^{2+}, with an optimum adsorption pH of 5.5 and an overall negative surface charge facilitating the adsorption process [19]. The adsorption capacities of OSP towards Cu^{2+} and Ni^{2+} could reach 49.26–103.1 mg g^{-1} and 48.75–94.3 mg g^{-1}, respectively, through physical adsorption mechanisms [7].

The present study aims to find mixed amendments with the best comprehensive benefits for immobilizing potential toxic metals in contaminated sediment. Based on the previous works, kaolinite, CB, FeS, OSP, and HAP were selected as potentially ideal amendments to immobilize metals in sediment. According to the survey report of Taiwan in 2012 [23] on main rivers with contaminated sediment, Ni, Cr, Cu, Zn, and Hg are the five major potential toxic metals present in sediments, which were selected as the targets of remediation in this study. Furthermore, methylmercury (MeHg) converted from Hg was measured in this study due to its high toxicity and bioaccumulative ability. Aqueous batch experiments were first conducted by using the five adsorbents to capture the five potential toxic metals individually at various concentrations. Then, five columns containing contaminated sediment covered by different mixed materials were set based on the obtained results of batch experiments to better comprehend the immobilization effectiveness of mixed capping for the potential toxic metals as compared to the controlled column without caps. Results from this study are helpful in designing an economically and technically feasible in–situ approach for sediment remediation, which could not only reduce the potential toxic metal release from sediment to overlying water, but also reduce the ecological and human health risk.

2. Materials and Methods

2.1. Adsorbents Preparation

The kaolinite and FeS used in this study were all reagent-grade chemicals purchased from Sigma-Aldrich. The CB used was obtained from Enrestec Inc. (Tainan, Taiwan), and was considered as a low-cost, recycled materials because it was a byproduct from waste rubber tire pyrolysis for oil production.

OSP was prepared by grinding oyster shells collected as food waste materials. The oyster shells were scrubbed carefully to remove impurities, then dried for 48 h at 100 °C [19] and ground to homogenized powder that could pass through a 30-mesh sieve. Then this sieved material was shaken over a 60-mesh sieve and any material that passed through was rejected (ASTM D2765). [24]

To synthesize HAP, the OSP passed though the 60-mesh sieve was collected as a raw material. The synthesis was operated according to the following reaction: $10Ca^{2+} + 6PO_4{}^{3-} + 2OH^- = Ca_{10}(PO_4)_6(OH)_2$, referring to previous studies [18,25–27] (Figure 1). In the synthesis process, an amount of Na_2HPO_4 and the corresponding amount of oyster shell powder (Ca/P molar ratio = 1.67) were used as P and Ca precursors, respectively. Firstly, 5 g of that oyster shell powder (<60 mesh) was dissolved in 1:3 hydrochloride acid/water solution and stirred thoroughly. The supernatant of the solution filtered by 0.45 μm filter was put into 500 mL of 0.1 M EDTA. Then reaction process was carried out by drop wise addition of 0.06 M Na_2HPO_4 solution under continuous stirring of 200 rpm. The pH value of the reaction mixture was regulated within the range of 10.5–11.5 by 5 M NaOH solution and maintained throughout the process the dripping velocity was controlled at 2 mL min^{-1}. After mixing, the resulting mixture was left to maturate for the next 24 h. The obtained white precipitate was washed with distilled water by centrifuge and dried in the hot air oven at 65 °C for 24 h to get the final product.

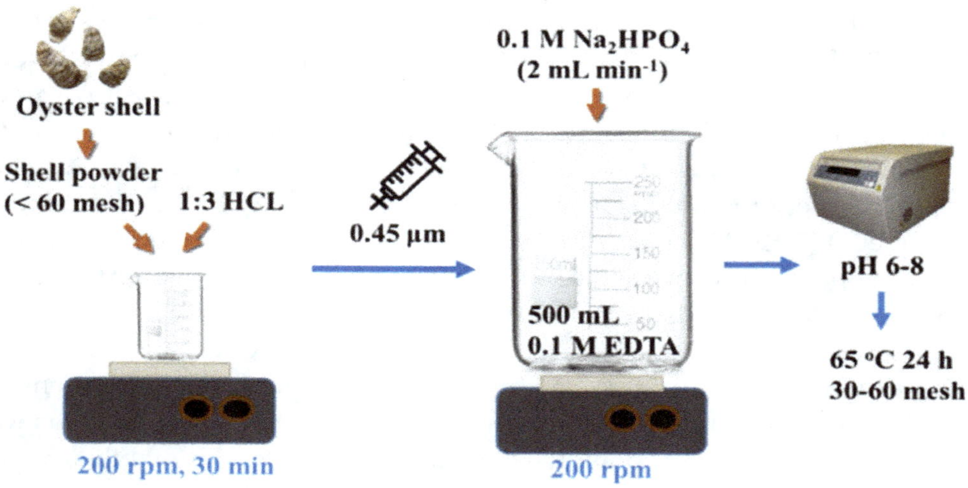

Figure 1. The process of synthesizing HAP.

All of the above materials were sieved to the range of 30–60 mesh (ASTM D2765) [24] and analyzed using a physisorption analyzer (Micromeritics Inc. ASAP 2420, Norcross, GA, USA) to obtain the 77 K N_2 adsorption isotherm. The total surface area was then obtained based on the Brunauer-Emmett-Teller (BET) equation and described as BET surface area (S_{BET}). The chemical compositions of the five materials were analyzed for the C/H/N/S contents (Elementar Vario EL cube, Langenselbold, Germany) and O content by automatic elemental analyzers (Flash 2000, Thermo Fisher Scientific, Waltham, MA, USA).

2.2. *Artificial Fresh Water and Sediment Incubation*

In order to simulate the river environment, artificial freshwater prepared based on the formula provided by USEPA [28] was used in this system, instead of using purified or deionized water. To prepare 20 L of synthetic, moderately hard, reconstituted fresh water, the reagent grade chemicals as follows were used: 1.20 g $MgSO_4$, 1.92 g $NaHCO_3$, 0.080 g KCl, and 1.06 g $CaSO_4$.

The metal incubation concentrations of artificial sediment and reagent grade chemicals were determined with reference to the survey report of Taiwan in 2012 [23] on main rivers with contaminated sediment. The sediment was designed to consist of 400 mg kg^{-1} Ni, 400 mg kg^{-1} Cr, 1000 mg kg^{-1} Cu, 1000 mg kg^{-1} Zn, and 50 mg kg^{-1} Hg by adding $Ni(NO_3)_2{\cdot}6H_2O$, $Cr(NO_3)_3{\cdot}9H_2O$, $Cu(NO_3)_2{\cdot}3H_2O$, $Zn(NO_3)_2{\cdot}6H_2O$, and $HgCl_2$. The sediment (2000 g) was placed in a 3 L glass bottle, filled with artificial freshwater, mixed thoroughly, sealed tightly, and put aside for incubation. The time of incubation lasted up to 60 days.

2.3. *Aqueous Batch Experiments*

The aqueous adsorption experiment was performed according to Bouhamed et al. [1]. To determine the adsorption isotherms of adsorbents, 10, 30, and 50 mg L^{-1} of Ni, Cr, Cu, and Zn were tested by aqueous batch experiments. For Hg, 0.2, 0.6, and 1.0 mg L^{-1} were tested. A serial dilution of standard solution was made to the intended concentrations; the pH value of the solution was controlled to be 5 ± 0.1 by using 0.01 M NaOH and 0.01 M HCl, which simulates the real contaminated sediment environment.

Batch adsorption experiments were carried out in a rotary shaker at 125 rpm using 50 mL capped glass bottles containing 20 mL of metal ion solutions and 25 mg of adsorbents, similar to our previous study [29]. The temperature was controlled at 30 °C and the contact time was up to 24 h to achieve an adsorption equilibrium. All the experimental operations were triplicated.

After shaking, samples were filtered with 0.45 μm filters and the supernatant was kept for metal analyses. For a long-time preservation, each sample of Ni, Cr, Cu, and Zn was preserved with 0.5% HNO_3 and measured by flame atomic absorption spectroscopy (FAAS; Perkin Elmer AAnalyst 800, Waltham, MA, USA); each sample of Hg was preserved with 0.5% BrCl and estimated by cold vapor atomic fluorescence spectroscopy (CVAFS; Brooks Rand Automated Total Mercury System, Seattle, WA, USA). QA/QC were regularly checked by analyzing duplicate samples and quality control samples from each batch.

2.4. Microcosm Experiments

2.4.1. Microcosm Design

The microcosm shown in Figure 2 was designed by modifying the system used in Ting et al. [29]. Multi-columns containing contaminated sediment with mixed caps on the top were set to stimulate the release of metal compounds and examine the efficiencies of capping materials. Water entered the system from the bottom of the column, vertically moved upwards to fill the column, and then discharged to the outside of the system.

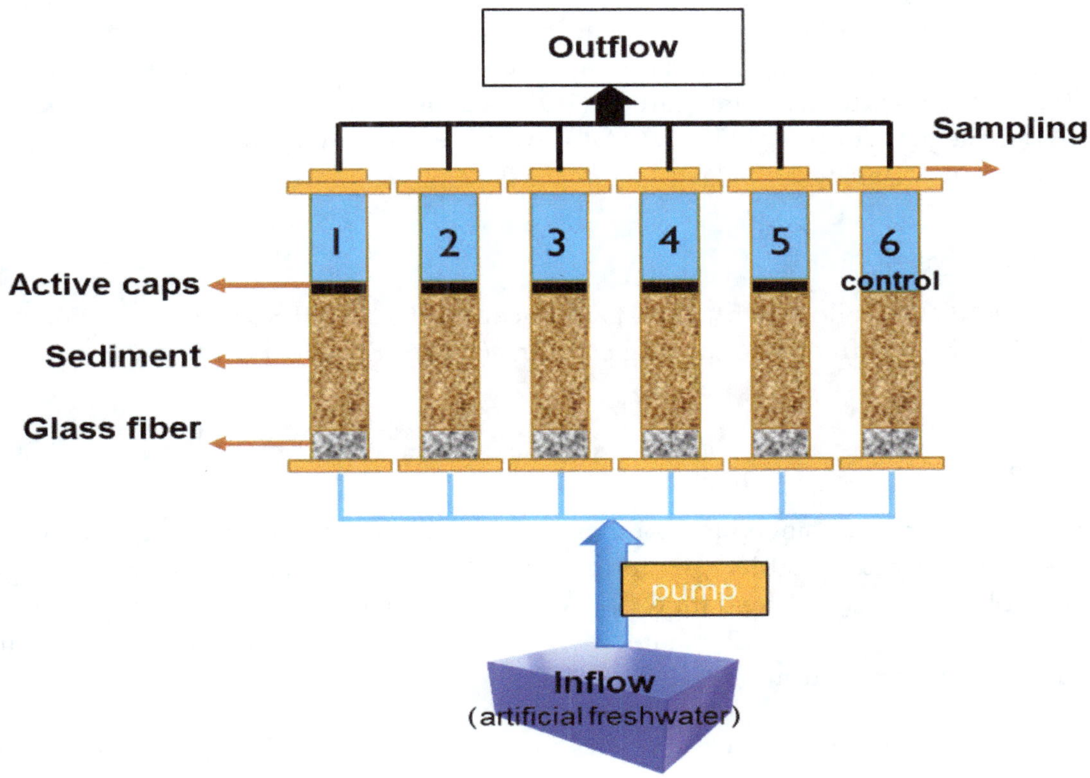

Figure 2. The lab-scale microcosm system.

2.4.2. Microcosm Operation

Six freshwater columns were set up to investigate the efficiencies and stabilities of different mixed caps. Dried incubated sediment was added into the columns and each column contained 200 g of sediment. Five columns were capped by the mixed adsorbents with five different ratios, which were determined according to the results of batch experiments. The sixth column was not capped as the controlled unit. To start up the system, each column was filled with water and allowed to settle for 24 h. Next, the microcosm was activated to start inflow and this day was counted as the operation day 1. The flow rate was maintained to be 0.2 L day^{-1} constantly by using a peristaltic pump (Gamma ST100SV2, Shakopee, MN, USA).

Notably, in order to investigate the stabilization of different mixed caps, the pH of artificial water was adjusted to 3 ± 0.1 by adding HNO_3 during day 101–114 to simulate the extreme condition with acid influent.

2.4.3. Water Sampling and Analyses

Periodic water sampling was conducted continuously once a week for up to two months. At each sampling time, the pH, dissolved oxygen (DO), electrical conductivity (EC), oxidation-reduction potential (ORP) were measured one by one initially. DO of the overlying water was directly measured by a DO meter (EXStik DO600, Nashua, NH, USA). pH value was measured by a pH meter (SunTex SR-2300, Hsi-Chih, Taipei, Taiwan). The electrical conductivity (EC) was measured by a conductivity meter (Taina EZDO-6021, Taiwan). The oxidation–reduction potential (ORP) of sediment within the depth of 1–3 cm of caps was measured by an ORP meter (SunTex SR-2300, Hsi-Chih, Taipei, Taiwan). After the basic measurement, 100 mL of water was collected from each column by disposable syringes and filtered with 0.45 μm filter membranes. The supernatant was kept in 20 mL glass bottles in a refrigerator at 4 °C before analyses.

The samples prepared for Hg analysis were preserved with 0.5% BrCl. For Ni, Cr, Cu, and Zn analyses, samples were preserved with 0.5% HNO_3 and determined by inductively coupled plasma optical emission spectrometry (Agilent ICP-OES 700 series, Santa Clara, CA, USA). For MeHg analysis, samples were put in brown bottles in the shadows to avoid light and analyzed immediately by ethylation, purge and trap, and gas chromatography/CVAFS (Brooks Rand Automated Total Mercury System, Seattle, WA, USA). Total organic carbon (TOC) was analyzed by a TOC analyzer (OI Analytical Aurora 1030W, College Station, TX, USA). QA/QC were regularly checked by analyzing the quality control samples from each batch.

2.4.4. Statistical Analysis

A one-way ANOVA, followed by a least significant difference (LSD) test ($p<0.05$), was used to determine the significance differences among columns (IBM SPSS statistics).

3. Results

3.1. Adsorbents Properties

Five capping adsorbents, sieved to 30–60 mesh, were analyzed and their physical and chemical properties are summarized in Table 1, including the BET surface areas (S_{BET}), average pore sizes, and total pore volumes (V_{total}). The data revealed that HAP had the absolutely largest BET surface area and pore volume among these five materials, and CB had a relatively larger specific surface area and the second largest pore volume.

Table 1. BET surface areas, average pore sizes and total pore volumes (V_{total}) of five materials.

Adsorbent	S_{BET} (m^2 g^{-1})	Pore Size (nm)	V_{total} (cm^3 g^{-1})
Kaolinite	23.3	14.0	0.082
FeS	106.2	60.2	0.042
CB	93.0	17.3	0.376
HAP	367.4	9.2	0.501
OSP	0.06	4.8	0.011

The results of element analyses of the five materials are given in Table 2. FeS was not analyzed and theoretically it is composed of 63.64 wt% of iron and 36.36 wt% of sulfur. It can be seen that CB was rich in C (77.87 wt%) and had a great amount of S (2.53 wt%) because it originated from waste tire. OSP and HAP were rich in O (32.29 and 13.24 wt%, respectively). The proportions of C in OSP roughly corresponded to the proportion of $CaCO_3$ in oyster shell.

Table 2. Elemental analyses of the five test materials.

Adsorbent	C (wt%)	N (wt%)	H (wt%)	O (wt%)	S (wt%)
Kaolinite	0.03	0.15	1.43	10.88	0.05
FeS			-		
CB	77.87	0.46	1.02	1.6	2.53
HAP	0.00	0.11	1.71	13.24	0.00
OSP	11.61	0.20	0.55	32.29	0.07

3.2. *Adsorption Efficiency*

The aqueous metal removal efficiency by the adsorbents was calculated with the following equation:

$$R = \frac{C_0 - C_t}{C_0} \times 100\%, \quad (1)$$

where R (%) is the removal efficiency of adsorbents, C_0 (mg L^{-1}) is the initial metal concentration detected in blank solution, and C_t (mg L^{-1}) is the concentration of remaining sorbate at any time.

Figure 3 and Table 3 show the removal efficiencies of five materials towards five metals at various initial concentrations, which illustrated roughly that OSP and HAP had the best affinities to Cr, Zn, Cu, and Ni, and CB took the third place. As for Cu, HAP performed the largest removal efficiency than the others with a reduction of 98.39%. For Ni, OSP presented a significant removal rate of 76.47% at a low initial concentration and, contrarily, poor efficiency at high initial metal concentrations. For the adsorption of Hg, inversely, FeS showed the best removal efficiency, which was up to 100%. CB was the second excellent one with a removal rate of 86.4%.

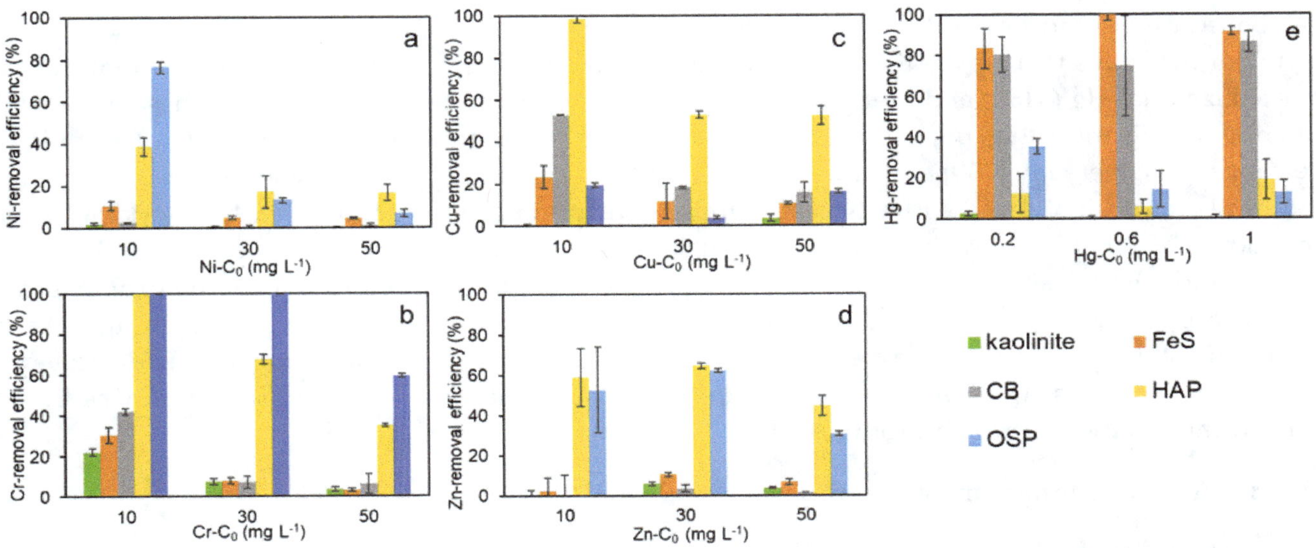

Figure 3. Removal efficiencies of five materials towards (**a**) Ni, (**b**) Cr, (**c**) Cu, (**d**) Zn, and (**e**) Hg at various initial concentrations.

Table 3. Removal efficiencies (in percentage) of five adsorbents for five metals ($n = 3$).

Metal	C_0 (mg L^{-1})	Kaolinite	FeS	CB	HAP	OSP
Ni	10	2.04 ± 0.50	10.73 ± 1.98	2.60 ± 0.27	38.85 ± 4.36	76.47 ± 2.73
	30	0.86 ± 0.12	4.96 ± 0.91	0.60 ± 0.53	17.11 ± 7.56	12.99 ± 1.19
	50	0.41 ± 0.15	4.63 ± 0.48	1.37 ± 0.72	16.62 ± 3.98	6.85 ± 1.81
Cr	10	22.15 ± 1.66	30.65 ± 3.85	42.00 ± 1.41	102.58 ± 0.15	103.13 ± 0.03
	30	7.38 ± 1.61	7.88 ± 1.63	6.98 ± 3.11	67.81 ± 2.30	105.18 ± 0.07
	50	3.73 ± 1.10	3.07 ± 0.77	6.19 ± 5.09	35.06 ± 0.73	59.58 ± 0.90
Cu	10	−1.13 ± 0.95	23.45 ± 5.41	52.80 ± 0.21	98.39 ± 1.65	19.43 ± 1.17
	30	−0.20 ± 0.11	11.89 ± 8.29	18.27 ± 0.41	52.90 ± 1.85	3.82 ± 0.76
	50	3.69 ± 1.83	10.78 ± 0.69	15.91 ± 4.83	52.41 ± 4.46	16.05 ± 1.14
Zn	10	−6.63 ± 2.65	2.09 ± 6.73	−29.32 ± 10.30	58.88 ± 14.26	52.71 ± 21.53
	30	5.86 ± 0.81	10.55 ± 1.03	3.47 ± 1.62	64.56 ± 1.20	62.30 ± 0.98
	50	3.69 ± 0.45	6.71 ± 1.33	1.44 ± 0.00	44.38 ± 5.02	30.62 ± 1.21
Hg	0.2	2.93 ± 1.23	83.37 ± 9.65	80.30 ± 8.55	12.48 ± 9.35	35.32 ± 3.81
	0.6	−2.36 ± 1.15	101.12 ± 3.15	74.70 ± 24.47	5.92 ± 3.47	14.43 ± 8.85
	1	−0.50 ± 1.75	91.51 ± 2.10	86.40 ± 5.09	18.86 ± 9.43	12.86 ± 5.70

3.3. Microcosm

3.3.1. Mixed Caps Design

Five columns covered by different mixed materials and one controlled column were set as shown in Table 4 for conducting the microcosm experiments. The design of capping mixing ratio was based on the integrating consideration of removal efficiencies, cost of preparation, and utilization of recycling resources.

Table 4. Six columns covered by different mixed materials.

Column	1	2	3	4	5	6
kaolinite (wt%)	10	10	10	10	-	
FeS (wt%)	2.5	-	5	2.5	2.5	
CB (wt%)	2.5	5	-	2.5	2.5	-
HAP (wt%)	35	35	35	5	35	
OSP (wt%)	50	50	50	80	50	
Total (wt%)	100	100	100	100	90	

Based on the aforementioned results from batch experiments, HAP and OSP were selected as the major materials to adsorb Ni, Cr, Cu, and Zn; FeS and CB were selected as the major materials to immobilize total Hg (THg) and MeHg. Although the adsorption effect of kaolinite on these five metals was not significant, it has been shifted that kaolinite may probably be more capable of stabilizing the caps than other clays [30] and was therefore added for the purpose of resisting flow disturbance. Column 1 was designed as the ideally best column with appropriate amounts of each material. Column 2 replaced FeS completely with CB and Column 3 did the opposite for comparing FeS with CB. As far as costs and complexity of preparation were concerned, Column 4 tried to reduce the proportion of HAP to grope for a cheaper mixed amendment with equal ability. The cap ratio of Column 5 was almost the same as Column 1, except for the absence of 10 wt% kaolinite, which was designed to investigate the necessity of existence of kaolinite. The last column was composed of only sediment without caps as the controlled group.

3.3.2. Results of Sediment Incubation

The concentrations of the five metals before and after 60 days of incubation and the concentrations of the five metals in the supernatant after incubation are shown in Table 5. Concentrations of metals in the sediment after incubation had almost met the experimental requirements. It is worth mentioning that the concentrations of Ni and Zn of supernatant were much higher than those of the other three metals.

Table 5. Concentrations of the five metals in sediment before and after 60 days of incubation and the concentrations of the five metals in the supernatant after incubation.

Metal	Sediment (mg kg^{-1})			Supernatant (mg L^{-1})
	Before	Design	After	
Ni	61.35 ± 1.46	400	519.59 ± 9.52	133.46 ± 0.98
Cr	102.16 ± 0.63	400	593.96 ± 2.81	0.16 ± 0.002
Cu	94.64 ± 2.38	1000	1383.30 ± 6.58	17.71 ± 0.12
Zn	373.70 ± 12.18	1000	1579.13 ± 14.31	358.79 ± 1.06
THg	0.20 ± 0.03	50	72.32 ± 1.96	0. 03 ± 0.008
MeHg	$1.50 \times 10^{-4} \pm 2.54 \times 10^{-5}$	–	$8.56 \times 10^{-3} \pm 3.23 \times 10^{-3}$	–

3.3.3. pH and ORP

Figure 4 shows the changes of oxidation-reduction potential (ORP) and water pH in the microcosm during the operation. The data of changes of electrical conductivity (EC), pH, dissolved oxygen (DO), and oxidation-reduction potential (ORP) are also shown in Tables S1–S4. Higher pH values and lower ORP in columns capped with materials were attained as compared to the one without caps. However, the differences tended to be slight during the later stage.

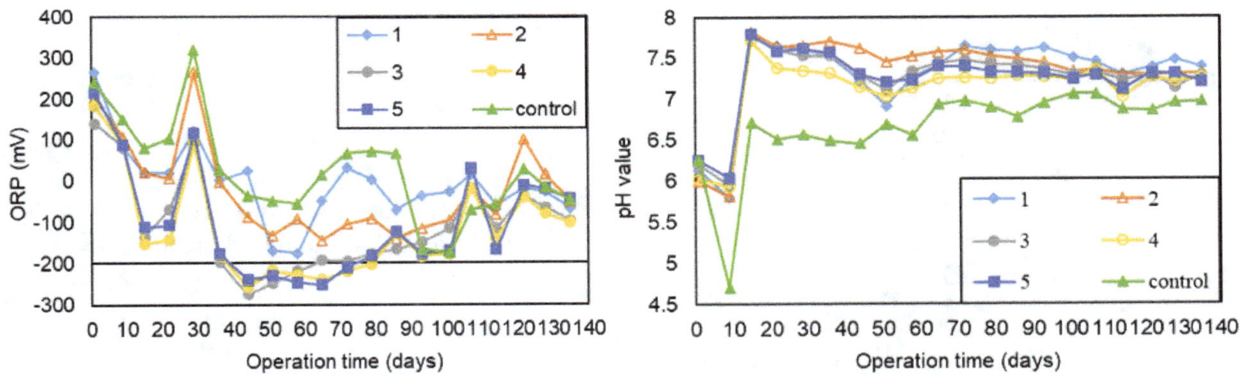

Figure 4. The changes of oxidation-reduction potential (ORP) and water pH values in microcosm during the operation.

3.3.4. Metal Immobilization

The concentrations of five metals and MeHg are shown in Figure 5 and Tables S5–S10 (in Supplementary Material). The results of one-way ANOVA and LSD test ($p < 0.05$) are shown in Tables 6 and 7. In consistence with the expectation, Column 1 showed the most considerable comprehensive consequences of inhibiting the release of metals to the overlying water. It was unexpected that even without 10 wt% of kaolinite, Column 5 achieved almost the same results as Column 1, which is also proved by no significant ($p > 0.05$) correlation in ANOVA (Table 6) between the two columns for all metals, indicating that kaolinite did not show greater stabilization ability for mixed caps as compared to the other test materials.

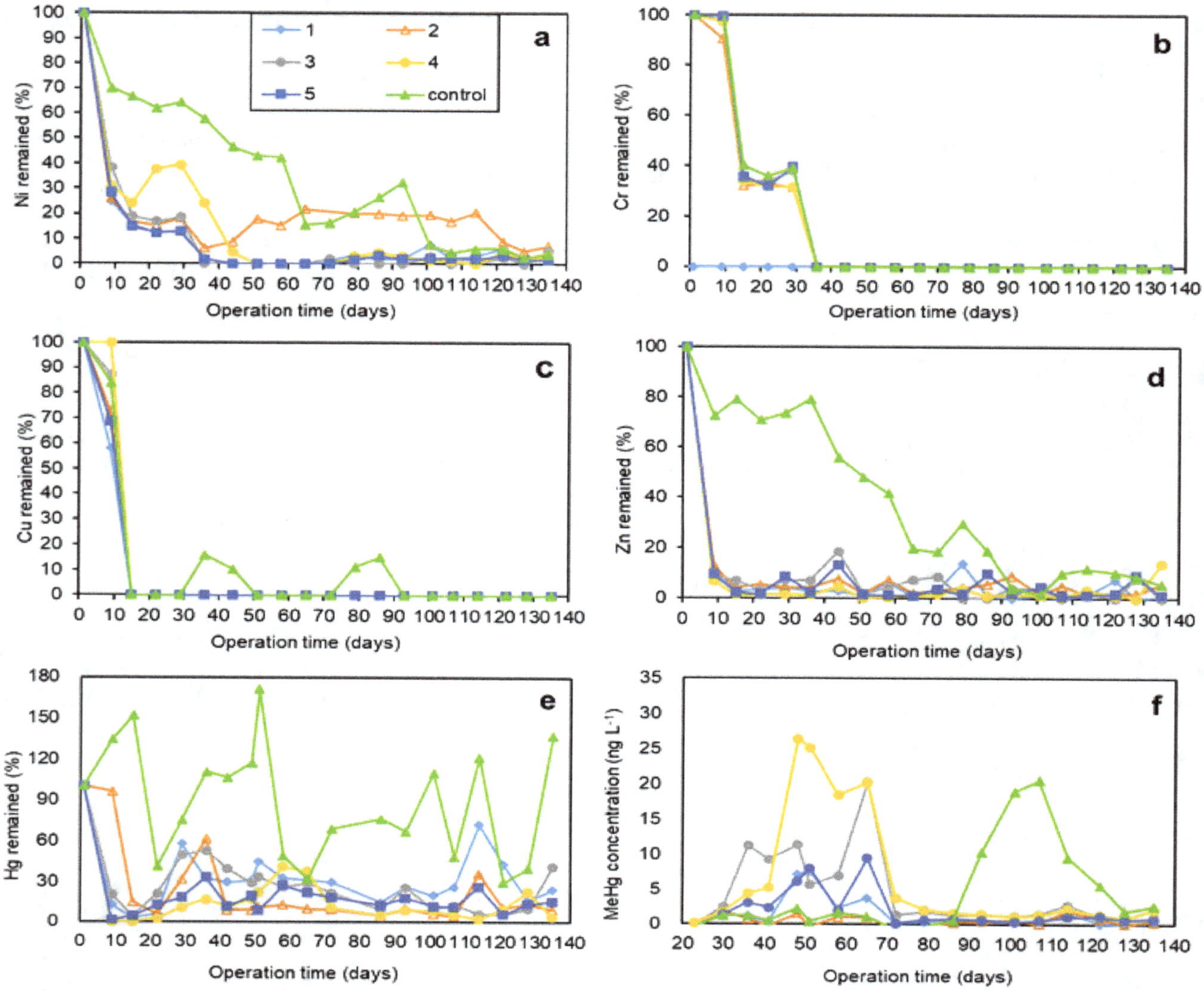

Figure 5. The remaining proportion (%) of (**a**) Ni, (**b**) Cr, (**c**) Cu, (**d**) Zn, (**e**) Hg, and (**f**) the concentration of MeHg in six columns. The remaining proportion in aqueous phase was calculated through dividing the concentration of each contaminant by the concentration on the first day.

Table 6. Metal concentrations in six columns before adding acid (day 1–100).

Column	Ni (n = 14) (mg L^{-1})	Cr (n = 14) (mg L^{-1})	Cu (n = 14) (mg L^{-1})	Zn (n = 14) (mg L^{-1})	THg (n = 14) (μg L^{-1})	MeHg (n = 12) (ng L^{-1})
1	0.091 ± 0.189^{b}	0.000 ± 0.000^{a}	0.013 ± 0.036^{a}	0.150 ± 0.354^{b}	0.145 ± 0.109^{c}	2.09 ± 2.68^{bc}
2	0.162 ± 0.157^{b}	0.015 ± 0.025^{a}	0.012 ± 0.030^{a}	0.140 ± 0.289^{b}	0.382 ± 0.463^{ab}	0.535 ± 0.622^{c}
3	0.064 ± 0.126^{b}	0.015 ± 0.025^{a}	0.010 ± 0.027^{a}	0.090 ± 0.182^{b}	0.233 ± 0.166^{bc}	6.06 ± 6.09^{ab}
4	0.140 ± 0.200^{b}	0.016 ± 0.026^{a}	0.012 ± 0.031^{a}	0.131 ± 0.385^{b}	0.288 ± 0.373^{abc}	9.18 ± 10.2^{a}
5	0.079 ± 0.165^{b}	0.015 ± 0.025^{a}	0.012 ± 0.031^{a}	0.135 ± 0.311^{b}	0.178 ± 0.194^{bc}	2.85 ± 3.25^{bc}
6	0.315 ± 0.161^{a}	0.015 ± 0.025^{a}	0.018 ± 0.035^{a}	0.498 ± 0.288^{a}	0.496 ± 0.222^{a}	1.56 ± 2.85^{c}

Different letters for metal concentrations in six columns indicate a significant difference at $p < 0.05$. For example, the concentrations of Ni in Column 1 with letter b are significantly different from the concentrations of Ni in Column 6 with letter a.

When comparing Column 1 (50% OSP + 35% HAP) with Column 4 (80% OSP + 5% HAP), it is demonstrated by Figure 5 that the releases of Ni, Cr, Cu, and Zn were inhibited well in both columns. However, Column 4 with less proportion of HAP could not achieve an equal effect to Column 1 for Ni, Cr and MeHg immobilization (Figure 5), and a significant correlation was observed by ANOVA ($p < 0.05$) (Table 6) in MeHg concentrations of the two columns (Table 6).

Table 7. Metal concentrations in six columns after adding acid (day 101–135)

Column	Ni (n = 6) (mg L^{-1})	Zn (n = 6) (mg L^{-1})	THg (n = 6) (μg L^{-1})	MeHg (n = 6) (ng L^{-1})
1	0.029 ± 0.017^{ab}	0.024 ± 0.038^{b}	0.151 ± 0.094^{b}	0.453 ± 0.544^{b}
2	0.029 ± 0.017^{ab}	0.033 ± 0.022^{ab}	0.182 ± 0.162^{b}	0.605 ± 0.672^{b}
3	0.009 ± 0.010^{c}	0.017 ± 0.023^{b}	0.100 ± 0.096^{b}	1.29 ± 0.729^{b}
4	0.016 ± 0.012^{bc}	0.049 ± 0.076^{ab}	0.121 ± 0.106^{b}	1.42 ± 0.532^{b}
5	0.015 ± 0.005^{bc}	0.038 ± 0.039^{ab}	0.113 ± 0.057^{b}	0.721 ± 0.311^{b}
6	0.035 ± 0.011^{a}	0.078 ± 0.036^{a}	0.433 ± 0.251^{a}	9.80 ± 8.20^{a}

Different letters for metal concentrations in six columns indicate a significant difference at $p < 0.05$. For example, the concentrations of Ni in Column 1 with letter a and b are significantly different from the concentrations of Ni in Column 3 with letter c.

To compare FeS with CB, 5% FeS in Column 3 made it more effective in immobilizing Ni than 5% CB in Column 2. On the other hand, Column 3 did worse than Column 2 for lowering the concentration of MeHg in the long term, which was demonstrated doubly by Figure 5 and statistical analysis (Table 6).

The stabilization of different mixed caps under extreme condition with acid influent during day 101–114 is also shown in Figure 5. It was observed that the concentrations of Hg and MeHg in every column increased during the days with acid influent (Figure 5f), especially for the controlled column without capping (Table 7). Also, Column 6 released the most amount of Zn during day 101–114 (Figure 5d). The experimental results also suggested that columns with caps were less affected by acid influent than the controlled one, which could be supported by strong significant correlations observed by ANOVA between Column 6 and the other columns.

3.3.5. Total Organic Matter (TOC)

Figure 6 shows that the amount of TOC was positively related to the concentration of MeHg. Column 4, with the largest percentage of OSP, released maximum MeHg during day 20–80. In a later stage from approximately day 85, however, both MeHg and TOC concentrations in Column 6 increased suddenly.

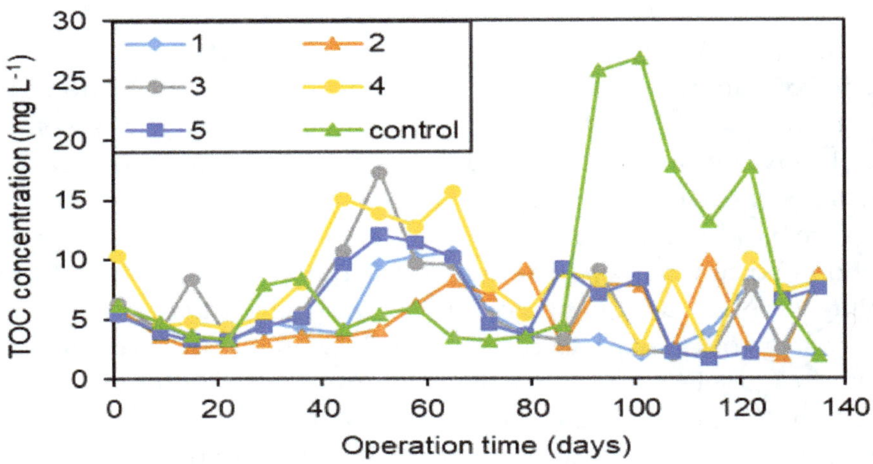

Figure 6. Total organic carbon (TOC) in six columns during the operation

4. Discussion

The data of material properties showed the absolutely largest BET surface area and pore volume of HAP, CB, and FeS, which may lead to its excellent performance during adsorption due to more adsorption sites. In contrast, OSP is a non-porous material with insignificant surface area and pore volume, but OSP still had appreciable adsorption performance for several metals, suggesting that the adsorption behavior of OSP is different with its derived HAP. Abundant C content in CB indicated its additional benefits as an active material [31]. HAP was rich in H and O, probably as a result of the richness of hydroxyl groups on its surface [32], contributing to its high ability as a metal scavenger [33].

In the part of batch experiments, OSP and HAP had the optimum removal efficiencies for Cr, Zn, Cu, and Ni. HAP is regarded as an ideal material with large specific surface and high stability under both reducing and oxidizing conditions [16]. It can release phosphate to interact with metals, forming metal phosphates of low water solubility [34]. OSP has been proven to be able to raise the pH value when applied to soil [35]. If the pH is above the value that promotes metal precipitation, the removal mechanisms will be related to precipitation [36]. However, it is known that the hydroxides of Ni have the highest solubility product (K_{sp}) among those of the five metals, making it comparatively difficult for Ni to be removed by precipitation. As for Hg, FeS and CB showed the best immobilizing ability due to their high Hg affinity. Notably, the CB used in this study was a waste tire recycled product, which contains a significant amount of S that can form chemical bonding with Hg.

The metal potential leachability could determine their environmental risks, and it has been reported to decrease in the order of Zn > Ni > Cu > Cr at pH 4 [37], which corresponded to the concentrations of the five metals in the supernatant after incubation, revealed stronger leachability of Ni and Zn, and was related to their behaviors in the microcosm.

It is generally accepted that sediment pH is an important factor in the adsorption of adsorbate on adsorbent. Under low pH condition, most metal ions are in the cationic state in the solution. The hydrogen ions can compete for the adsorption sites with the metal ions, influencing the exchange adsorption of potential toxic metals, promoting the desorption of metal ions and causing a higher release rate [38,39]. The ability of OSP to raise the concentration of hydroxide ions mentioned before may account for the increase of pH in the columns and its high removal capacity of metals. As one of the most important factors influencing the mobility of metals, the increase of ORP in sediment will correspondingly promote the oxidization of metal sulfides and the degradation of organic compounds, both accelerating the release of the adsorbed/complexing metals [40]. It appeared that ORP of Column 6 was higher than the other five capped in general, mainly due to the decrease of dissolved oxygen caused by caps in the other five columns.

In the part of the lab-scale microcosm experiments, the data from Column 5 indicated the unnecessary existence of kaolinite whether for adsorption or for stabilization in such a condition, which is inconsistent with the suggestions from some earlier studies. The reason may be due to that HAP and OSP could play the same role as kaolinite to stabilize the caps. The difference between Column 1 and Column 4 shown in Figure 5 was related to different performances of HAP and OSP to remove Ni as Figure 3a illustrated. Nevertheless, using caps mainly composed of OSP can yet be regarded as a good choice when considering costs. In batch adsorption, FeS showed the third best adsorption ability for Ni by a reduction of 10.73%, and yet CB hardly worked (Figure 3a). Thus, the release in the later stage of the experiment of Ni in Column 2 may be due to the ineffectiveness of CB. In contrast, Column 3 (5% FeS) did worse than Column 2 (5% CB) for lowering the concentration of MeHg in the long term (Figure 3f). FeS can reduce the concentration of soluble Hg species, resulting in less methylation of Hg [41]. Moreover, part of FeS was likely to be converted into insoluble HgS and taken out by the overlying water, which may explain the unsuccessful results of Column 3 to immobilize MeHg.

The relationship between TOC and MeHg during day 20–80 was likely caused by the remaining organic matter existing in OSP. It has been observed that organic matter content seemed to play a critical role for MeHg formation, acting as electron donor for Hg methylation bacteria [42,43]. The sediment in Column 4 with more OSP would likely release a greater amount of MeHg. Without capping, TOC was possibly easier to release from sediment, which then caused the uprush of MeHg in Column 6 from day 85.

5. Conclusions

Based on the results obtained from the aqueous batch experiments, OSP and HAP had the optimum removal efficiencies to Cr, Zn, Cu, and Ni with CB taking the third place. As for Hg, FeS and CB showed the best immobilizing ability. Kaolinite presented the weakest removal performance towards all of these five metals, and lack of the presence of kaolinite did not show significant influence on the hydraulic stability of mixed caps.

Based on the lab-scale microcosm experiments, the mixing ratio of 10% kaolinite + 2.5% FeS + 2.5% CB + 35% HAP + 50% OSP (i.e., Column 5) had the most prominent effect to immobilize the five metals present in the test sediment. Although Column 4 (the lowest cost one) with 80% OSP showed unsuccessful results for reducing Ni and MeHg, it performed well in inhibiting the release of the other metals. When considering cost effectiveness and environmental impact, using caps mainly composed of OSP can yet be a good choice. But when applied to real sites, decisions should be made after comprehensive evaluations based on actual conditions.

This study helps to construct guidelines of using mixed materials mostly prepared from recycled materials to remediate multi-contaminated sediments and provide some references for active capping application and development. Results from this study would also be helpful in reducing the human health and ecological risks by reducing the potential toxic metal release from sediment to overlying water.

Supplementary Materials
Figure S1: title, Table S1: Oxidation reduction potential (ORP) during operation, Table S2: pH values of microcosm during operation, Table S3: Dissolved oxygen (DO) (mg L^{-1}) of the overlying water in the microcosm, Table S4: Electrical conductivity (EC) of the overlying water in the microcosm, Table S5: THg concentration of the overlying water in the microcosm, Table S6: MeHg concentration of the overlying water in the microcosm, Table S7: Ni concentration of the overlying water in the microcosm, Table S8: Cr concentration of the overlying water in the microcosm, Table S9: Cu concentration of the overlying water in the microcosm, Table S10: Zn concentration of the overlying water in the microcosm.

Author Contributions: Conceptualization, M.-Y.O., Y.T., T.-C.C., and H.-C.H.; methodology, M.-Y.O., Y.T., B.-L.C., and C.C.; formal analysis, M.-Y.O., Y.T., B.-L.C., and Y.-H.C.; investigation, M.-Y.O.; data curation, M.-Y.O. and Y.-H.C.; writing—original draft preparation, M.-Y.O.; writing—review and editing, T.-C.C. and H.-C.H.; visualization, M.-Y.O.; funding acquisition, H.-C.H. All authors have read and agreed to the published version of the manuscript.

Acknowledgments: We are very grateful to all members of the project team and the Groundwater Pollution Remediation Funds of Taiwan Environmental Protection Administration for their funding support.

References

1. Bouhamed, F.; Elouear, Z.; Bouzid, J.; Ouddane, B. Multi-component adsorption of copper, nickel and zinc from aqueous solutions onto activated carbon prepared from date stones. *Environ. Sci. Pollut. Res.* **2016**, *23*, 15801–15806. [CrossRef]
2. Kutuniva, J.; Mäkinen, J.; Kauppila, T.; Karppinen, A.; Hellsten, S.; Luukkonen, T.; Lassi, U. Geopolymers as active capping materials for in situ remediation of metal (loid)-contaminated lake sediments. *J. Environ. Chem. Eng.* **2019**, *7*, 102852. [CrossRef]
3. Zhang, C.; Zhu, M.Y.; Zeng, G.M.; Yu, Z.G.; Cui, F.; Yang, Z.Z.; Shen, L.Q. Active capping technology: A new environmental remediation of contaminated sediment. *Environ. Sci. Pollut. Res.* **2016**, *23*, 4370–4386. [CrossRef] [PubMed]
4. Race, M.; Nabelkova, J.; Fabbricino, M.; Pirozzi, F.; Raia, P. Analysis of heavy metal sources for urban creeks in the Czech Republic. *Water Air Soil Pollut.* **2015**, *226*, 322. [CrossRef]
5. Salomons, W.; De Rooij, N.M.; Kerdijk, H.; Bril, J. Sediments as a source for contaminants? *Hydrobiologia* **1987**, *149*, 13–30. [CrossRef]

6. Zoumis, T.; Schmidt, A.; Grigorova, L.; Calmano, W. Contaminants in sediments: Remobilisation and demobilisation. *Sci. Total Environ.* **2001**, *266*, 195–202. [CrossRef]
7. Hsu, T.C. Experimental assessment of adsorption of Cu^{2+} and Ni^{2+} from aqueous solution by oyster shell powder. *J. Hazard. Mater.* **2009**, *171*, 995–1000. [CrossRef]
8. Jiang, M.Q.; Jin, X.Y.; Lu, X.Q.; Chen, Z.L. Adsorption of Pb (II), Cd (II), Ni (II) and Cu (II) onto natural kaolinite clay. *Desalination* **2010**, *252*, 33–39. [CrossRef]
9. Latifi, Z.; Jalali, M. Measuring and simulating Co (II) sorption on waste calcite, zeolite and kaolinite. *Nat. Resour. Res.* **2019**, 1–15. [CrossRef]
10. Jalees, M.I.; Farooq, M.U.; Basheer, S.; Asghar, S. Removal of Heavy Metals from Drinking Water Using Chikni Mitti (Kaolinite): Isotherm and Kinetics. *Arabian J. Sci. Eng.* **2019**, *44*, 6351–6359. [CrossRef]
11. Alasadi, A.; Khaili, F.; Awwad, A. Adsorption of Cu (II), Ni (II) and Zn (II) ions by nano kaolinite: Thermodynamics and kinetics studies. *Chem. Int.* **2019**, *5*, 258–268.
12. Jeong, H.Y.; Klaue, B.; Blum, J.D.; Hayes, K.F. Sorption of mercuric ion by synthetic nanocrystalline mackinawite (FeS). *Environ. Sci. Technol.* **2007**, *44*, 7699–7705. [CrossRef] [PubMed]
13. Xiong, Z.; He, F.; Zhao, D.; Barnett, M.O. Immobilization of mercury in sediment using stabilized iron sulfide nanoparticles. *Water Res.* **2009**, *43*, 5171–5179. [CrossRef] [PubMed]
14. Feng, Y.; Gong, J.L.; Zeng, G.M.; Niu, Q.Y.; Zhang, H.Y.; Niu, C.G.; Yan, M.; Deng, J.-H. Adsorption of Cd (II) and Zn (II) from aqueous solutions using magnetic hydroxyapatite nanoparticles as adsorbents. *Chem. Eng. J.* **2010**, *162*, 487–494. [CrossRef]
15. Jang, S.H.; Min, B.G.; Jeong, Y.G.; Lyoo, W.S.; Lee, S.C. Removal of lead ions in aqueous solution by hydroxyapatite/polyurethane composite foams. *J. Hazard. Mater.* **2008**, *152*, 1285–1292. [CrossRef]
16. Krestou, A.; Xenidis, A.; Panias, D. Mechanism of aqueous uranium (VI) uptake by hydroxyapatite. *Miner. Eng.* **2004**, *17*, 373–381. [CrossRef]
17. Narwade, V.N.; Khairnar, R.S.; Kokol, V. In situ synthesized hydroxyapatite—Cellulose nanofibrils as biosorbents for heavy metal ions removal. *J. Polym. Environ.* **2018**, *26*, 2130–2141. [CrossRef]
18. Kumar, G.S.; Girija, E.K.; Venkatesh, M.; Karunakaran, G.; Kolesnikov, E.; Kuznetsov, D. One step method to synthesize flower-like hydroxyapatite architecture using mussel shell bio-waste as a calcium source. *Ceram. Int.* **2017**, *43*, 3457–3461. [CrossRef]
19. Wu, Q.; Chen, J.; Clark, M.; Yu, Y. Adsorption of copper to different biogenic oyster shell structures. *Appl. Surf. Sci.* **2014**, *311*, 264–272. [CrossRef]
20. Moon, D.H.; Wazne, M.; Cheong, K.H.; Chang, Y.Y.; Baek, K.; Ok, Y.S.; Park, J.H. Stabilization of As-, Pb-, and Cu-contaminated soil using calcined oyster shells and steel slag. *Environ. Sci. Pollut. Res.* **2015**, *22*, 11162–11169. [CrossRef]
21. Ahmad, M.; Lee, S.S.; Yang, J.E.; Ro, H.M.; Lee, Y.H.; Ok, Y.S. Effects of soil dilution and amendments (mussel shell, cow bone, and biochar) on Pb availability and phytotoxicity in military shooting range soil. *Ecotoxicol. Environ. Saf.* **2012**, *79*, 225–231. [CrossRef] [PubMed]
22. Ahmad, M.; Lee, S.S.; Lim, J.E.; Lee, S.E.; Cho, J.S.; Moon, D.H.; Hashimoto, Y.; Ok, Y.S. Speciation and phytoavailability of lead and antimony in a small arms range soil amended with mussel shell, cow bone and biochar: EXAFS spectroscopy and chemical extractions. *Chemosphere.* **2014**, *95*, 433–441. [CrossRef] [PubMed]
23. EPA Taiwan. *Investigation of Sediment Pollution Sources and Transmission Mode-Taking Key Rivers as Examples*; Taiwan Environmental Protection Administration: Taipei, Taiwan, 2011; EPA-100-GA102-02-A232.
24. ASTM D2765-16. *Standard Test Methods for Determination of Gel Content and Swell Ratio of Crosslinked Ethylene Plastics*; ASTM International: West Conshohocken, PA, USA, 2016; Available online: www.astm.org (accessed on 1 July 2020).
25. Wu, S.C.; Tsou, H.K.; Hsu, H.C.; Hsu, S.K.; Liou, S.P.; Ho, W.F. A hydrothermal synthesis of eggshell and fruit waste extract to produce nanosized hydroxyapatite. *Ceram. Int.* **2013**, *39*, 8183–8188. [CrossRef]
26. Trinkunaite–Felsen, J.; Prichodko, A.; Semasko, M.; Skaudzius, R.; Beganskiene, A.; Kareiva, A. Synthesis and characterization of iron-doped/substituted calcium hydroxyapatite from seashells Macoma balthica (L.). *Adv. Powder Technol.* **2015**, *26*, 1287–1293. [CrossRef]
27. Sobczak–Kupiec, A.; Malina, D.; Kijkowska, R.; Wzorek, Z. Comparative study of hydroxyapatite prepared by the authors with selected commercially available ceramics. *Digest J. Nanomater. Biostructures* **2012**, *7*, 385–391.

28. Lewis, P.A. *Short-Term Methods for Estimating the Chronic Toxicity of Effluents and Receiving Waters to Freshwater Organisms*; EPA, Environmental Monitoring Systems Laboratory: Cincinnati, OH, USA, 1994.
29. Ting, Y.; Chen, C.; Ch'ng, B.L.; Wang, Y.L.; Hsi, H.C. Using raw and sulfur-impregnated activated carbon as active cap for leaching inhibition of mercury and methylmercury from contaminated sediment. *J. Hazard. Mater.* **2018**, *354*, 116–124. [CrossRef]
30. Ting, Y.; Ch'ng, B.L.; Chen, C.; Ou, M.Y.; Cheng, Y.H.; Hsu, C.J.; Hsi, H.C. A simulation study of mercury immobilization in estuary sediment microcosm by activated carbon/clay-based thin-layer capping under artificial flow and turbation. *Sci. Total Environ.* **2020**, *708*, 135068. [CrossRef]
31. Kwapinski, W.; Byrne, C.M.; Kryachko, E.; Wolfram, P.; Adley, C.; Leahy, J.J.; Hayes, M.H.; Novotny, E. Biochar from biomass and waste. *Waste Biomass Valoriz.* **2010**, *1*, 177–189. [CrossRef]
32. Kay, M.I.; Young, R.A.; Posner, A.S. Crystal structure of hydroxyapatite. *Nature* **1964**, *204*, 1050–1052. [CrossRef]
33. Wang, X.; Yu, R.; Wang, P.; Chen, F.; Yu, H. Co-modification of F− and Fe (III) ions as a facile strategy towards effective separation of photogenerated electrons and holes. *Appl. Surf. Sci.* **2015**, *351*, 66–73. [CrossRef]
34. Fuller, C.C.; Bargar, J.R.; Davis, J.A.; Piana, M.J. Mechanisms of uranium interactions with hydroxyapatite: Implications for groundwater remediation. *Environ. Sci. Technol.* **2002**, *36*, 158–165. [CrossRef] [PubMed]
35. Lee, C.H.; Lee, D.K.; Ali, M.A.; Kim, P.J. Effects of oyster shell on soil chemical and biological properties and cabbage productivity as a liming materials. *Waste Manag.* **2008**, *28*, 2702–2708. [CrossRef] [PubMed]
36. Glatstein, D.A.; Francisca, F.M. Influence of pH and ionic strength on Cd, Cu and Pb removal from water by adsorption in Na-bentonite. *Appl. Clay Sci.* **2015**, *118*, 61–67. [CrossRef]
37. Singh, S.P.; Ma, L.Q.; Tack, F.M.; Verloo, M.G. Trace metal leachability of land-Disposed dredged sediments. *J. Environ. Qual.* **2000**, *29*, 1124–1132. [CrossRef]
38. Li, H.; Shi, A.; Li, M.; Zhang, X. Effect of pH, temperature, dissolved oxygen, and flow rate of overlying water on heavy metals release from storm sewer sediments. *J. Chem.* **2013**, 434012. [CrossRef]
39. Zhai, X.; Li, Z.; Huang, B.; Luo, N.; Huang, M.; Zhang, Q.; Zeng, G. Remediation of multiple heavy metal-contaminated soil through the combination of soil washing and in situ immobilization. *Sci. Total Environ.* **2018**, *635*, 92–99. [CrossRef]
40. Calmano, W.; Hong, J.; Förstner, U. Binding and mobilization of heavy metals in contaminated sediments affected by pH and redox potential. *Water Sci. Technol.* **1993**, *28*, 223–235. [CrossRef]
41. Mehrotra, A.S.; Horne, A.J.; Sedlak, D.L. Reduction of net mercury methylation by iron in Desulfobulbus propionicus (1pr3) cultures: Implications for engineered wetlands. *Environ. Sci. Technol.* **2003**, *37*, 3018–3023. [CrossRef]
42. Parks, J.M.; Johs, A.; Podar, M.; Bridou, R.; Hurt, R.A.; Smith, S.D.; Brandt, C.C.; Palumbo, A.V. The genetic basis for bacterial mercury methylation. *Science* **2013**, *339*, 1332–1335. [CrossRef]
43. Bravo, A.G.; Bouchet, S.; Tolu, J.; Björn, E.; Mateos-Rivera, A.; Bertilsson, S. Molecular composition of organic matter controls methylmercury formation in boreal lakes. *Nat. Commun.* **2017**, *8*, 1–9. [CrossRef]

3

AQUALIFE Software: A New Tool for a Standardized Ecological Assessment of Groundwater Dependent Ecosystems

Giovanni Strona [1], Simone Fattorini [2], Barbara Fiasca [2], Tiziana Di Lorenzo [3], Mattia Di Cicco [2], Walter Lorenzetti [4], Francesco Boccacci [4] and Diana M. P. Galassi [2,*]

[1] Research Centre for Ecological Change, University of Helsinki, P.O. Box 4 FI-00014 Helsinki, Finland; giovanni.strona@helsinki.fi

[2] Department of Life, Health and Environmental Sciences, University of L'Aquila, 67100 L'Aquila, Italy; simone.fattorini@univaq.it (S.F.); barbara.fiasca@univaq.it (B.F.); diciccomattiauni@yahoo.it (M.D.C.)

[3] Italian National Research Council, Research Institute on Terrestrial Ecosystems, Via Madonna del Piano 10, Sesto Fiorentino, 50019 Florence, Italy; tiziana.dilorenzo@cnr.it

[4] Gis3W S.r.l, Viale Verdi 24, Montecatini Terme, 51016 Pistoia, Italy; lorenzetti@gis3w.it (W.L.); boccacci.francesco@gmail.com (F.B.)

* Correspondence: dianamariapaola.galassi@univaq.it

Abstract: We introduce a suite of software tools aimed at investigating multiple bio-ecological facets of aquatic Groundwater Dependent Ecosystems (GDEs). The suite focuses on: (1) threats posed by pollutants to GDE invertebrates (Ecological Risk, ER); (2) threats posed by hydrological and hydromorphological alterations on the subsurface zone of lotic systems and groundwater-fed springs (Hydrological-Hydromorphological Risk, HHR); and (3) the conservation priority of GDE communities (Groundwater Biodiversity Concern index, GBC). The ER is assessed by comparing tolerance limits of invertebrate species to specific pollutants with the maximum observed concentration of the same pollutants at the target site(s). Comparison is based on an original, comprehensive dataset including the most updated information on tolerance to 116 pollutants for 474 freshwater invertebrate species. The HHR is assessed by accounting for the main direct and indirect effects on both the hyporheic zone of lotic systems and groundwater-fed springs, and by scoring each impact according to the potential effect on subsurface invertebrates. Finally, the GBC index is computed on the basis of the taxonomical composition of a target community, and allows the evaluation of its conservation priority in comparison to others.

Keywords: AQUALIFE software; groundwater dependent ecosystems; threats; biodiversity

---

1. Introduction

Groundwater dependent ecosystems (GDEs) are ecosystems whose biological structure and ecological processes are directly or indirectly influenced by groundwater [1,2]. They may include springs, wetlands, streams and rivers, hyporheic zones, lakes, coastal lagoons, and saturated aquifers (also indicated as Subsurface Groundwater Dependent Ecosystems, SGDEs). Aquatic GDEs are known to host a high biodiversity, including highly specialized, rare, and often narrow endemic species [3–5]. Preserving their biodiversity is therefore identified as a conservation priority [5–11]. Many GDEs, and hence their biodiversity, are known to be at risk of impairment worldwide [4,7,10,12,13].

Some regulations for the protection of GDEs have recently entered in force in the Australian subcontinent [14,15] and, to a lesser extent, in Europe [10]. Although these ecosystems are cited in Annex II, Section 2, "Groundwater", pt. 2.2. of the EU Directive 2000/60/EC [16], and later in the Directive 2006/118/EC, also called Groundwater Daughter Directive [17], only a few GDEs have been included in Nature 2000 network or designated as Ramsar areas [18,19]. In the United States, a mosaic of laws acts directly or indirectly towards the protection of groundwater and GDEs by controlling contamination, managing conflicts among users, and quantifying abstraction but a comprehensive governance is still missing [20,21]. Proposals for management and recovery of GDEs have been advanced for particular situations, such as mining [12] and groundwater withdrawal [22]. However, current proposals [10] for management actions remain quite vague and protocols to assess the impacts of chemical and hydrological-hydromorphological threats on aquatic GDEs and their conservation priorities are still lacking.

To cope with one of the major challenges of integrating scientific knowledge and environmental management policy, the European Union Life + project "LIFE12 BIO/IT/000231—AQUALIFE" developed a set of procedures and indicators for the assessment of aquatic GDE biodiversity and its threats. The project implemented this set of procedures and indicators into the software suite called "AQUALIFE Software" (hereafter, AS), which we introduce here.

The AS consists of a user-friendly online interface, and it is free to use (after registration) at: http://app.aqualifeproject.eu.

2. Software Capabilities

AS allows users to compute a set of different metrics to evaluate the ecological risk, the hydrological-hydromorphological risk, and the conservation priority of aquatic GDEs. The software has been mainly developed using the Python programming language [23] in combination with the Django web framework [24].

The target for the application of the software is, in principle, a set of monitoring sites and their associated biological communities. The software was generated by assembling a rich biodiversity dataset for selected GDEs in the Abruzzo region (Italy), which was chosen as a pilot area for AS development. Users can use their own biodiversity datasets, together with the concentration of the pollutants measured in their sites and, if working on the hyporheic habitat of streams and rivers or groundwater-fed springs, the hydrological-hydromorphological impacts detected in their study sites. All the AS outputs can be saved online, accessed, and downloaded at any later stage.

AS comprises three sections, namely: Ecological Risk (ER), Hydrological-Hydromorphological Risk (HHR), and Conservation Priority Index (GBC). The first two focus on a particular aspect of the GDEs' vulnerability while the last one on the conservation value of GDE biodiversity. Each section includes three subsections. The first subsection, I—Info, provides general information about the rationale of the main section, as well as technical details regarding the associated computation procedures. The second subsection, R—Risk for the first two sections and I—Index for the third section, allows users to perform the analyses. Finally, the third subsection, H—History, allows users to explore and access saved results from previous analyses.

In the following, we provide a detailed description of the different software sections. In doing that, we mainly focus on the technical aspects of computation procedures and on practical aspects of the user experience. Our aim is to provide stakeholders with enough information to start actively using the software in a fully-conscious manner.

2.1. Section 1: Ecological Risk

The Ecological Risk (ER) quantifies the probability that chemicals produced by human activity (hereafter referred as "pollutants") harm the health status of a biological community inhabiting an aquatic GDE [5,25–28]. Since chemical pollutants usually occur in mixtures in GDEs, ER is quantified as in Equation (1) [29,30]:

$$ER = RQ_{mix} = \Sigma RQ_i = \Sigma \frac{MEC_i}{PNEC_i} \quad (1)$$

where RQ_i is the risk posed by the ith pollutant in the mixture, MEC_i stands for the Measured Environmental Concentration and $PNEC_i$ for the Predicted No-Effect Concentration of the ith pollutant in the mixture. If more environmental concentrations are available for the *i*th pollutant, MEC_i of Equation (1) is equal to the highest concentration measured by the user, to account for the worst-case scenario [31]. $PNEC_i$ is the concentration at which the *i*th pollutant is predicted to have no significant detrimental effect on the GDE community [32]. $PNEC_i$ is computed as follows [31]:

$$PNEC_i = \sqrt[n]{PNEC_1 \times PNEC_2 \times \ldots PNEC_n} \tag{2}$$

where

$$PNEC_n = \frac{LC50(EC50)}{AFa} \tag{3}$$

or

$$PNEC_n = \frac{NOEC}{AFb} \tag{4}$$

$PNEC_i$ is the geometric mean of $PNEC_n$ where $PNEC_n$ is the predicted no-effect concentration of the *i*th pollutant on the *n*th taxon of a GDE community represented by *n* taxa [31,33,34]. LC_{50} (Lethal Concentration 50%) and EC_{50} (Effect Concentration 50%) are the concentrations of the *i*th pollutant producing, respectively, a lethal effect or a maximal effect, on 50% of individuals tested in short-term trials (≤96 h) and NOEC (No-Observed Effect Concentration) is the concentration producing no significant effects under long-term exposures (>96 h). AF is an assessment factor that reflects the uncertainty in extrapolating $PNEC_n$ values from laboratory toxicity tests [27], being AF*a* equal to 1000 and AF*b* equal to 100 [35,36]. If more than one LC_{50} and/or EC_{50} and/or NOEC values are available for the *n*th taxon, the toxicity data are aggregated by computing their geometric mean [31].

AS provides an internal database of PNECs for 116 pollutants known to be potential threats to GDEs; they were derived from 8806 toxicity values related to 474 taxa either found in GDEs or in other freshwater habitats according to information availability (Tables S1 and S2). Toxicity data were retrieved from the US EPA database ECOTOX "Aquatic Data" [37]. When unavailable, LC_{50} values were estimated by US EPA ECOSAR v.2.0 [38]. We used the most updated literature to amend errors (for example, wrong habitat attribution of the records, wrong attribution of the species to the freshwater environment, and wrong species names) as well as to expand the AS database with additional information (e.g., new toxicity data for freshwater species).

In freshwater studies, it is common to focus on tolerance limits of algae, macrophytes, invertebrates, and small fish; however, to assemble the dataset used by AS, we focused only on invertebrates. The rationale behind this choice stems from the circumstantial evidence that in aquatic GDEs there is virtually no primary productivity (except for rare cases of chemoautotrophy [39]) and, as for Europe, no underground fish are known.

AS assigns a risk magnitude and a color to ER based on five classes (Table 1), following the assumption that, if $RQ_{mix} > 1$, an ER is expected [27,28].

Table 1. ER rating, magnitude and color per rank.

ER Rating	ER Magnitude	ER Color Ranking
$0 \leq RQ_{mix} < 1$	no risk	
$1 \leq RQ_{mix} < 10$	low risk	
$10 \leq RQ_{mix} < 100$	medium risk	
$100 \leq RQ_{mix} < 1000$	high risk	
$RQ_{mix} \geq 1000$	very high risk	

Because of the dominance of the crustaceans in certain GDEs (e.g., SGDEs), AS allows users to select one out of two subsets of the toxicity database: (i) "*crustacean species*" (choice suggested for SGDEs such as the saturated aquifers where crustaceans are the dominant group); or (ii) "*freshwater invertebrates*" (choice suggested for springs and the hyporheic zone of streams and rivers, where also non-crustacean species are present, and often with high abundances).

The computation of RQ_{mix} by AS is straightforward. After loading the Ecological Risk page, the user can select the pollutants from a drop-down list, enter the MECs for the pollutants detected in the GDE sites under study, select one out the two subsets of toxicity database, and start the calculation of RQ_{mix} by pressing the "RUN" button. After that, AS will provide the user with both the individual PNEC values of the selected pollutants, and the RQ_{mix} value along with the ER magnitude and color.

AS allows the user to save the analyses and access them at a later moment. To do this, the user will have to associate the analysis to a geographical location (either a location indicated in previous analyses or a new one; in the latter case, the user will have to assign an arbitrary name to the locality and will have the possibility to provide its geographical coordinates, in form of latitudinal/longitudinal decimal degrees). An interactive map will show the position of the target site(s). After that, a "SAVE" button will become available. The last step consists in attributing an identifier name to the analysis, allowing its later retrieval. Saved analyses are accessible by the user in the History Page of the Ecological Risk section. Clearly, users have personalized histories, and do not have access to other user pages. Data from the History subsection can be directly exported to spreadsheet documents for being processed outside the software environment.

2.2. *Section 2: Hydrological-Hydromorphological Risk*

The second section of AS aims at identifying and scoring potential risks posed by man-induced alterations of the hydrological connectivity (Hydrological-Hydromorphological Risk, HHR) of specific GDE types such as hyporheic zones of streams and rivers and groundwater-fed springs. We designed our procedure taking as a model the Index of Morphological Quality and the Morphological Alteration Index, two operational tools of the Italian "Hydromorphological evaluation system, analysis and monitoring of rivers" named IDRAIM [40].

Several factors can potentially alter the hydrological connectivity of the subsurface habitats from surface waters as well as from the underlying aquifer [41,42], critically influencing hyporheic invertebrate communities [43–45]. Among them, important ones are alteration of streamflow dynamics, changes in streambed and river banks structure and morphology, and reduction of streambed sediments heterogeneity (i.e., colmation or clogging [46]), mainly arising from the presence of dams [42,47–49], barriers, water derivation works, and groundwater overexploitation [21] (HHYC = altered habitat due to hydrological changes [50]).

The presence of solid walls and/or embankments limiting permeability with the alluvial plain of streams and rivers, concreted riverbed, presence of knick points in rivers, artificial riverbed aggradation, channelization [51], clogging or colmation [52–54], artificial elevation of the riverbed, and removal of perifluvial vegetation [55,56] may lead to an alteration of the vertical connectivity river/aquifer [48,57] (HMOC = altered habitats due to morphological changes [50]).

Concerning groundwater-fed springs, alteration of the discharge in the eucrenal-hypocrenal of the spring environment, and in the epirhytron (mountain streams) and alteration of the spring morphology affect severely the crenocenosis and may also cause a disconnection between the spring and the stream [58,59]. If the aquifer feeding the spring is overexploited, an alteration of the spring discharge is expected [58,60–62]. Spring casing and/or concretion (piping, impoundment, and diversion of spring flows), removal of riparian vegetation [63] for maintenance works are reflected in HMOC impacts, such as alteration of the spring morphology with interruption of the spring/aquifer and spring/stream connections, habitat losses, and changes in the nutrient dynamics after vegetation removal [64].

The computation of the HHR is based on the pressures generating effective impacts on GDE invertebrate faunas, with a distinction between those insisting on the hyporheic zones (Table 2) and those affecting springs (Table 3). The list of pressures has been compiled based on expert judgment and is available in the INFO subsection of the main menu under the heading "Hydrological-Hydromorphological Risk" (Tables 2 and 3), which includes a description of the pressures and the indication of the impacts, along with the relative category (i.e., impacts altering habitats due to hydrological changes (HHYC) and impacts altering habitats due to morphological changes

(HMOC)) [50]. A cutoff of 10 ("at risk" vs. "no risk") was identified as the threshold value of the overall impact, which corresponds to the maximum partial impact score attributed to the alteration of the river/aquifer connection exerted by different pressures, considering this alteration sufficient to produce the disappearance of a GDE. All the scores assigned to each impact are stored into the AS.

Table 2. Relevant pressures and related impacts for hyporheic habitats. Pressures and impacts are as so in the Guidance No. 35 of the Water Framework Directive Reporting Guidance 2016, Annexes 1b and 8n [50].

<table>
<tr><th colspan="2">Relevant Pressures: Names and Descriptions</th><th colspan="2">Impacts: Names and Categories</th></tr>
<tr><td>Hydrological alteration
Abstraction or flow diversion
Dams, barriers, and locks</td><td>Presence of transversal works with reduction of discharge flow rates (detention and/or diversion, dams, weirs)</td><td>Alteration of discharge flow rates</td><td>HHYC-Altered habitats due to hydrological changes</td></tr>
<tr><td>Hydromorphological alteration</td><td>Presence of transversal works with reduction of discharge flow rates (detention and/or diversion, dams, weirs) limiting sediment and woody debris transport on riverbed</td><td>Alteration of the longitudinal connectivity in sediment and woody debris in streams/rivers benthic and hyporheic zones</td><td rowspan="6">HMOC-Altered habitats due to morphological changes</td></tr>
<tr><td>Abstraction or flow diversion</td><td>Presence of solid walls and/or embankments limiting permeability with the alluvial plain of streams and rivers</td><td>Alteration of the lateral connectivity in streams/rivers benthic and hyporheic zones</td></tr>
<tr><td>Dams, barriers, and locks</td><td>Concreted riverbed, presence of knick points in rivers or channels where the longitudinal (lengthways) slope of the bed suddenly steepens (incision), artificial riverbed aggradation, clogging, overexploitation of aquifers</td><td>Alteration of the vertical connectivity river/aquifer</td></tr>
<tr><td>Physical alteration of channel/bed/riparian area/shore</td><td>Artificially straightened channel</td><td>Morphological alteration of the river stretch and riverbed sediment composition</td></tr>
<tr><td>Groundwater alteration of water level or volume</td><td>Removal of perifluvial or riparian vegetation</td><td>Increase of riverbank erosion</td></tr>
<tr><td></td><td>Large woody debris removal</td><td>Large woody debris removal</td></tr>
</table>

For hyporheic habitats, most of the pressures are to be evaluated in a river stretch of at least 100 m upstream of the target site. As the effects of dams or river derivations can occur several tens of kilometers downstream, the user will assess their effectiveness at the target site.

A similar set of criteria was adopted to assess the HHR for groundwater-fed springs, where the main pressures are represented by hydrological alteration due to abstraction or flow diversion, dams, or barriers downstream from the spring, which may determine HHYC and HMOC impacts (Table 3 and Table 5).

A score was assigned to each impact based on expert judgment, considering the deviation from reference conditions of unmodified rivers or spring systems. Therefore, the score reflects the degree of alteration of a GDE site related to a given pressure (Tables 4 and 5).

Table 3. Relevant pressures and related impacts for groundwater-fed springs are as so in the Guidance No. 35 of the Water Framework Directive Reporting Guidance 2016, Annexes 1b and 8n [50].

Relevant Pressures		**Impacts**	
Hydrological alteration Abstraction or flow diversion Dams, barriers, and locks	Presence of transversal works with reduction of discharge flow rates (detention and/or diversion, dams, weirs) downstream the spring emergence	Alteration of discharge flow rate	HHYC–Altered habitats due to hydrological changes
	Overexploitation of aquifers	Alteration of the vertical connectivity	
Physical alteration of channel/bed/riparian area/shore	Spring used for different purposes or concreted at its emergence	Morphological alteration of spring habitats with loss of the ecotonal nature of GW-fed springs	HMOC–Altered habitats due to morphological changes
	Riparian vegetation removal	Habitat and microhabitat losses, nutrient dynamics alteration, current velocity alteration	

Table 4. Impact scores for hyporheic habitats.

Impacts	**Impact Score**
Alteration of discharge flow rates	6
Alteration of the longitudinal connectivity limiting sediment and woody debris transport on riverbed	4
Alteration of the lateral connectivity	3
Alteration of the vertical connectivity river/aquifer	10
Morphological alteration of the river stretch and riverbed sediment composition	6
Alteration of riparian vegetation	2
Lowering of woody debris storage in streams/rivers	1

Table 5. Impact scores for groundwater-fed springs.

Impacts	**Impact Score**
Alteration of discharge flow rates	6
Alteration of the ecological features of the spring (e.g., change from rheocrenic to limnocrenic spring as effect of damming)	4
Alteration of the vertical connectivity	10
Morphological alteration of spring habitats with loss of the ecotonal nature of GW-fed springs	5
Alteration of the riparian vegetation, habitat and microhabitat losses, nutrient dynamics alteration, current velocity alteration	3

The computation of HHR is performed by AS in two simple steps. In the first step, users are asked to select the target habitat (either hyporheic or spring habitats).

Depending on this first selection, a pre-defined list of pressures and impacts becomes available, that is either Table 2 or Table 3. The second step consists in choosing from this list the impact(s) recorded at the target site (Table 4 or Table 5). After that, the "RUN" button leads the user to the results window, which provides a risk score (Table 6), and a summary of the user selections.

Table 6. HHR rating, risk ranking, and color per rank.

HHR Rating	**Risk Ranking**	**HHR Color Ranking**
HHR < 10	no risk	
HHR ≥ 10	at risk	

As for ER, results can be associated with a geographic location and stored in an internal database that the user can access at any moment through the "History" subsection.

2.3. Section 3: Conservation Priority Index

The third section of AS allows the computation of three conservation indices: Sum of Species Scores (SSS), Biodiversity Conservation Concern (BCC), and Groundwater Biodiversity Concern (GBC). These indices aim at prioritizing GDE communities on the basis of their conservation importance. Although species richness is an obvious criterion to select conservation priority areas, not all species have the same conservation importance. For example, areas that host more imperiled species should be considered more important [65,66]. However, estimating species extinction risk may be extremely difficult, especially for less known invertebrates [67]. For this reason, there is increasing interest in the use of rarity measures as a proxy for the extinction risk of invertebrates [68–70]. Rarity is a relative and multidimensional concept, because a species can be more or less rare compared to other species for various characteristics [71–73]. Once the rarity of a species has been assessed for the various dimensions that are considered, this information can be used to obtain scores that express the overall species' conservation priority. Under these assumptions, AS uses a two-step protocol to assess the conservation priority of GDEs based on invertebrate species. The first step consists in computing the conservation priority of each species occurring in GDE sites based on selected species traits. The second step is to compute site conservation indices using species scores.

The computation of the indices requires users to fill and upload: (1) a dataset where each species (but other taxonomic units can be used) is associated with a set of traits that expresses its conservation importance (the "Dataset" file must contain species in rows, traits in columns); and (2) a matrix of each species abundance across the compared sites (if abundance data are not available, presence/absence can be used) (the "Sampling" file must contain species in rows and sites in columns). Both need to be provided to AS as spreadsheets (.xls). By processing these two input files, AS will compute both a conservation score for each species in the dataset, and three conservation indices for each site (with different scores of the GBC index being associated with different conservation priority categories).

Trait scores are attributed by the user to the collected species for five dimensions (geography, ecology, biology, population, and evolutionary history), each divided into different traits (Table 7). For each trait, the higher is the score, the higher is the species' conservation importance. Different traits have different score ranges. For example, in the "Geography" dimension, the trait "Degree of Endemicity" can obtain a score from 1 to 4, while the trait "Extent of occurrence" can obtain a score from 1 to 7 [74].

Table 7. Traits and their scores used to calculate the Conservation Priority Index.

Dimension	Trait	Score Range
1. GEOGRAPHY	Extent of occurrence	1–7
	Degree of endemicity	1–4
2. ECOLOGY	Degree of GW dependence	1–5
	Ecological specialization to GW	1–3
	Size of ecological niche	1–5
	Tolerance to variation of climatic parameters	1–5
	Microhabitat preferences	1–5
3. BIOLOGY	Trophic role	1–5
	Life time span	1–3
4. POPULATION	Frequency	1–4
	Abundance	1–4
5. EVOLUTIONARY HISTORY	Evolutionary origin	1–3
	Phylogenetic rarity	1–5

Except for "Frequency" and "Abundance", which are calculated by AS, scores for the other traits are assigned to each species present in the dataset by the user on the basis of the best available information. Thus, these scores are based on expert judgment. Frequency and Abundance scores are calculated by AS using site data because these traits are not species-specific, but site-specific (the same species may show different population characteristics at different sites). "Frequency" is given by AS calculating first the frequency of occurrence of each species across the total number of studied sites, and then by attributing to each species a score (from 1 to 4) based on quartiles (with score 1 assigned to species falling above the third quartile, which are the most widely distributed species, and score 4 assigned to species with frequency below the first quartile, which are the species with the narrowest distribution). The "Abundance" trait is computed by AS in a similar way: relative abundance is computed for all species, and then to each species is attributed a score (from 1 to 4) based on quartiles. If abundance data are not available, and only presence/absence data are used, this trait is automatically excluded from calculations.

Trait scores (assigned to each species by the user) belonging to the same dimension are multiplied by AS to obtain a synthetic score for each dimension. Once this operation has been made for all the species in the dataset, a distribution of synthetic scores becomes available for each dimension. These synthetic scores have different ranges. For example, the score for Dimension 1 (Geography) may vary between 1 and 28 (for a species that ranked highest for the two traits included in this dimension), while the score for Dimension 2 (Ecology) may vary between 1 and 1875 (for a species that obtained the maximum score for each of the five traits included in this dimension). For this reason, scores resulting from multiplications are divided using quartiles, and values falling within the same interquartile range receive the same score, ranging from 1 (values below the first quartile) to 4 (values above the third quartile). The use of quartiles to rescale original products into four classes with integer values (i.e., 1, 2, 3, and 4) was adopted in AS to make dimensions comparable. After the quartilization of products, the new scores are summed up to obtain the overall species score. This sum of scores (which can theoretically range from 5, for a species which scored 1 for all five dimensions, to 20, for a species which scored 4 for all dimensions) is rescaled to vary between 1 and 4 by simply dividing by 5. In this way, each species receives a score represented by a value varying between 1 and 4 (SS, Species Scores).

AS generates and uses the SS scores to compute three distinct site-specific indices, and particularly:

(1) Sum of Species Scores (SSS) normalized between 0 and 1 using the formula (Equation (5)):

$$\text{SSSrescaled} = (\text{SSS} - \text{SSS}_{\text{min}})/(\text{SSS}_{\text{max}} - \text{SSS}_{\text{min}}) \quad (5)$$

with SSS_{min} and SSS_{max} being, respectively, the minimum and maximum SSS scores recorded across all studied sites.

(2) Biodiversity Conservation Concern (also varying between 0 and 1) (Equation (6)):

$$\text{BCC} = (\text{SSS}/\text{N} - 1)/(\text{SS}_{\text{max}} - 1) \quad (6)$$

where N is the total number of species present in a site and SS_{max} is the maximum value of the SS values.

(3) Groundwater Biodiversity Concern (GBC) computed as the mean of SSSrescaled and BCC. GBC values are finally used to identify different priority classes (Table 8).

Table 8. GBC rating, priority ranking and color per rank.

GBC Rating	Priority Ranking	GBC Color Ranking
$0 \leq \text{GBC} \leq 0.2$	very low priority	(blue)
$0.2 < \text{GBC} \leq 0.4$	low priority	(green)
$0.4 < \text{GBC} \leq 0.6$	medium priority	(yellow)
$0.6 < \text{GBC} \leq 0.8$	high priority	(orange)
$0.8 < \text{GBC} \leq 1$	very high priority	(red)

After uploading the files including the table of species scores for each trait, and the matrix of species occurrences/abundances, by pressing the "RUN" button users will be prompted to the results page, which will include both the individual species scores of all species included in the analysis, as well as the site-specific scores, including the raw and rescaled SSS, the BCC, and the GBC (together with the corresponding conservation priority class).

As for the previous sections, users can save all their results, access them at any moment, and export them for additional analysis outside the AS.

3. Example of Application: The Hyporheic Zone of the River Sagittario

To understand the main outputs of AS, namely ER, HHR, and GBC, six hyporheic sites (ABRO86/3S, ABRO87/3S, ABRO88/3S, ABRO89/3S, ABRO90/3S, and ABRO91/3S; Figure 1) of the River Sagittario (Italy) were selected from the AQUALIFE network of sites (Tables S3–S5). The River Sagittario in the Sulmona plain is among the most altered rivers in central Italy [75], thus representing a model environmental system for assessing the effects of different types of pressures on hyporheic biodiversity. At each site, we performed two temporal surveys, namely in December 2014 (winter) and in June 2015 (summer).

Hyporheic samples were collected pumping 10 L of interstitial water by a membrane pump connected to steel pipes screened at −40 cm deep and filtered through a 60 μm hand net. Faunal samples were preserved in a 70% alcohol solution. In the laboratory, specimens were sorted under a stereomicroscope at 16× magnification and identified to the lowest taxonomic level possible using a LEICA M205C stereomicroscope. Based on the human activities carried out in the catchment of the River Sagittario, 108 chemicals (including N-compounds, phosphates, metals, volatile organic compounds, and pesticides) were analyzed per each sample taking a 2 L volume. The concentrations of most chemicals (101 out of 108) were below the limit of detection. The MECs of the seven chemical compounds out of 108 (namely, nitrites, nitrates, sulfates, chloride, phosphates, ammonium, and αHCH) exceeding the limits of detection are reported in Table S3. The hydrological-hydromorphological impacts affecting the GDE sites are reported in Table S4. Overall, 58 invertebrate taxa were identified (Table S5) and the trait scores, used by AS to calculate the GBC index, were assigned (Table S5).

The invertebrate assemblages were mostly composed by stygobionts or species highly dependent on groundwater (Table S5; see the column "Ecological specialization to GW"). AS was run for each site and the results are summarized in Table 9 and Figure 1. The ER outputs revealed that all sites were at medium risk along the whole river sector (Table 9). No hydrological-hydromorphological risk was reported for the upstream sites ABRO86/3S and ABRO87/3S (HHR < 10) while the four remaining downstream sites showed an HHR > 10 (Table 9). Finally, the highest GBC value (GBC = 0.667) was reached in ABRO86/3S where medium ER (RQ_{mix} = 58.85) and HHR = 6 (no risk) were calculated by AS. In the remaining sites the GBC was in the range very low–medium. The lowest GBC (GBC = 0.172) was assigned by AS to the site ABRO89/3S with low species richness, medium ER (RQ_{mix} = 72.10), and HHR = 13 (at risk) (Table 9 and Figure 1).

The application of AS to the River Sagittario has highlighted that the hyporheic zone has only one site with a significant conservation importance. The GDE assemblages experience a medium risk due to pollution of a limited number of substances in all sites, while the hydrological-hydromorphological alteration of the GDE sites under study occurs only in the most downstream sites. This scenario requires management choices that are discussed in detail in the next paragraph.

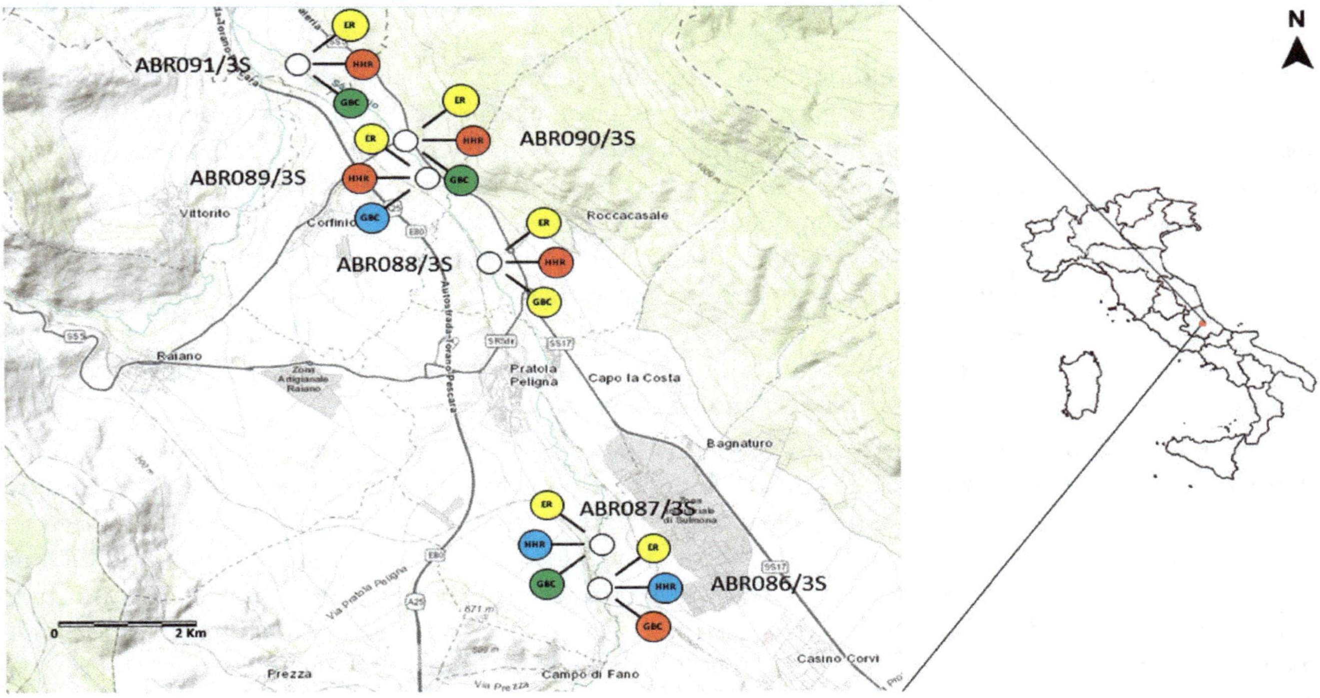

Figure 1. Sampling sites (white circles) of the hyporheic zone of the River Sagittario. Color-coded circles are used to indicate the ranks of the ecological risk (ER), the hydrological-hydromorphological risk (HHR), and the groundwater biodiversity concern index (GBC).

Table 9. AS outputs for ER (Environmental Risk), HHR (Hydrological-Hydromorphological Risk) and the conservation priority indices (SSS, Sum of Species Scores; SSS rescaled, Sum of Species Scores rescaled; BCC, Biodiversity Conservation Concern; GBC, Groundwater Biodiversity Concern) for the example of application in the hyporheic zone of the River Sagittario (Italy).

Sampling Site	ER (RQ_{mix})	ER Ranking	HHR	HHR Ranking	SSS	SSS Rescaled	BCC	GBC	Priority Ranking
ABRO86/3S	58.85	medium	6	no risk	16	1	0.333	0.667	high
ABRO87/3S	83.75	medium	6	no risk	13.8	0.421	0.324	0.372	low
ABRO88/3S	32.11	medium	28	at risk	15	0.737	0.381	0.559	medium
ABRO89/3S	72.10	medium	20	at risk	12.2	0	0.344	0.172	very low
ABRO90/3S	51.22	medium	16	at risk	13.6	0.368	0.314	0.341	low
ABRO91/3S	49.05	medium	15	at risk	13.8	0.421	0.324	0.372	low

4. Discussion

The introduction of the concept of GDE in the context of the Water Framework Directive and the subsequent issuing of specific guidelines have certainly represented a considerable advancement in the protection of aquatic and, in particular, groundwater dependent ecosystems [10]. Due to their communicative effectiveness, scores and color-coded maps have been strongly recommended by legislations worldwide to illustrate the ecological status, the risk, and the monitoring network of water bodies [16]. In fact, scores and colors are intuitive and of immediate understanding even for non-experts and have the advantage of making people aware of information that has been long confined to technicians.

AS provides a user-friendly toolkit to assess (i) the risk effectively or potentially posed to the GDE communities by chemical pollutants; (ii) the hydrological-hydromorphological risk (available for ecotonal GDEs, such as hyporheic zones and groundwater-fed springs); and (iii) the conservation priority of GDE sites based on the community composition. Due to its structure and, above all, thanks to the outputs that are given in scores and color-coded maps, AS fully meets the requirements of intuitive communication of the legislations.

In AS, each index may work either alone or in conjunction with the others, without the limitations of single-function applications. AS creatively addresses also a risk scale for the ecological risk, based on the RQ_{mix} rating, thus allowing the user to measure how far each site is from the condition of "no risk". This graded scale of risk, even if not provided for by any international regulation, allows the user to make an informed choice on which sites require priority recovery and which do not. ER has more classes (five) than HHR (two). This is due to the higher level of knowledge that is available for the ecological risk compared to the hydrological-hydromorphological risk for which scoring is available only for surface water bodies, thus preventing a fine-tuning of the HHR scoring for the GDEs. On a regulatory basis, water bodies are still divided in at risk vs. not at risk, but a further division in classes of the ecological risk is possible ([27] and references therein).

The major advantage of the software is that of providing integrated information. The integration of the most commonly assessed ecological risk with the hydrological-hydromorphological risk is certainly a key-point of AS. In fact, through the HHR index, it is immediately evident to the user that chemical pollution is not always the main factor of GDE degradation. In fact, the greatest impact on GDEs is often due to human activities such as river bed reshaping, grinding, bank defenses, embankments, coatings, burial, vegetation cutting, etc. The attention paid by AS to the physical components of GDEs makes it possible to broaden the horizon of the management measures of protection and recovery [56]. We recommend computing ER along with HHR, because hydrological and hydromorphological impacts occurring in the river channels or groundwater-fed springs can lead to an increasing pollutant concentration both in the hyporheic zone or in spring water and in the feeding aquifers. For instance, areas with intensive agriculture, where a massive riparian vegetation removal has been practiced along river banks, are characterized by higher concentrations of nitrates and ammonium, mainly in the hyporheic zone [26,76] and in the SGDEs, i.e., the saturated aquifers [31,76,77]. The third tool of AS allows the assessment of the conservation priority of sites on the basis of the uniqueness of the species collected in aquatic GDE (and SGDE) sites. This set of indicators (SSS, BBC, and GBC) may work in combination with ER and HHR, based on the assumption that the higher the ER and HHR scores, the higher the risk of biodiversity loss, especially for the most sensitive species. This assumption is not necessarily true, since many variables come into play to define whether a given site is permanently disturbed or not. Furthermore, individual sites are often interconnected, for example if they belong to the same aquifer, or to the hyporheic zone of the same river, thus ensuring a recolonization of perturbed sites from unpolluted aquifer sectors or hyporheic stretches [78].

AS can support European Member States in the preparation of the GDE management plans under the EU Directive 2000/60/EC [16]. Every six years (duration of a management plan) all Member States must indicate the GDEs at risk or not at risk and the related management measures. The actions concern: (i) a new census of anthropic pressures; (ii) a monitoring program, which can be distinguished in surveillance monitoring (which applies exclusively to not-at-risk GDEs) and operational monitoring (which applies to at-risk GDEs); and (iii) protection measures involving interdictions of impactful human activities in the entire GDE or part of it. The list of pressures must be updated every six years, regardless of the risk conditions of the GDE. The choice between the types of monitoring programs has profound economic impact. Surveillance monitoring is largely sustainable because it can be performed only in one point of a given GDE, on basic physico-chemical (e.g., pH, electrical conductivity, and nitrates) and hydrological-hydromorphological parameters (e.g., flow rate, water level, damming, etc.) and with a frequency of one time every six years. On the other hand, the operational monitoring is much more expensive because it involves a significant number of sites and parameters and a frequency of at least once a year for the entire duration of the management plan. In addition, protection measures are often required in at-risk GDEs. To facilitate decision-makers in the choice of the type of monitoring program, we propose to refer to the individual values of the indicators, as shown in Table 10, where ER and HHR provide indications for the type of monitoring and GBC for the protection measures. Based on this rationale, if ER and HHR indicate that the GDE site is not at risk, the monitoring must be based on surveillance pursuant to the EU Directive 2000/60/EC [16] and no protective action is required regardless of the conservation value of the GDE. If, instead, ER and HHR indicate that the GDE is at risk, the monitoring must always be operational, both for the chemical and hydrological-hydromorphological parameters, but the protection actions may be necessary only at certain risk values. For example, protection measures are not necessary if GBC = 0.7 (High Priority) and ER = 9 and HHR < 10. The cases in which AS shows its full utility is when ER does not indicate risk but HHR does (or vice versa). In this case, the monitoring must be operational only for the chemical (or hydrological-hydrogeological) parameters, with a relative cost containment. The study case of the hyporheic zone of the River Sagittario allowed the application of the ER and the HHR together with the GBC (Figure 1). Based on the obtained results, the River Sagittario basin management plan should take into account primarily the restoration of the connectivity between river flowing waters and the underlying aquifer via the hyporheic zone to preserve benthic and hyporheic fauna and the ecosystems services they provide. Four sites out of six were at hydrological-hydromorphological risk for the presence of a dam, water diversion for agriculture, presence of solid walls and/or embankments limiting permeability with the alluvial plain, channelization, artificially straightened channel, and artificial elevation of the riverbed. According to Table 10, the monitoring of the chemical-physical parameters should be operational in all sampling sites while that of the hydrological-hydromorphological variables should be operational in four sites out of six, being based on surveillance in the remaining two sites. Finally, the GBC index indicated that protection measures are required for just one site out of six, namely ABRO86/3S.

Table 10. Decision-making table. Surveillance (SMPs) and operational monitoring program (OMPs) and protection measures (PMs) are indicated according to the values of ER, HHR and GBC indices.

ER Values	ER Monitoring Measures	HHR Values	HHR Monitoring Measures	GBC Values	Measures
$0 \leq RQ_{mix} < 1$	SMPs	HHR < 10	SMPs	$0 \leq GBC \leq 0.2$	No PMs
$1 \leq RQ_{mix} < 10$	OMPs	HHR ≥ 10	OMPs	$0.2 < GBC \leq 0.4$	PMs if RQ ≥ 1000 and HHR ≥ 10
$10 \leq RQ_{mix} < 100$	OMPs			$0.4 < GBC \leq 0.6$	PMs if RQ > 100 and HHR ≥ 10
$100 \leq RQ_{mix} < 1000$	OMPs			$0.6 < GBC \leq 0.8$	PMs if RQ > 10 and HHR ≥ 10
$RQ_{mix} \geq 1000$	OMPs			$0.8 < GBC \leq 1$	PMs if RQ > 1 and/or HHR ≥ 10

From the 2000s until today, other valid tools for assessing the ecological and environmental health in GDEs (but not their conservation values) were presented. For instance, the GW-Fauna-Index [79] was the first attempt to classify GDEs (hyporheic zones and alluvial aquifers) in terms of ecologically relevant conditions using factors which best correlated with stygofauna, namely detritus, temperature, and dissolved oxygen. In 2019, an approach based on three microbiological parameters (prokaryotic cell density, prokaryotic intracellular ATP concentrations, and assimilable organic carbon) was proposed as an integration of the existing routine of GDE monitoring [80]. A single index value obtained from multivariate analyses of the microbial parameters proved to best discriminate between disturbed (organic contamination with hydrocarbons, surface water intrusion, and agricultural activities) and undisturbed alluvial aquifers [80]. A more structured and comprehensive framework for measuring GDE health is the Groundwater Health Index (GHI) [81,82] that was presented in 2011 and successively refined in 2017. Relying upon a combination of microbial, stygofaunal, water chemistry, and environmental indicators, GHI can discriminate three distinct ecosystem health classifications, namely "similar to reference", "mild deviation from reference", and "major deviation from reference".

Other tools were presented to preserve biodiversity of GDEs based on their conservation value (but not on their quality status). A conservation strategy to minimize the risk of groundwater biodiversity loss due to human activities in European GDEs was proposed by designing a network of reserve areas that collectively include most groundwater species [8]. Three area selection methods (species richness hotspots, endemism hotspots, and complementarity) were proposed to preserve 1059 groundwater species of hyporheic zones, springs, and alluvial and karst aquifers in six European regions. Compared to the aforementioned tools, AS offers high flexibility, and can be used, depending on the user needs, in different GDEs and at different spatial scales. In addition, AS is the only available tool that calculates the chemical and hydrological-hydromorphological risks faced by GDEs along with their conservation priority.

However, along with the merits, AS faces also some methodological limitations such as those related, for example, to the number of pollutants for which AS provides the PNECs. Emerging contaminants, such as pharmaceutical compounds, are not represented in the internal database. The HHR index is currently applicable only to ecotonal GDEs, such as hyporheic zones and springs. A HHR for the aquifers (that takes into account quarries, for example, waste materials from stone processing, and the widespread phenomenon of fracking) is still missing. Finally, the scores assigned to each impact are based on expert judgment, which will be potentially tuned with the advancement of knowledge on the GDE functioning and the relative impacts. For this reason, AS will likely be useful for the authorities in charge of GDE management; however, the version presented here should be considered a preliminary workbench toward an appealing refinement of the software itself.

5. Conclusions

The purpose of this paper is to present the AQUALIFE software, which is the main product of the AQUALIFE project. AS consists of a user-friendly online interface, and is free to use (after registration) at: http://app.aqualifeproject.eu. AS is a relatively fast, economical tool of investigation that can be used by operators with basic knowledge of aquatic ecosystems dependent on groundwater and that, through training courses, can complete their knowledge and acquire the necessary experience. AS allows the user to assess the conservation priority of GDEs based on species conservation importance along with chemical and hydrological-hydromorphological risks. Thanks to its novelty, AS has the potential to become a user-friendly standard tool not only for researchers, but also for practitioners and decision makers. The scientific awareness of the existence of a unique biodiversity in the GDEs (and SGDEs) and the role it plays in terms of ecosystem services has not had any impact on the perception and cognition of this reality at the regulatory level; in this regard, AS acts as a tool for ensuring communication between scientists and end-users of environmental information. Being effective both for the calculation of conservation priorities and in terms of potential impact assessment, AS represents a profitable tool to set up appropriate policies to achieve concrete GDE improvement results.

Author Contributions: Conceptualization, D.M.P.G., G.S., S.F., B.F., and T.D.L.; methodology, D.M.P.G., S.F., and B.F.; software, G.S., W.L., and F.B.; validation, D.M.P.G., S.F., B.F., and M.D.C.; analysis, D.M.P.G. and S.F.; data curation, D.M.P.G. and B.F.; writing—original draft preparation, D.M.P.G., G.S., S.F., T.D.L., and B.F.; writing—review and editing, D.M.P.G., T.D.L., B.F., S.F., and M.D.C.; visualization, M.D.C.; supervision, D.M.P.G.; project administration, B.F.; and funding acquisition, D.M.P.G.

Acknowledgments: The Stygobiology Laboratory (University of L'Aquila, Department of Life, Health and Environmental Sciences) is greatly acknowledged for the field activity and sorting of the invertebrates during the AQUALIFE project.

References

1. Hatton, T.J.; Evans, R. *Dependence of Ecosystems on Groundwater and Its Significance to Australia*; CSIRO (Land and Water): Canberra, Australia, 1998.
2. Eamus, D.; Froend, R. Groundwater-dependent ecosystems: The where, what and why of GDEs. *Aust. J. Bot.* **2006**, *54*, 91–96. [CrossRef]

3. Gibert, J.; Deharveng, L. Subterranean ecosystems: A truncated functional biodiversity. *Bioscience* **2002**, *52*, 473–481. [CrossRef]
4. Humphreys, W.F. Aquifers: The ultimate groundwater-dependent ecosystems. *Aust. J. Bot.* **2006**, *54*, 115–132. [CrossRef]
5. Mammola, S.; Cardoso, P.; Culver, D.C.; Deharveng, L.; Ferreira, R.L.; Fišer, C.; Galassi, D.M.P.; Griebler, C.; Halse, S.; Humphreys, W.F.; et al. Scientists' Warning on the Conservation of Subterranean Ecosystems. *Bioscience* **2019**, *69*, 641–650. [CrossRef]
6. Ferreira, D.; Malard, F.; Dole-Olivier, M.-J.; Gibert, J. Obligate groundwater fauna of France: Diversity patterns and conservation implications. *Biodivers. Conserv.* **2007**, *16*, 567. [CrossRef]
7. Deharveng, L.; Stoch, F.; Gibert, J.; Bedos, A.; Galassi, D.; Zagmajster, M.; Brancelj, A.; Camacho, A.; Fiers, F.; Martin, P.; et al. Groundwater biodiversity in Europe. *Freshw. Biol.* **2009**, *54*, 709–726. [CrossRef]
8. Michel, G.; Malard, F.; Deharveng, L.; Di Lorenzo, T.; Sket, B.; De Broyer, C. Reserve selection for conserving groundwater biodiversity. *Freshw. Biol.* **2009**, *54*, 861–876. [CrossRef]
9. Galassi, D.M.P.; Fiasca, B.; Tosto, D. Patterns of Copepod diversity (Copepoda: Cyclopoida, Harpacticoida) in springs of central Italy: Implications for conservation issues. In *Studies on Freshwater Copepoda: A Volume in Honour of Bernard Dussart*; Brill: Buckinghamshire, UK, 2011; Volume 16, pp. 199–226.
10. European Commission. *Technical Report on Groundwater Associated Aquatic Ecosystems*; Technical Report No. 9; European Commission: Luxembourg, 2015.
11. Marmonier, P.; Maazouzi, C.; Baran, N.; Blanchet, S.; Ritter, A.; Saplairoles, M.; Dole-Olivier, M.J.; Galassi, D.M.P.; Eme, D.; Doledec, S.; et al. Ecology-based evaluation of groundwater ecosystems under intensive agriculture: A combination of community analysis and sentinel exposure. *Sci. Total Environ.* **2018**, *613*, 1353–1366. [CrossRef]
12. Jaffé, R.; Prous, X.; Zampaulo, R.; Giannini, T.C.; Imperatriz-Fonseca, V.L.; Maurity, C.; Oliveira, G.; Brandi, I.V.; Siqueira, J.O. Reconciling Mining with the Conservation of Cave Biodiversity: A Quantitative Baseline to Help Establish Conservation Priorities. *PLoS ONE* **2016**, *11*. [CrossRef]
13. EPA. *EPA's Draft Report on the Environment Technical Document*; EPA: Washington, DC, USA, 2003.
14. Tomlinson, M.; Boulton, A.J.; Hancock, P.J.; Cook, P.G. Deliberate omission or unfortunate oversight: Should stygofaunal surveys be included in routine groundwater monitoring programs? *Hydrogeol. J.* **2007**, *15*, 1317. [CrossRef]
15. Serov, P.; Kuginis, L.; Williams, J.P. *Risk Assessment Guidelines for Groundwater Dependent Ecosystems, Volume 1—The Conceptual Framework*; NSW Department of Primary Industries, Office of Water: Sydney, Australia, 2012.
16. Council of the European Communities. Directive 2000/60/EC of the European Parliament and the Council of 23 October 2000 establishing a framework for Community action in the field of water policy. *Off. J. Eur. Commun.* **2000**, *327*, 1–73.
17. Council of the European Communities. Directive 2006/118/EC of the European Parliament and of the Council of 12 December 2006 on the protection of groundwater against pollution and deterioration. *Off. J. Eur. Union* **2006**, *372*, 19–31.
18. *Millennium Ecosystem Assessment, Ecosystems and Human Well-being: Synthesis*; Island Press: Washington, DC, USA, 2005.
19. Kløve, B.; Ala-Aho, P.; Bertrand, G.; Boukalova, Z.; Erturk, A.; Goldscheider, N.; Ilmonen, J.; Karakaya, N.; Kupfersberger, H.; Kvaerner, J.; et al. Groundwater dependent ecosystems. Part I: Hydroecological status and trends. *Environ. Sci. Policy* **2011**, *14*, 770–781. [CrossRef]
20. Megdal, S.B.; Gerlak, A.K.; Varady, R.G.; Huang, L.Y. Groundwater Governance in the United States: Common Priorities and Challenges. *Groundwater* **2015**, *53*, 677–684. [CrossRef] [PubMed]
21. Saccò, M.; Blyth, A.; Bateman, P.W.; Hua, Q.; Mazumder, D.; Nicole, W.; Humphreys, W.F.; Laini, A.; Griebler, C.; Grice, K. New light in the dark—A proposed multidisciplinary framework for studying functional ecology of groundwater fauna. *Sci. Total Environ.* **2019**, *662*, 963–977. [CrossRef]
22. Taylor, R.G.; Scanlon, B.; Doll, P.; Rodell, M.; van Beek, R.; Wada, Y.; Longuevergne, L.; Leblanc, M.; Famiglietti, J.S.; Edmunds, M.; et al. Ground water and climate change. *Nat. Clim. Chang.* **2013**, *3*, 322–329. [CrossRef]
23. Phyton Software Foundation (US). Available online: https://www.python.org/ (accessed on 3 October 2019).
24. Django Project. Available online: https://www.djangoproject.com (accessed on 3 October 2019).

25. Avramov, M.; Schmidt, S.I.; Griebler, C. A new bioassay for the ecotoxicological testing of VOCs on groundwater invertebrates and the effects of toluene on *Niphargus inopinatus*. *Aquat. Toxicol.* **2013**, *130*, 1–8. [CrossRef]
26. Di Lorenzo, T.; Borgoni, R.; Ambrosini, R.; Cifoni, M.; Galassi, D.M.P.; Petitta, M. Occurrence of volatile organic compounds in shallow alluvial aquifers of a Mediterranean region: Baseline scenario and ecological implications. *Sci. Total Environ.* **2015**, *538*, 712–723. [CrossRef]
27. Di Lorenzo, T.; Castano-Sanchez, A.; Di Marzio, W.D.; Garcia-Doncel, P.; Nozal Martinez, L.; Galassi, D.M.P.; Iepure, S. The role of freshwater copepods in the environmental risk assessment of caffeine and propranolol mixtures in the surface water bodies of Spain. *Chemosphere* **2019**, *220*, 227–236. [CrossRef]
28. Di Lorenzo, T.; Di Marzio, W.D.; Fiasca, B.; Galassi, D.M.P.; Korbel, K.; Iepure, S.; Pereira, J.L.; Reboleira, A.; Schmidt, S.I.; Hose, G.C. Recommendations for ecotoxicity testing with stygobiotic species in the framework of groundwater environmental risk assessment. *Sci. Total Environ.* **2019**, *681*, 292–304. [CrossRef]
29. Calamari, D.; Vighi, M. A proposal to define quality objectives for aquatic life for mixtures of chemical-substances. *Chemosphere* **1992**, *25*, 531–542. [CrossRef]
30. Vighi, M.; Altenburger, R.; Arrhenius, Å.; Backhaus, T.; Bödeker, W.; Blanck, H.; Consolaro, F.; Faust, M.; Finizio, A.; Froehner, K.; et al. Water quality objectives for mixtures of toxic chemicals: Problems and perspectives. *Ecotoxicol. Environ. Saf.* **2003**, *54*, 139–150. [CrossRef]
31. Di Lorenzo, T.; Cifoni, M.; Fiasca, B.; Di Cioccio, A.; Galassi, D.M.P. Ecological risk assessment of pesticide mixtures in the alluvial aquifers of central Italy: Toward more realistic scenarios for risk mitigation. *Sci. Total Environ.* **2018**, *644*, 161–172. [CrossRef] [PubMed]
32. Pennington, D. Extrapolating ecotoxicological measures from small data sets. *Ecotoxicol. Environ. Saf.* **2003**, *56*, 238–250. [CrossRef]
33. Backhaus, T.; Faust, M. Predictive Environmental Risk Assessment of Chemical Mixtures: A Conceptual Framework. *Environ. Sci. Technol.* **2012**, *46*, 2564–2573. [CrossRef]
34. Gustavsson, M.; Kreuger, J.; Bundschuh, M.; Backhaus, T. Pesticide mixtures in the Swedish streams: Environmental risks, contributions of individual compounds and consequences of single-substance oriented risk mitigation. *Sci. Total Environ.* **2017**, *598*, 973–983. [CrossRef]
35. ECHA. *Chapter, R.10: Characterisation of Dose [Concentration]-Response for Environment. Guidance for the Implementation of REACH—Guidance on Information Requirements and Chemical Safety Assessment*; European Chemical Agency: Helskinki, Finland, 2008.
36. European Commission. *Technical Guidance for Deriving Environmental Quality Standard*; European Commission: Brussels, Belgium, 2018.
37. ECOTOX Database. Available online: https://cfpub.epa.gov/ecotox/ (accessed on 3 October 2019).
38. ECOSAR Software. Available online: http://www.epa.gov/oppt/newchems/tools/21ecosar.htm (accessed on 3 October 2019).
39. Galassi, D.M.P.; Fiasca, B.; Di Lorenzo, T.; Montanari, A.; Porfirio, S.; Fattorini, S. Groundwater biodiversity in a chemoautotrophic cave ecosystem: How geochemistry regulates microcrustacean community structure. *Aquat. Ecol.* **2017**, *51*, 75–90. [CrossRef]
40. Rinaldi, M.; Surian, N.; Comiti, F.; Bussettini, M. *IDRAIM—Sistema di Valutazione Idromorfologica, Analisi E Monitoraggio dei Corsi d'acqua—Versione Aggiornata 2016*; ISPRA: Rome, Italy, 2016.
41. Brunner, P.; Cook, P.G.; Simmons, C.T. Disconnected Surface Water and Groundwater: From Theory to Practice. *Ground Water* **2011**, *49*, 460–467. [CrossRef]
42. Siergieiev, D.; Ehlert, L.; Reimann, T.; Lundberg, A.; Liedl, R. Modelling hyporheic processes for regulated rivers under transient hydrological and hydrogeological conditions. *Hydrol. Earth Syst. Sci.* **2015**, *19*, 329–340. [CrossRef]
43. Hancock, P.J. The response of hyporheic invertebrate communities to a large flood in the Hunter River, New South Wales. *Hydrobiologia* **2006**, *568*, 255–262. [CrossRef]
44. Arroita, M.; Flores, L.; Larranaga, A.; Martinez, A.; Martinez-Santos, M.; Pereda, O.; Ruiz-Romera, E.; Solagaistua, L.; Elosegi, A. Water abstraction impacts stream ecosystem functioning via wetted-channel contraction. *Freshw. Biol.* **2017**, *62*, 243–257. [CrossRef]
45. Liu, D.S.; Zhao, J.; Chen, X.B.; Li, Y.Y.; Weiyan, S.P.; Feng, M.M. Dynamic processes of hyporheic exchange and temperature distribution in the riparian zone in response to dam-induced water fluctuations. *Geosci. J.* **2018**, *22*, 465–475. [CrossRef]

46. Mathers, K.L.; Hill, M.J.; Wood, P.J. Benthic and hyporheic macroinvertebrate distribution within the heads and tails of riffles during baseflow conditions. *Hydrobiologia* **2017**, *794*, 17–30. [CrossRef]
47. Sawyer, A.H.; Cardenas, M.B.; Bomar, A.; Mackey, M. Impact of dam operations on hyporheic exchange in the riparian zone of a regulated river. *Hydrol. Process.* **2009**, *23*, 2129–2137. [CrossRef]
48. Boano, F.J.; Harvey, W.; Marion, A.; Packman, A.I.; Revelli, R.; Ridolfi, L.; Wörman, A. Hyporheic flow and transport processes: Mechanisms, models, and biogeochemical implications. *Rev. Geophys.* **2014**, *52*, 603–679. [CrossRef]
49. Magliozzi, C.; Grabowski, R.; Packman, A.; Krause, S. Scaling down hyporheic exchange flows: From catchments to reaches. *Hydrol. Earth Syst. Sci. Discuss.* **2017**. [CrossRef]
50. European Commission. *WFD Reporting Guidance 2016. Final Draft 6.0.6*; European Commission: Brussels, Belgium, 2016.
51. Loheide, S.P.; Booth, E.G. Effects of changing channel morphology on vegetation, groundwater, and soil moisture regimes in groundwater-dependent ecosystems. *Geomorphology* **2011**, *126*, 364–376. [CrossRef]
52. Pacioglu, O.; Shaw, P.; Robertson, A. Patch scale response of hyporheic invertebrates to fine sediment removal in two chalk rivers. *Fundam. Appl. Limnol.* **2012**, *181*, 283–288. [CrossRef]
53. Jones, I.; Growns, I.; Arnold, A.; McCall, S.; Bowes, M. The effects of increased flow and fine sediment on hyporheic invertebrates and nutrients in stream mesocosms. *Freshw. Biol.* **2015**, *60*, 813–826. [CrossRef]
54. Pacioglu, O.; Robertson, A. The invertebrate community of the chalk stream hyporheic zone: Spatio-temporal distribution patterns. *Knowl. Manag. Aquat. Ecosyst.* **2017**, *418*, 10. [CrossRef]
55. Hancock, P.J. Human impacts on the stream-groundwater exchange zone. *Environ. Manag.* **2002**, *29*, 763–781. [CrossRef] [PubMed]
56. Merill, L.; Tonjes, D.J. A Review of the Hyporheic Zone, Stream Restoration, and Means to Enhance Denitrification. *Crit. Rev. Sci. Technol.* **2014**, *44*, 2337–2379. [CrossRef]
57. Wondzell, S.M. Effect of morphology and discharge on hyporheic exchange flows in two small streams in the Cascade Mountains of Oregon, USA. *Hydrol. Process.* **2006**, *20*, 267–287. [CrossRef]
58. Fumetti von, S.; Dmitrovic, D.; Pesic, V. The influence of flooding and river connectivity on macroinvertebrate assemblages in rheocrene springs along a third-order river. *Fundam. Appl. Limnol.* **2017**, *190*, 251–263. [CrossRef]
59. Pesic, V.; Dmitrovic, D.; Savic, A.; Milosevic, D.; Zawal, A.; Vukasinovic-Pesic, V.; Fumetti von, S. Application of macroinvertebrate multimetrics as a measure of the impact of anthropogenic modification of spring habitats. *Aquat. Conserv. Mar. Freshw. Ecosyst.* **2019**, *29*, 341–352. [CrossRef]
60. Salameh, E. Over-exploitation of groundwater resources and their environmental and socio-economic implications: The case of Jordan. *Water Int.* **2008**, *33*, 55–68. [CrossRef]
61. Tal, A.; Katz, D. Rehabilitating Israel's streams and rivers. *Int. J. River Basin Manag.* **2012**, *10*, 317–330. [CrossRef]
62. Siwek, J.; Pociask-Karteczka, J. Springs in South-Central Poland—Changes and threats. *Episodes* **2017**, *40*, 38–46. [CrossRef]
63. Dwire, K.A.; Mellmann-Brown, S.; Gurrieri, J.T. Potential effects of climate change on riparian areas, wetlands, and groundwater-dependent ecosystems in the Blue Mountains, Oregon, USA. *Clim. Serv.* **2018**, *10*, 44–52. [CrossRef]
64. Sabater, F.; Butturini, A.; Mart, E.; Oz, I.; Romaní, A.; Wray, J.; Sabater, S. Effect of Riparian Vegetation Removal on Nutrient Retention in a Mediterranean Stream. *J. N. Am. Benthol. Soc.* **2000**, *19*, 609–620. [CrossRef]
65. Possingham, H.P.; Wilson, K.A. Biodiversity—Turning up the heat on hotspots. *Nature* **2005**, *436*, 919–920. [CrossRef] [PubMed]
66. Funk, S.M.; Fa, J.E. Ecoregion Prioritization Suggests an Armoury Not a Silver Bullet for Conservation Planning. *PLoS ONE* **2010**, *5*, e8923. [CrossRef] [PubMed]
67. Cardoso, P.; Erwin, T.L.; Borges, P.A.V.; New, T.R. The seven impediments in invertebrate conservation and how to overcome them. *Biol. Conserv.* **2011**, *144*, 2647–2655. [CrossRef]
68. Fattorini, S. Use of insect rarity for biotope prioritisation: The tenebrionid beetles of the Central Apennines (Italy). *J. Insect Conserv.* **2010**, *14*, 367–378. [CrossRef]
69. Fattorini, S. Biotope prioritisation in the Central Apennines (Italy): Species rarity and cross-taxon congruence. *Biodivers. Conserv.* **2010** *19*, 3413–3429. [CrossRef]

70. Fattorini, S.; Cardoso, P.; Rigal, F.; Borges, P.A.V. Use of Arthropod Rarity for Area Prioritisation: Insights from the Azorean Islands. *PLoS ONE* **2012**, *7*. [CrossRef] [PubMed]
71. Kattan, G.H. Rarity and vulnerability: The birds of the Cordillera Central of Colombia. *Conserv. Biol.* **1992**, *6*, 64–70. [CrossRef]
72. Rabinowitz, D. Seven forms of rarity. In *The Biological Aspects of Rare Plant Conservation*; Synge, H., Ed.; Sinauer Associates: Chichester, UK, 1981.
73. Rabinowitz, D.; Cairns, S.; Dillon, T. Seven forms of rarity and their frequency in the flora of the British Isles. In *Conservation Biology: The Science of Scarcity and Diversity*; Soulè, M.E., Ed.; British Ecological Society: London, UK, 1986; pp. 182–204.
74. Fattorini, S.; Fiasca, B.; Di Lorenzo, T.; Di Cicco, M.; Galassi, D.M.P. A new protocol for assessing the conservation priority of groundwater dependent ecosystems. *Aquat. Conserv. Mar. Freshw. Ecosyst.* submitted for publication.
75. Caschetto, M.; Galassi, D.M.P.; Petitta, M.; Aravena, R. Evaluation of the sources of nitrogen compounds and their influence on the biological communities in the hyporheic zone of the Sagittario River, Italy: An isotopic and biological approach. *Ital. J. Geosci.* **2017**, *136*, 145–156. [CrossRef]
76. Di Lorenzo, T.; Cannicci, S.; Spigoli, D.; Cifoni, M.; Baratti, M.; Galassi, D.M.P. Bioenergetic cost of living in polluted freshwater bodies: Respiration rates of the cyclopoid *Eucyclops serrulatus* under ammonia-N exposures. *Fundam. Appl. Limnol.* **2016**, *188*, 147–156. [CrossRef]
77. Di Lorenzo, T.; Cifoni, M.; Lombardo, P.; Fiasca, B.; Galassi, D.M.P. Ammonium threshold values for groundwater quality in the EU may not protect groundwater fauna: Evidence from an alluvial aquifer in Italy. *Hydrobiologia* **2015**, *743*, 139–150. [CrossRef]
78. Scharf, B.; Brunke, M. The recolonization of the river Elbe with benthic and hyporheic Ostracoda (Crustacea) after the reunion of Germany in 1989. *Int. Rev. Hydrobiol.* **2013**, *98*. [CrossRef]
79. Hahn, H.J. The GW-Fauna-Index: A first approach to a quantitative ecological assessment of groundwater habitats. *Limnologica* **2006**, *36*, 119–137. [CrossRef]
80. Fillinger, L.; Hug, K.; Trimbach, A.M.; Wang, H.; Kellermann, C.; Meyer, A.; Bendinger, B.; Griebler, C. The D-A-(C) index: A practical approach towards the microbiological-ecological monitoring of groundwater ecosystems. *Water Res.* **2019**, *163*, 114902. [CrossRef] [PubMed]
81. Korbel, K.; Hose, G.C. The weighted groundwater health index: Improving the monitoring and management of groundwater resources. *Ecol. Indic.* **2017**, *75*, 164–181. [CrossRef]
82. Korbel, K.L.; Hose, G.C. A tiered framework for assessing groundwater ecosystem health. *Hydrobiologia* **2011**, *661*, 329–349. [CrossRef]

4

Reduction of Nutrient Leaching Potential in Coarse-Textured Soil by using Biochar

Yu-Lin Kuo [1], Chia-Hisng Lee [2] and Shih-Hao Jien [3,*]

1 Department of Civil Engineering, National Pingtung University of Science and Technology, Pingtung 91201, Taiwan; q3489505@gmail.com
2 Center for Sustainability Science, Academia Sinica, Taibei 11529, Taiwan; d91623402@ntu.edu.tw
3 Department of Soil and Water Conservation, National Pingtung University of Science and Technology, Pingtung 91201, Taiwan
* Correspondence: shjien@mail.npust.edu.tw

Abstract: *Background*: Loss of nutrients and organic carbon (OC) through leaching or erosion may degrade soil and water quality, which in turn could lead to food insecurity. Adding biochar to soil can effectively improve soil stability, therefore, evaluating the effects of biochar on OC and nutrient retention and leaching is critical. *Methods*: We conducted a 42-day column leaching experiment by using sandy loam soil samples mixed with 2% of biochar pyrolyzed from Honduran mahogany (*Swietenia macrophylla*) wood sawdust at 300 °C (WB300) and 600 °C (WB600) and a control sample. Leaching was achieved by flushing the soil column on day 4 and every week during the 42-day experiment and adding a water volume for each flushing equivalent to the field water capacity. *Results*: Biochar application increased the final soil pH and OC, NH_4^+-N, NO_3^--N, available P concentrations but not exchangeable K concentrations. In particular, WB600 exhibited superior performance in alleviating soil acidification; WB300 engendered high NO_3^--N concentrations. Biochar application effectively retained water in soil and inhibited the leaching of the aforementioned nutrients and dissolved OC. WB300 reduced NH_4^+-N and K leaching by 30%, and WB600 reduced P leaching by 68%. *Conclusions*: Biochar application can improve nutrient retention and reduce the leaching potential of soils and connected water bodies.

Keywords: biochar; organic carbon; nutrients; leaching; nitrogen; phosphorus; potassium

---

1. Introduction

Nutrients (nitrogen, phosphorus, and potassium) and soil organic carbon (SOC) are critical components of a healthy soil, which is the foundation of a strong food system [1]. Tropical ecosystems are particularly susceptible to the loss of nutrients through soil erosion or leaching processes [2]. Intense rainfall in tropical or subtropical areas results in the leaching of fertilizer containing N, P, and K from soil bodies. Nutrient leaching could diminish soil fertility, accelerate soil acidification, increase fertilizer costs for farmers, and reduce crop yields [3]. The deposition of leached nutrients into water bodies adversely affects aquatic environments because of potential risks such as eutrophication. Leaching of N and P and agricultural runoffs are among the leading contributors to non-point source (NPS) pollution, which has a detrimental effect on drinking and ground water, aquatic habitats, and other water resources. Agricultural runoffs often contain several contaminants, including nutrients, pesticides, pathogens, sediment, salts, trace metals, and other substances, which contribute to biological oxygen demand [4]. Moreover, SOC, comprising nutrient and soil biota, leaches out over time [5], which could aggravate nutrient losses and water pollution. An enormous quantity of fertilizers must be applied to counter the dwindling fertility of agricultural soil.

Biochar is a solid bioresource obtained through the pyrolysis of organic waste. Residues from agricultural and forestry production processes are suitable raw materials for the production of high-quality biochar [6,7]. Biochar is a porous substance containing high levels of carbon and various functional groups. Accordingly, the addition of biochar to agricultural soil has emerged as a feasible strategy to enhance soil water retention capacity [8–10], soil quality [11–14], soil organic matter stability and nutrient retention [15,16], organic carbon (OC) sequestration [17], and greenhouse gases emission reduction [18–21]. Furthermore, biochar can affect soil microbial properties, including microbial activity [22] and microbial diversity [23]. However, the interactions between biochar and microbial properties in soil are not fully understood [24]. The application of biochar to soil could increase soil fertility and crop productivity by reducing leaching or even supplying nutrients [25–27]. However, the effects of biochar on nutrient leaching and OC retention has been reported to vary with the applied biochar pyrolysis temperature, raw material, and soil type [28,29]. Biochar produced from secondary forest residues could reduce fertilizer leaching and increase plant growth and nutrition [26]. Furthermore, the addition of biochar produced from hardwood to a typical Midwestern agricultural soil in the United States considerably reduced the leaching of total N and P by 11% and 69%, respectively [3]. Yao et al. [29] reported that the effect of biochar on nutrient retention and release varied with the nutrient and biochar type.

In this study, we conducted a 42-day column leaching experiment by using loamy sand soil samples that were obtained from a tropical/subtropical area and treated with two types of wood dust biochar pyrolyzed at 300 and 600 °C. The objective of this study was to determine the effects of biochar application on water, nutrient, and OC retention and leaching from the observed soil. The results are expected to be valuable for assessing the potential of biochar for the retention and immobilization of nutrients in soils and inhibition of water body contamination.

2. Materials and Methods

2.1. Collection of Soil Samples and Preparation of Biochars

Surface soil samples (0–15 cm) were collected from a field in Pingtung, Southern Taiwan (22°31′57.9″ N 120°33′38.1″ E). As of April 2016, pineapple (*Ananas comosus* (L.) Merr.) was the dominant crop on this land. The soil samples were air-dried, sieved through a 2-mm screen, and stored at room temperature. The biochar used in this study comprised Honduran mahogany (*Swietenia macrophylla*) wood sawdust obtained from the Department of Wood Design, National Pingtung University of Science and Technology. Two biochar materials were used in this study, namely WB300 and WB600, that were produced at pyrolysis temperatures of 300 and 600 °C, respectively. The biochar used in this study was supplied by the Industrial Technology Research Institute (ITRI) of Taiwan. Before being charred, the wood sawdust was dried at 60 °C for 24 h to < 10% moisture and cut to a particle size of 2 cm. For pyrolysis, the samples were placed in a tubular furnace (ITRI, Tainan, Taiwan) equipped with a corundum tube (diameter, 32 mm; length, 700 mm) and a N_2 purging mechanism (flow rate, 1 L/min) to ensure an oxygen-free atmosphere. Heat treatments were performed at temperatures of 300 and 600 °C, with the heating rate being 5 °C min^{-1}. The temperature was maintained for 2 h before cooling to an ambient temperature under an N_2 flow. After the pyrolysis, the biochar materials were ground to pass through a 2-mm sieve, followed by homogenization through stirring.

2.2. Preparations of Leaching Column

Similar to the procedures applied by Lo [30], the biochar materials were thoroughly mixed with the collected sandy loam soil at application rates of 0% (Control, 0 tons ha^{-1}) and 2% (40 tons ha^{-1}) w/w for Bok choy (*Brassica rapa chinensis*) cultivation in Taiwan. Briefly, nutrient solutions of ammonium sulfate ($(NH_4)_2SO_4$), calcium dihydrogen phosphate ($Ca(H_2PO_4)_2$), and potassium chloride (KCl) were added to the soil at application rates of 2076, 227, and 191 kg ha^{-1}, respectively (approximately 220, 30, and 100 kg ha^{-1} for N, P, and K, respectively). The fertilizers were dissolved in deionized (DI) water and then mixed thoroughly with the soil samples. The volume of the nutrient solution applied was 60% of the water retention capacity of the treated soil samples. The treatment samples are outlined as follows: (1) control, comprising soil only (CK); (2) WB300, comprising soil to which 2% of the biochar pyrolyzed at 300 °C was added; and (3) WB600, comprising the soil to which 2% of the biochar pyrolyzed at 600 °C was added. A leaching experiment was conducted for each treatment in three replicates. As illustrated in Figure 1, a soil column with an internal diameter of 20.6 cm was constructed. The column was composed of two polyvinyl chloride (PVC) tubings of equal length, which were connected through a PVC fitting with a 5-cm interval. A nylon mesh (1 mm^2) with filter paper (Whatman grade no. 42) above was placed between the joints to separate the soil in the upper part from the quartz sand (~2 mm in diameter) filled in the center PVC fitting. At a soil depth of 15 cm, the volume of the soil column was approximately 5000 cm^3. All the columns were packed with the tested soil samples to obtain an initial bulk density of 1.2 g cm^{-3}.

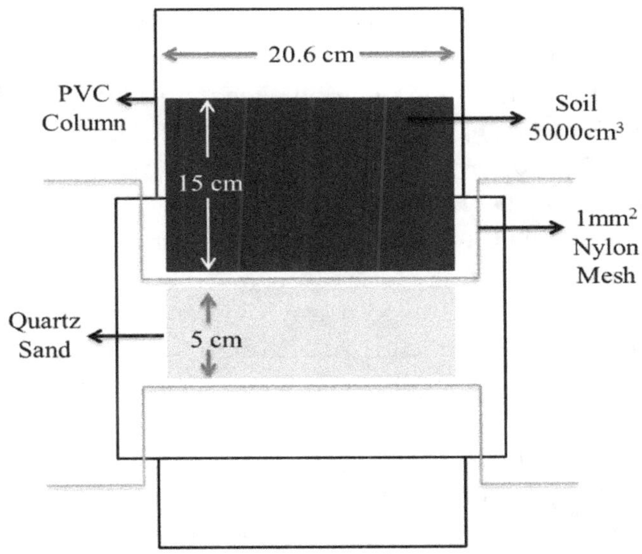

Figure 1. Schematic of the soil column constructed for the leaching experiment.

2.3. *Soil Column Incubation and Leaching*

The soil columns were subjected to a 42-day incubation process conducted at room temperature (25–28 °C) and humidity (60%–80%) with repeated leaching in order to investigate the effects of biochar application on (1) the physicochemical properties of, (2) the hydraulic properties of, and (3) nutrient retention and leaching from the soils. Short-term duration of incubation period was chosen based on Yoo et al. (2013) [31], where the leaching experiment was finished within 60 days. Likewise, based on our previous studies, the variation of chemical and physical properties [13] and dynamic changes of N and P [7] after biochar addition might occur and finish within 8 weeks, therefore, a short-term experiment period (42 days) was selected in this study. Table 1 lists the analysed items.

According to the soil porosity and volume determined for the studied soil column, we used a leaching volume of 700 mL for each flushing process. A fine sieve was placed above the columns to minimize water disturbance to the soil surfaces during flushing. Throughout the experimental period, all columns were leached seven times (on days 4, 7, 14, 21, 28, 35, and 42) using DI water. The leachates were collected using 1000-mL measuring cylinders, and the volumes of the leachates were recorded. Leachates were then subjected to chemical analyses. After the final leaching event, the soil of each column was collected, air-dried, and ground to pass through a 2-mm sieve before further chemical analysis.

Table 1. Analysis items and relevant abbreviations for the leachates and soils.

Properties		Leachate	Soil
Leachate Volume	V_L	✓	
pH	pH		✓
Bulk Density	D_B		✓
Organic Carbon	OC	✓	✓
Ammonium Nitrogen	NH_4^+-N	✓	✓
Nitrate-Nitrogen	NO_3^--N	✓	✓
Available Phosphorus	Ava. P	✓	✓
Exchangeable Potassium	Ex. K	✓	✓

2.4. Analytical Methods

The bulk density (D_B) was determined using the core method [32]. The pH values of the soil samples and biochar materials mixed with DI water (1:2.5 and 1:20 *w/v*, respectively) were determined using a Horiba F-74 BW meter [33]. We performed electrical conductivity measurements on saturated paste extracts of the soil samples by using a Horiba F-74 BW meter [34]. The soil particle size distribution was determined using the pipette method [35]. Cation exchange capacity (CEC) was determined using the ammonium acetate method (pH 7.0) [36]. Exchangeable K was extracted using 1 mol L^{-1} NH_4OAc (1:10 *w/v* for the soil samples; 1:20 *w/v* for the biochar materials), and the extract was analyzed through atomic absorption spectrometry (Z-2300, Hitachi, Tokyo, Japan). The OC concentration was determined through wet oxidation [37]. Available P was determined using the Bray P-1 extract test [38]. Inorganic N was extracted using 2 M KCl (1:10 *w/v*), and the concentrations of NH_4^+-N and NO_3^--N were determined though steam distillation conducted using MgO and Devarda's alloy [39]. The microscale structure of the biochar materials was characterized through optical microscopy using reflected light, followed by scanning electron microscopy (SEM; Hitachi, S-3000N, Japan). A backscattered electron image representing the mean atomic abundance in a black-and-white image was observed on the surface of the samples coated with Au. The C components of biochar horizons were examined through solid-state CPMAS ^{13}C nuclear magnetic resonance (DSX 400-MHz solid-state NMR, Bruker, Karlsruhe, Germany). Data acquisition was executed under the following conditions: spectrometer frequency, 100.46 MHz; spinning speed, 7000 Hz; contact time, 1 ms; and pulse delay time, 1 s. We determined the total signal intensity and the proportion contributed by each C functional group by integrating the spectra in the chemical-shift region: 0–50 ppm (aliphatic C), 50–110 ppm (O-alkyl-C), 110–165 ppm (aromatic C), and 165–190 ppm (carboxyl C). Methoxyl C contributed a wide shoulder between 50 and 60 ppm within the O-alkyl-C range (Alpha-T, Bruker).

2.5. Statistical Analysis

Data were analyzed using IBM SPSS Statistics 22 for Windows (IBM Corp., Armonk, NY, USA). Data sets were subjected to mean separation analysis using one-way analysis of variance, with significance being set to a p value of 0.05. The differences between mean values under different treatments were identified using Duncan's test.

3. Results

3.1. Properties of the Soil and Biochar Materials

Table 2 lists the properties of the soil samples and biochar materials. The texture of the studied soil was sandy loam; the soil was determined to have a neutral pH and low OC content. The porosity, bulk density, and particle density were in the normal ranges for the coarse-textured soil samples. The soil used in this study was sourced from an intensively cultivated field with high human input, which may result in high nutrient concentrations. The pH values of WB300 and WB600 (Honduran mahogany wood sawdust pyrolyzed at 300 and 600 °C) were 6.5 (neutral) and 10.4 (alkaline), respectively. The OC content in WB300 was 6.8%, which was higher than that in WB600 (2.0%). By contrast, the total carbon content was 69% in WB300, which was lower than that in WB600 (79.5%). These results indicate that WB600 contained a higher level of inorganic carbon than W300 did. The higher pyrolysis temperature reduced the concentrations of oxygen, nitrogen, ammonium–nitrogen, nitrate–nitrogen, available phosphorus, and exchangeable potassium in the biochar materials. Figure 2 depicts SEM images of both biochar materials. WB300 exhibited coarser pores than WB600 did but had a lower number of pores for the same volume. Because of its more porous structure—signifying a larger surface area—WB600 could have distinct effects on the physicochemical properties of soil and groundwater when compared with WB300.

Table 2. Properties of the studied soil and Honduran mahogany (*Swietenia macrophylla*) wood sawdust biochar samples pyrolyzed at 300 °C (WB300) and 600 °C (WB600).

Properties	Soil	Wood Biochar (WB)	
		300 °C (WB300)	600 °C (WB600)
pH	6.2	6.5	10.4
EC (dS m^{-1})	0.35	-	-
Sand (%)	72.5	-	-
Silt (%)	18.1	-	-
Clay (%)	9.4	-	-
Texture	SL	-	-
D_B (g cm^{-3})	1.44	-	-
D_P (g cm^{-3})	2.69	-	-
Porosity (%)	42.8	-	-
CEC (cmol(+)/kg)	10.2	55.1	20.4
OC (%)	0.33	6.8	2.0
TC (%)	-	69.0	79.5
H (%)	-	4.5	2.9
O (%)	-	24.7	14.6
N (%)	-	0.92	0.82
H/C	-	0.06	0.04
O/C	-	0.36	0.18
NH_4^+-N (mg/kg)	82.4	56.7	41.7
NO_3^--N (mg/kg)	131	548	341
Ava. P (mg/kg)	6.63	5.29	3.30
Ex. K (mg/kg)	259	449	323

EC: electrical conductivity; SL: sandy loam; D_B: bulk density; D_P: particle density; OC: organic carbon; TC: total carbon; H: hydrogen; O: oxygen; N: nitrogen; NH_4^+-N: ammonium-nitrogen; NO_3^--N: nitrate-nitrogen; Ava. P: available phosphorous; Ex. K: exchangeable potassium; -: Not determined.

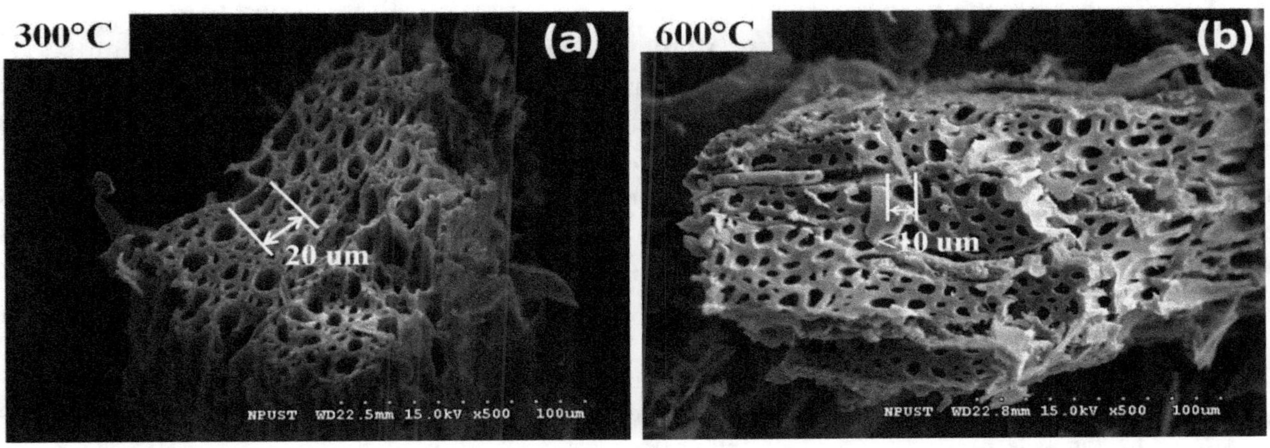

Figure 2. Scanning electron microscope (SEM) images of Honduran mahogany (*Swietenia macrophylla*) wood sawdust biochar pyrolyzed at (**a**) 300 °C and (**b**) 600 °C.

Figure 3 illustrates the functional groups of C within the structures of the WB300 and WB600. O-alkyl-C is the major C group in the natural composition of Honduran mahogany. By pyrolyzing at 300 °C, the WB300 consisted of more aromatic-C, less O-alkyl-C, more alkyl-C, and more carboxylic-C than the raw wood dust. At 600 °C, the pyrolysis process resulted in the predominating aromatic-C in the WB600, and the other C groups became less observable. The results of the physical and chemical properties of the biochars as affected by pyrolysis temperature are consistent with previous studies [7,13,40,41].

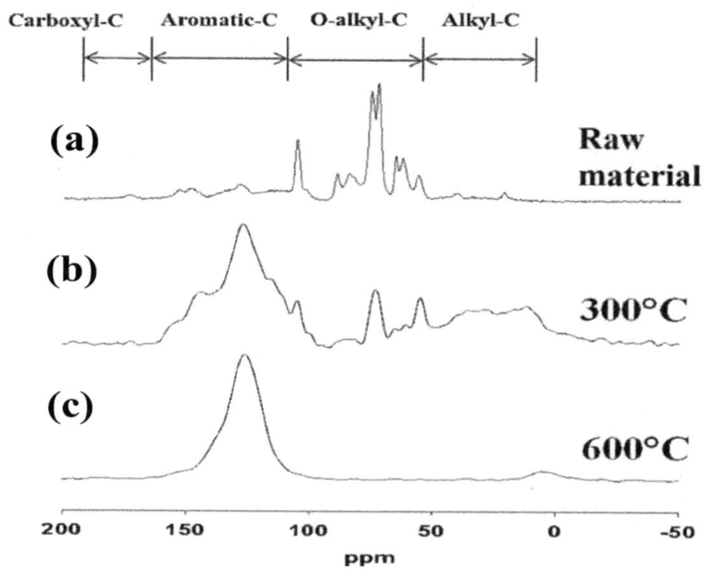

Figure 3. Solid-state ^{13}C cross-polarization magic-angle spinning nuclear magnetic resonance spectra for Honduran mahogany (*Swietenia macrophylla*) wood sawdust (**a**) and its biochar materials pyrolyzed at 300 °C (**b**) and 600 °C (**c**).

3.2. Soil Physicochemical Properties

Table 3 shows the major soil properties under different treatments before and after the 42-days experiment. The pH value of the untreated soil (before mixing with fertilizer) was 6.8, as shown in Table 2. After fertilization, the pH value under control dropped to 6.1. The pH of soil treated with WB300 was ~6.2, whereas WB600 had the highest pH value of 6.5. At day 42, the soil pH values of both biochar treatments were significantly higher than that of the control (pH 4.6), as shown in Table 3. The WB600 treated soil still revealed the highest pH value (5.8) at day 42 among all treatments. Although the soil under WB300 treatment had the similar pH value (6.2) with the control on day 0, it revealed a higher pH than the control at day 42.

The D_B values observed on day 42 for all the treated samples were lower than the initial D_B of 1.20 g cm^{-3} (achieved when the soil columns were packed). The control exhibited the highest D_B (1.11 g cm^{-3}); the WB600-treated sample had the lowest value (1.05 g cm^{-3}). However, the differences in D_B between the treated samples were not significant ($p = 0.05$) (Table 3). Because biochar typically contains low levels of OC, the SOC levels observed for all treatments were low (0.21%–0.46%). The WB300-treated sample had the highest SOC content levels on both day 0 and day 42 (0.46% and 0.39%, respectively), indicating an SOC content loss of only 0.7% throughout the experiment (Table 3). The SOC levels observed for the WB600-treated sample did not differ significantly from that observed for the control (Table 3).

The NH_4^+-N concentration did not differ significantly between any of the treated samples on day 0 (Table 3). On day 42, the NH_4^+-N concentration in all treated samples decreased drastically from approximately 205 mg kg^{-1} to less than 6.5% of the initial concentrations. On day 42, the control had the lowest concentration (5.87 mg kg^{-1}), and the WB300- and WB600-treated samples had significantly higher concentrations (12.7 mg kg^{-1}). The NO_3^--N and inorganic N concentrations in the treated samples exhibited similar trends to the NH_4^+-N concentrations. On day 42, all treated samples exhibited considerably lower NO_3^--N concentrations when compared with the initial concentrations; the NO_3^--N concentrations were high in the samples treated with the two biochar materials, particularly the WB300-treated sample.

Table 3. The Soil physicochemical properties on day 0 and day 42 (n = 3).

Properties	Day	Treatments		
		CK	WB300	WB600
pH	0	6.1 ± 0.1 [a]	6.2 ± 0.1 [a]	6.5 ± 0.1 [b]
	42	4.6 ± 0.1 [a]	5.2 ± 0.2 [b]	5.8 ± 0.1 [c]
D_B (g cm^{-3})	42	1.11 ± 0.24 [a]	1.09 ± 0.11 [a]	1.05 ± 0.10 [a]
SOC (%)	0	0.33 ± 0.05 [a]	0.46 ± 0.03 [b]	0.36 ± 0.05 [a]
	42	0.21 ± 0.05 [a]	0.39 ± 0.05 [b]	0.31 ± 0.07 [a,b]
NH_4^+-N (mg kg^{-1})	0	206 ± 6.28 [a]	205 ± 6.16 [a]	205 ± 6.35 [a]
	42	5.87 ± 1.19 [a]	12.7 ± 1.15 [b]	12.7 ± 1.21 [b]
NO_3^--N (mg kg^{-1})	0	131 ± 8.35 [a]	138 ± 8.98 [a]	135 ± 8.14 [a]
	42	7.52 ± 1.14 [a]	33.0 ± 1.31 [c]	28.0 ± 3.63 [b]
Inorganic N (mg kg^{-1})	0	336 ± 8.61 [a]	343 ± 9.60 [a]	339 ± 8.38 [a]
	42	13.4 ± 0.82 [a]	45.7 ± 0.11 [c]	40.7 ± 3.42 [b]
Ava. P (mg kg^{-1})	0	19.6 ± 0.27 [a]	19.6 ± 0.26 [a]	19.8 ± 0.28 [a]
	42	4.08 ± 0.45 [a]	5.10 ± 0.28 [b]	5.36 ± 0.75 [b]
Ex. K (mg kg^{-1})	0	488 ± 46.2 [a]	457 ± 30.6 [a]	497 ± 45.2 [a]
	42	302 ± 3.02 [a]	315 ± 0.24 [b]	351 ± 0.13 [c]

D_B: bulk density; SOC: soil organic carbon; NH_4^+-N: ammonium–nitrogen; NO_3^--N: nitrate–nitrogen; N: nitrogen; Ava. P: available phosphorous; Ex. K: exchangeable potassium. The values followed by the same superscript letters within a row are not significantly different ($p > 0.05$) between relevant treatments.

On day 0, the Ava. P concentrations did not differ significantly between the three treated soil samples (19.6–19.8 mg kg^{-1}). On day 42, the Ava. P concentration decreased to 4.08, 5.1, and 5.36 mg kg^{-1} in the control, WB300-treated, and WB600-treated samples, respectively. On day 0, the Ex. K concentrations in the control, WB300-treated, and WB600-treated samples were 302, 315, and 351 mg kg^{-1}, respectively. After the experiment, the Ex. K concentrations increased to 457–488 mg kg^{-1} in all treated samples and did not differ significantly between the samples.

3.3. Properties of Leachate

Both biochar-treated samples exhibited significantly smaller leachate volumes than that of the control for each flushing event (Table 4; Figure 4). On day 4 (the day of the first flushing event), the soil column with the control sample had a leachate volume of 530 mL, and both WB300- and WB600-treated samples retained approximately 150 mL more water than the control did (i.e., the leachate volume decreased by 28%). At the end of the experiment, the cumulative leachate volumes observed for the WB300- and WB600-treated samples were lower than that observed for the control by 9.2% and 13.7%, respectively.

Figure 5 displays the cumulative level of dissolved OC (DOC) in the leachate. This was highest in the control and lowest in the WB600-treated sample after the experiment (188 and 154 mg, respectively). After 42 days of incubation, the level of DOC leached from the soil column decreased by 6.50% and 20.0% in the WB300- and WB600-treated samples, respectively, compared with the control. The biochar materials contained low levels of OC. Accordingly, biochar introduces a negligible level of DOC into soils.

Table 4. Volume of the leachate from the soil columns after each flushing with DI water (n = 3).

Treatments	Volume of the Leachate (mL)						
	Incubation Time (Days)						
	4	7	14	21	28	35	42
CK	530 ± 45 [b]	539 ± 55	579 ± 39	598 ± 287	625 ± 21	624 ± 82	633 ± 08 [b]
WB300	378 ± 63 [a]	459 ± 53	539 ± 23	577 ± 18	600 ± 22	598 ± 08	597 ± 11 [a]
WB600	377 ± 05 [a]	426 ± 55	488 ± 110	525 ± 64	570 ± 05	598 ± 10	578 ± 24 [a]

The values followed by the same superscript letters within a column are not significantly different ($p > 0.05$) between the relevant treatments.

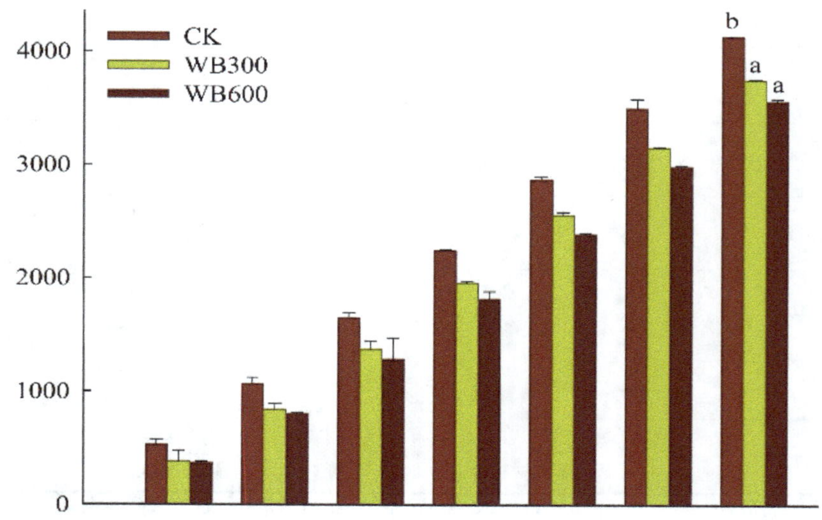

Figure 4. Cumulative leachate volume (V_L) of treated samples (n = 3). Different letters above the bars for day 42 indicate significant differences between the relevant treated samples ($p < 0.05$).

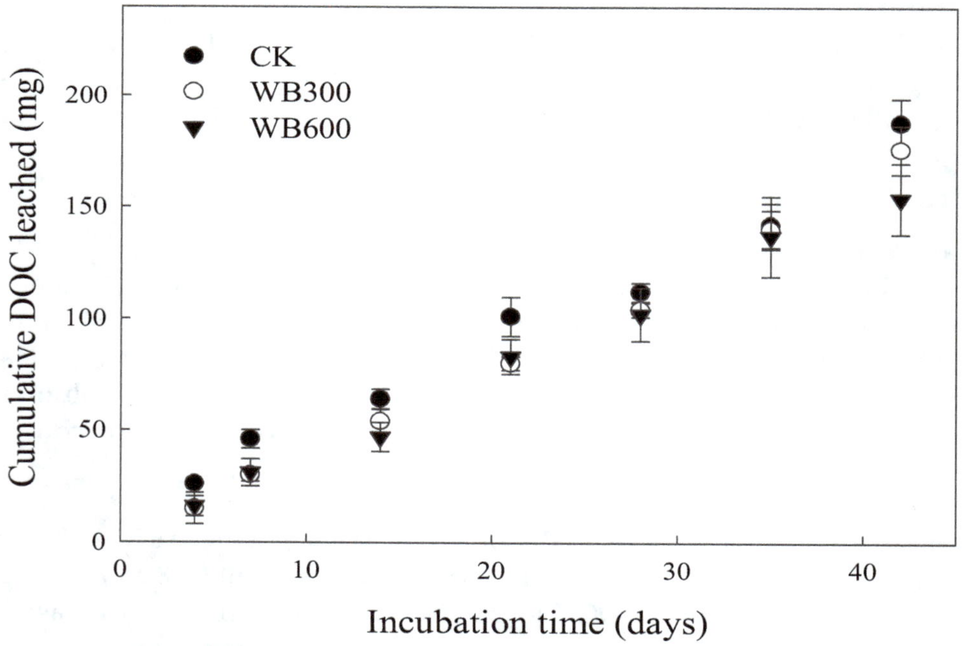

Figure 5. Cumulative concentration of dissolved organic carbon (DOC) for different treated samples (n = 3).

The cumulative quantities of NH_4^+-N and NO_3^--N leached from the soil columns are illustrated in Figure 6. WB300 remarkably reduced NH_4^+-N leaching by 30.5% relative to the control (69.6 mg). Although the inhibitory effect of WB600 on NH4+-N leaching was relatively weak, it still reduced the total quantity of NH_4^+-N leached from the soil by 10.6%, which was approximately one-third of that observed for the WC300-treated sample. The WB300- and WB600-treatments reduced the quantities of NO_3^--N leached from the soil samples by 13.8% and 16.4%, respectively, compared with the control (83.9 mg).

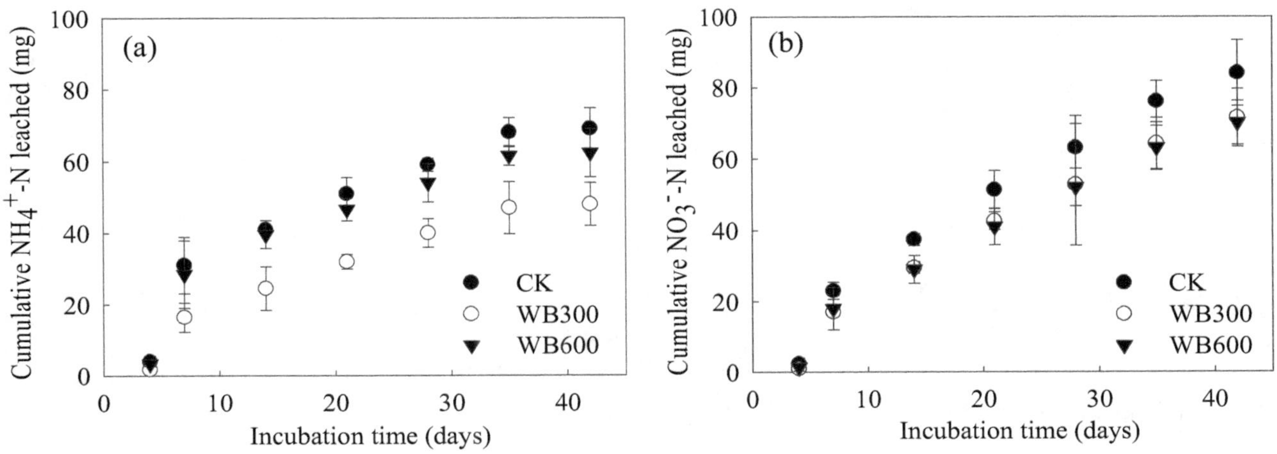

Figure 6. Cumulative quantities of (**a**) NH_4^+-N and (**b**) NO_3^--N leached from the soil columns (n = 3).

The cumulative quantity of inorganic N (summation of the quantities of NH_4^+-N and NO_3^--N) leached from the soil samples subjected to the different treatments differed significantly (Figure 7). The control exhibited the highest quantity of inorganic N leached from the soil (154 mg), and the WB300-treated sample exhibited the lowest quantity (33.3% lower than the control). Furthermore, WB600 treatment decreased the quantity of inorganic N leached from the soil by 13.7% only.

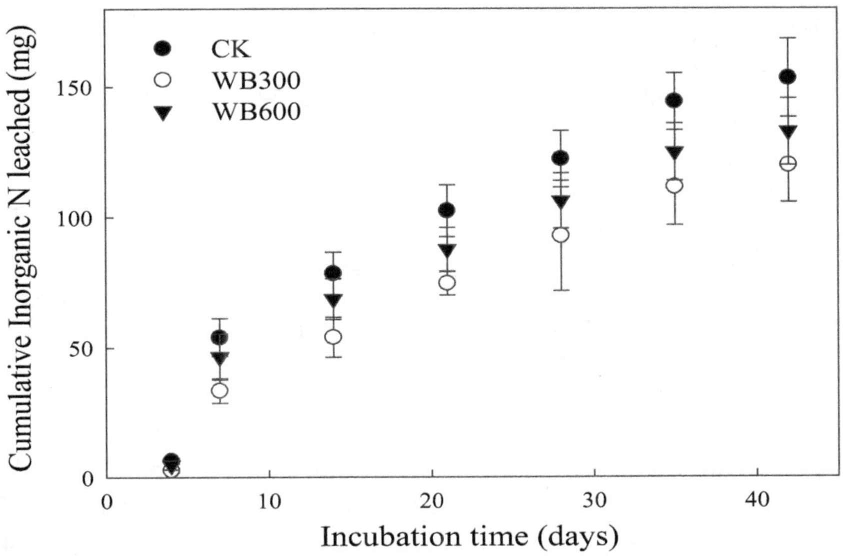

Figure 7. Cumulative quantity of inorganic N leached from the soil columns (n = 3).

Figure 8 displays the cumulative quantities of P leached from the soil columns. The total quantity of P leached from the WB600-treated sample decreased significantly (68.0%) compared with that from the control (12.2 mg). For the WB300-treated sample, the total quantity of P leached from the soil decreased by 45.2%. Compared with the control, the WB300 and WB600 treatments reduced the total quantities of P leached from the soil by 29.71% and 7.70% (156 and 210 mg leached), respectively (Figure 9).

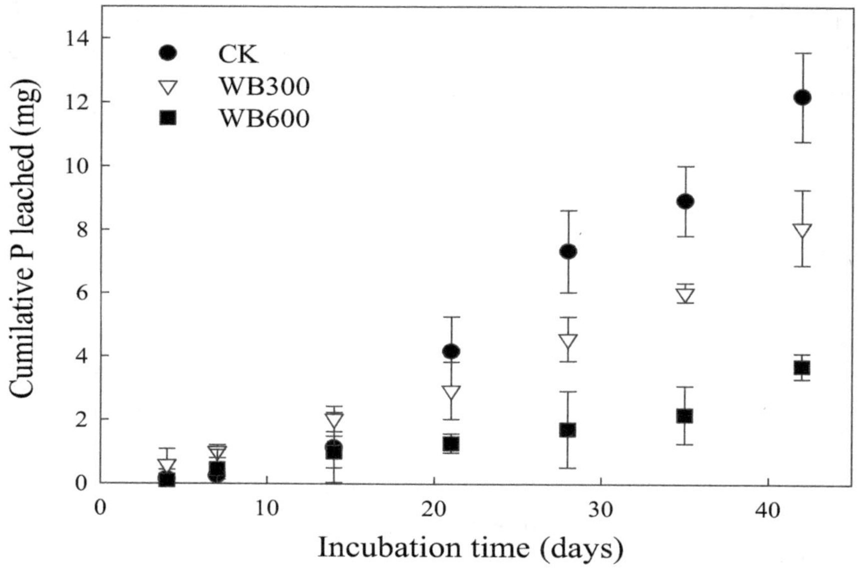

Figure 8. Cumulative quantity of phosphorus leached from the soil columns (n = 3).

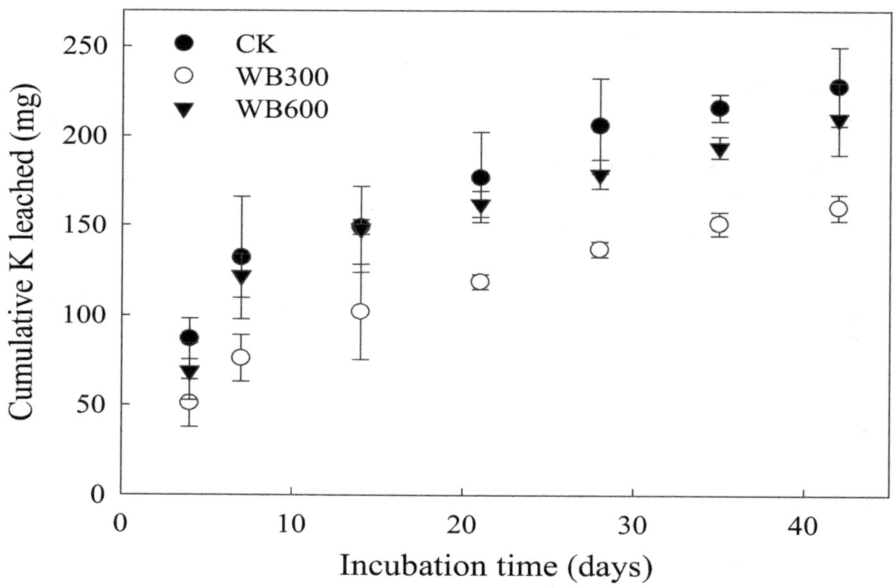

Figure 9. Cumulative quantity of potassium leached from the soil columns (n = 3).

4. Discussion

Compared with the original soil (control on day 0), the pH value of the fertilized soil decreased from 6.8 to 6.1, which could be attributed to the acidic properties of the two fertilizers, namely $(NH_4)_2SO_4$ and $Ca(H_2PO_4)_2$. The pH value of the control decreased to 4.6 on day 42, and those of the WB300- and WB600-treated samples were considerably higher (Table 3). The results indicate that both WB300 and WB600 could alleviate soil acidification. When biochar undergoes pyrolysis at a higher temperature, it generally has a higher pH [42]. Singh et al. [43] revealed that the $CaCO_3$ equivalence of biochar increased with the pyrolysis temperature. Accordingly, the application of biochar could engender liming effects. For biochar, a higher application rate or higher pyrolysis temperature could increase the pH or alkalinity of the biochar-treated soil [11]. The alleviation effects of biochar on soil

acidification could contribute to the retardation of the movement of several nutrients and pollutants from soils to groundwater and lower water bodies.

Applying biochar could decrease the D_B of a soil sample [13]. However, this phenomenon was not observed in this study; a possible reason is that the reorganization of the soil structure for mixing the materials and packing the soil columns neutralized the soil properties. A similar condition can be achieved in the field through intensive tilling. For a less- disturbed sandy loam soil, the application of biochar could improve the soil structure, increase the infiltration rate, and reduce runoff water and soil erosion, thus improving soil and water conservation [7,13,41]. The volumes of leachates increased with repeated leaching for all treatments in this study. Although the differences in leachate volume between the control and biochar-treated samples decreased gradually in this study, the biochar-treated samples still retained a significantly higher amount of water in the soil columns on day 42 (Table 4). The WB300 and WB600 treatments reduced the level of water loss by 9.2% and 13.7%, respectively. Our results demonstrate that the biochar materials, particularly WB600, exhibited a strong ability to conserve water in the soil samples when applied at a rate of 2%.

The OC contents in both of the biochars were much lower than that of other organic materials commonly used in farmland. The biochar applications displayed significantly affect the SOC in the WB300 treatment only in this study (Table 3). Accordingly, the effects of biochar on DOC leaching are the results of the sorption of organic carbon onto the biochar, either within the pores of the biochar or onto the external biochar surface [7]. The biochar-applied soils retained more water than the control (Figure 4), as illustrated earlier, which could also contribute to DOC retention since DOC move down and leach out from the soil column along with the soil water. Our results illustrated that the highest amounts of DOC and leachate were found in CK indicating biochar addition could effectively retain DOC in the soils. However, the highest DOC concentration (47.0 mg/L, DOC amount/leachate volume) was found in the leachate of WB300 treatment indicated that biochar might release solube C components into soil solution, particularly in biochar pyrolized with low temperatures [7,44]. The results implicated that soluable C components onto the biochar itself might increase DOC concentration in output water (runoff or eluate from soil pedon) duing rainfall events in the biochar-amended soils. The WB300 treatment revealed the efficiency of 9.2% for water retention and the lower efficiency of 6.5% for DOC retention (Figures 4 and 5). Under WB600 treatment, the retention efficiencies were 13.7% for water and a higher of 20.0% for DOC. This result indicates that WB600 had a stronger affinity to DOC than WB300. Kasozi et al. [45] reported that the organic matter sorption onto biochar surfaces is kinetically limited by slow diffusion into the subnanometer-sized pores dominating biochar surfaces. The various organo-mineral interactions lead to aggregations of soil and organic materials, which stabilizes both soil structure and the carbon compounds within the aggregates. Furthermore, the increase in the diversity and density of carbon groups within WB300 biochar may result in the slightly increased SOC but did not prevent the leaching of DOC as effective as the WB600 biochar.

Although the NH_4^+-N and NO_3^--N concentrations in the soil columns were low on day 42 (Table 3), the WB300- and WB600-treated samples, especially the WB300-treated sample, exhibited higher inorganic N concentrations than the control did. These results could be attributed to the high surface area and diverse functional groups, such as carboxyl C, O-alkyl-C, and alkyl C, of WB300 (Figure 3). Obvious higher CEC was also fould in WB300 than in WB600 in this study indicating that more NH_4^+-N and K could be retained in the soil treated with WB300, and which were demonstrated by our results in Figures 6 and 9 [46,47]. Addtionally, improvement in soil physical properties such as promotion of soil aggregation and increasing of water holding capacity might also bring positive effects in nutrient leaching. Yoo et al. [31] suggested that increasing formation of aggregates by biochar addition could effectively promoted retention of NO_3^-. Furthermore, they also indicated that increased water holding capacity after biochar addition was also a factor to reduce N leaching.

Overall, both biochars can effectively reduce N leaching and provide a potential N source of nutrient delivery to plants [48]. Agricultural non-point source (NPS) pollution is the leading source of water quality impacts to rivers and lakes. Nitrogen from fertilizers, manure, waste and ammonia turns into nitrite and nitrate. High levels of these toxins deplete waters of oxygen, killing all of the animals and fish. Nitrates also soak into the ground and end up in drinking water. Health problems can occur as a result of this and they contribute to methemeglopbinemia or blue baby syndrome which causes death in infants. Base on our results, application biochar might be a useful managememt practice to reduce NPS pollution in watersheds, particularly in tropical/subtropical climate regions. In this study, we measured the N losses through leaching. The N losses through denitrification and ammonia volatilization, which may form NH_3, N_2, or N_2O, were not accessed. As the soil was pH 6.8 and the soil columns were nearly water-saturated during the experiment, we used acidic fertilizer, $(NH_4)_2SO_4$ and $Ca(H_2PO_4)_2$ intentionally to minimize ammonium volatilization.

Applying the biochar materials engendered only a slight increase in the available P in the soil samples after the leaching experiment (Table 3). Both biochar materials, particularly WB600, could retain soil P, according to the leaching results (Figure 8). This effect also resulted in the higher concentrations of Ava. P in the soil samples after the experiment, as mentioned. P tends to precipitate with Fe under acidic conditions. However, the application of the biochar materials increased the soil pH, which could enhance the release of P. Therefore, the biochar materials were likely to have contained numerous bonding sites or other co-precipitation elements, resulting in considerably higher P retention efficiency levels compared with their water retention efficiency levels. During soil pH 4 to 6, increasing of soil pH might unlock P "adsorbed on soluble and hydrous Fe/Al oxides" into an available form, however, the unlocked P might be adsorbed again onto biochar to form a potential available form. Our results was consistence with the resuld of Laird et al. [3] who indicated that biochar addition could effectively reduce leaching of dissolved P in the soil column due to adsorption of orthophosphate and adsorption of organic P compounds by biochar.

The Ex. K concentration was higher in the biochar-treated soil samples, particularly the WB600-treated sample, than in the control at the beginning of the experiment; this could be attributed to the high mobility of K in plant ash. A higher pyrolysis temperature may result in a higher concentration of soluble K in biochar. However, the Ex. K concentrations did not differ significantly between the samples on day 42. Both biochar materials considerably reduced the total quantity of K leached from the soil samples. The efficiency levels of WB300 and WB600 in inhibiting K leaching were 29.7% and 7.7%, respectively. WB600 inhibited K leaching mainly through holding the soil solutions, and WB300 possibly had additional K sorption mechanisms mainly related to C functional groups.

Our results indicated that both WB300 and WB600 could effectively reduce the leaching of soil water, DOC, NH_4^+-N, NO_3^--N, P, and K. After incubation, not only nutrient concentration but also leachate volume were found lower in biochar-treated soil compared with those in control (Table 4). Reharding the strong water retention capacity of the biochar-treated soils, two possible reasons were speculated, which were (1) high degree of evaporation (about 30 °C in average during summer in southern Taiwan) for the soil column after water adding; therefore, the biochar treatments might still effectively retain water after water addition; (2) the volume of adding water (700 mL) to the soil column at each date still did not match the pore volume (~2000 cm^3) of soil column; therefore, the biochar treatments might still effectively retain water after water addition at day 7 and other dates. Nutrient leaching could be inhibited through increased water retention. Moreover, the inhibitory effect of the biochar materials on the mineralization of organic N, in terms of physical protection of organic matter [7,40], can reduce the quantity of nutrients released and thus reduce subsequent leaching. Our results reveal that WB300 exhibited a high NH_4^+-N and K (predominately cations) retention efficiency, and WB600 exhibited a high water, DOC, and P (predominately anions) retention efficiency. These results indicate that WB300 was negatively charged, which was most likely due to the distribution of various carbon functional groups. Overall, the biochar-induced retention of soil water, DOC, and nutrients could be considered to positively affect nutrient and water conservation and to improve soil quality. Reducing the leaching of water, DOC, and nutrients from soils could conserve groundwater and connected water bodies. Therefore, biochar application can benefit both soil and water conservation. Further research should be conducted to investigate whether these positive effects can be extended to the field and to downstream water bodies at the catchment level.

5. Conclusions

The results of this study demonstrate that the incorporation of Honduran mahogany (*Swietenia macrophylla*) wood sawdust biochar into sandy loam soil samples could improve soil health by increasing the capacity of the soil to retain nutrients and reduce nutrient leaching. The biochar materials applied in this study, particularly WB600, alleviated soil acidification. The incorporation of the biochar materials considerably increased SOC, inorganic N (NH_4^+-N and NO_3^--N), and ava. P concentrations in the soil samples. The biochar application did not significantly affect D_B or Ex. K in the soil. Furthermore, the total volume of leachates and cumulative quantity of DOC, inorganic N, P, and K leached from the soil samples decreased significantly in the biochar-treated samples. WB300 engendered the least quantities of NH_4^+-N and K, both of which are cations. Thus, WB300 could be more negatively charged than WB600 because of its inhibitory effects on cation leaching. By contrast, WB600 was likely to have a higher density of P-bonding sites, which resulted in a stronger inhibitory effect on P leaching. Although the ability to retain various nutrients in soil differed with pyrolysis temperatures, both biochar materials effectively contributed to the conservation of groundwater and subsequent downstream water bodies. Increased retention of these nutrients in soils can increase the probability of absorption by plant roots, thereby decreasing the risk of leaching into rivers or groundwater reservoirs. In future studies, the on-site effects of biochar application on underground and water bodies should be determined in terms of eutrophication and potential pollution, especially for intensively fertilized cropped fields.

Author Contributions: Conceptualization and methodology, S.-H.J.; sample analysis, Y.-L.K.; writing—original draft preparation, S.-H.J. and Y.-L.K.; writing—review and editing, S.-H.J. and C.-H.L. All authors have read and agreed to the published version of the manuscript.

Acknowledgments: We would like to thank the students of the Soil Survey and Conservation Laboratory, Department of Soil and Water Conservation, National Pingtung University of Science and Technology, for their assistance during the column leaching experiment and analysis.

References

1. Robertson, M. *Sustainability Principles and Practice*, 2nd ed.; Taylor & Francis: London, UK, 2017.
2. Jordan, C.F.; Herrera, R. Tropical rain forests: Are nutrients really critical? *Am. Nat.* **1981**, *117*, 167–180. [CrossRef]
3. Laird, D.; Fleming, P.; Wang, B.; Horton, R.; Karlen, D. Biochar impact on nutrient leaching from a Midwestern agricultural soil. *Geoderma* **2010**, *158*, 436–442. [CrossRef]
4. O'Green, A.T.; Budd, R.; Gan, J.; Manynard, J.J.; Parikh, S.J.; Dahlgren, R.A. Mitigating nonpoint source pollution in agriculture with constructed and restored wetlands. *Adv. Agron.* **2010**, *108*, 1–76.
5. Lal, R. Tillage and agricultural sustainability. *Soil Tillage Res.* **1991**, *20*, 133–146. [CrossRef]
6. Lu, K.; Yang, X.; Shen, J.; Robinson, B.; Huang, H.; Liu, D.; Bolan, N.; Pei, J.; Wang, H. Effect of bamboo and rice straw biochars on the bioavailability of Cd, Cu, Pb and Zn to Sedum plumbizincicola. *Agric. Ecosyst. Environ.* **2014**, *191*, 124–132. [CrossRef]
7. Jien, S.H.; Wang, C.C.; Lee, C.H.; Lee, T.Y. Stabilization of organic matter by biochar application in compost-amended soils with contrasting pH values and textures. *Sustainability* **2015**, *7*, 13317–13333. [CrossRef]
8. Busscher, W.J.; Novak, J.M.; Evans, D.E.; Watts, D.W.; Niandou, M.; Ahmedna, M. Influence of pecan biochar on physical properties of a Norfolk loamy sand. *Soil Sci.* **2010**, *175*, 10–14. [CrossRef]
9. Kammann, C.I.; Linsel, S.; Gößling, J.W.; Koyro, H.W. Influence of biochar on drought tolerance of Chenopodium quinoa Willd and on soil–plant relations. *Plant Soil* **2011**, *345*, 195–210. [CrossRef]

10. Basso, A.S.; Miguez, F.E.; Laird, D.A.; Horton, R.; Westgate, M. Assessing potential of biochar for increasing water-holding capacity of sandy soils. *Gcb Bioenergy* **2013**, *5*, 132–143. [CrossRef]
11. Yuan, J.H.; Xu, R.K.; Zhang, H. The forms of alkalis in the biochar produced from crop residues at different temperatures. *Bioresour. Technol.* **2011**, *102*, 3488–3497. [CrossRef]
12. Vaccari, F.; Baronti, S.; Lugato, E.; Genesio, L.; Castaldi, S.; Fornasier, F.; Miglietta, F. Biochar as a strategy to sequester carbon and increase yield in durum wheat. *Eur. J. Agron.* **2011**, *34*, 231–238. [CrossRef]
13. Jien, S.H.; Wang, C.S. Effects of biochar on soil properties and erosion potential in a highly weathered soil. *Catena* **2013**, *110*, 225–233. [CrossRef]
14. Zhao, X.; Wang, J.; Xu, H.; Zhou, C.; Wang, S.; Xing, G. Effects of crop-straw biochar on crop growth and soil fertility over a wheat-millet rotation in soils of China. *Soil Use Manag.* **2014**, *30*, 311–319. [CrossRef]
15. Clough, T.J.; Condron, L.M.; Kammann, C.; Müller, C. A review of biochar and soil nitrogen dynamics. *Agronomy* **2013**, *3*, 275–293. [CrossRef]
16. Ventura, M.; Sorrenti, G.; Panzacchi, P.; George, E.; Tonon, G. Biochar reduces short-term nitrate leaching from a horizon in an apple orchard. *J. Environ. Qual.* **2013**, *42*, 76–82. [CrossRef]
17. Laird, D.A. The charcoal vision: A win–win–win scenario for simultaneously producing bioenergy, permanently sequestering carbon, while improving soil and water quality. *Agron. J.* **2008**, *100*, 178–181.
18. Gaunt, J.L.; Lehmann, L. Energy balance and emissions associated with biochar sequestration and pyrolysis bioenergy production. *Environ. Sci. Technol.* **2008**, *42*, 4152–4158. [CrossRef]
19. Woolf, D.; Amonette, J.E.; Street-Perrott, F.A.; Lehmann, J.; Joseph, S. Sustainable biochar to mitigate global climate change. *Nat. Commun.* **2010**, *1*, 1–9. [CrossRef]
20. Zheng, J.; Stewart, C.E.; Cotrufo, M.F. Biochar and nitrogen fertilizer alters soil nitrogen dynamics and greenhouse gas fluxes from two temperate soils. *J. Environ. Qual.* **2012**, *41*, 1361–1370. [CrossRef]
21. Stewart, C.E.; Zheng, J.; Botte, J.; Cotrufo, M.F. Co-generated fast pyrolysis biochar mitigates green-house gas emissions and increases carbon sequestration in temperate soils. *GCB Bioenergy* **2013**, *5*, 153–164. [CrossRef]
22. Warnock, D.D.; Lehmann, J.; Kuyper, T.W.; Rillig, M.C. Mycorrhizal responses to biochar in soil–concepts and mechanisms. *Plant Soil* **2007**, *300*, 9–20. [CrossRef]
23. Jin, H. *Characterization of Microbial Life Colonizing Biochar and Biochar-Amended Soils*; Cornell University: New York, NY, USA, 2010.
24. Lehmann, J.; Rillig, M.C.; Thies, J.; Masiello, C.A.; Hockaday, W.C.; Crowley, D. Biochar effects on soil biota—A review. *Soil Biol. Biochem.* **2011**, *43*, 1812–1836. [CrossRef]
25. Glaser, B.; Lehmann, J.; Zech, W. Ameliorating physical and chemical properties of highly weathered soils in the tropics with charcoal—A review. *Biol. Fertil. Soils* **2002**, *35*, 219–230. [CrossRef]
26. Lehmann, J.; da Silva, J.P., Jr.; Steiner, C.; Nehls, T.; Zech, W.; Glaser, B. Nutrient availability and leaching in an archaeological Anthrosol and a Ferralsol of the Central Amazon basin: Fertilizer, manure and charcoal amendments. *Plant Soil* **2003**, *249*, 343–357. [CrossRef]
27. Major, J.; Rondon, M.; Molina, D.; Riha, S.J.; Lehmann, J. Maize yield and nutrition during 4 years after biochar application to a Colombian savanna oxisol. *Plant Soil* **2010**, *333*, 117–128. [CrossRef]
28. Mukherjee, A.; Zimmerman, A.R. Organic carbon and nutrient release from a range of laboratory-produced biochars and biochar–soil mixtures. *Geoderma* **2013**, *193*, 122–130. [CrossRef]
29. Yao, Y.; Gao, B.; Zhang, M.; Inyang, M.; Zimmerman, A.R. Effect of biochar amendment on sorption and leaching of nitrate, ammonium, and phosphate in a sandy soil. *Chemosphere* **2012**, *89*, 1467–1471. [CrossRef]
30. Lo, C.H. *Manual of Fertilizer Application*, 6th ed.; Agriculture and Food Agency, Council of Agriculture, Executive Yuan: Taipei, Taiwan, 2005.
31. Yoo, G.Y.; Kim, H.J.; Chen, J.J.; Kim, Y.S. Effects of biochar addition on nitrogen leaching and soil structure following fertilizer application to rice paddy soil. *Soil Sci. Soc. Am. J.* **2013**, *78*, 852–860. [CrossRef]
32. Blake, G.R.; Hartge, K.H. Bulk density. In *Methods of Soil Analysis, Part 1. Physical and Mineralogical Methods*; Klut, A., Ed.; ASA and SSSA: Madison, WI, USA, 1986; Agronomy monograph No. 9; pp. 363–375.
33. McLean, E. Soil pH and lime requirement. In *Methods of Soil Analysis. Part 2. Chemical and Microbiological Properties*; Page, A.L., Miller, R.H., Keeney, D.R., Eds.; ASA and SSSA: Madison, WI, USA, 1982; pp. 199–224.
34. Rhoades, J. 1982. Cation exchange capacity. In *Methods of Soil Analysis. Part 2. Chemical and Microbiological Properties*; Page, A.L., Miller, R.H., Keeney, D.R., Eds.; ASA and SSSA: Madison, WI, USA, 1982; pp. 149–157.

35. Gee, G.W.; Bauder, J.W.; Klute, A. Particle-size analysis. In *Methods of Soil Analysis. Part 1. Physical and Mineralogical Methods*; Klut, A., Ed.; ASA and SSSA: Madison, WI, USA, 1986; Agronomy monograph No. 9; pp. 383–411.
36. Sumner, M.E.; Miller, W.P. Cation exchange capacity and exchange coefficients. In *Methods of Soil Analysis: Soil Science Society of America Book Series 5 Part 3—Chemical Methods*; Sparks, D.L., Page, A.L., Helmke, P.A., Loeppert, R.H., Eds.; ASA and SSSA: Madison, WI, USA, 1996; pp. 1201–1229.
37. Nelson, D.W.; Sommers, L.E.; Sparks, D.; Page, A.; Helmke, P.; Loeppert, R.; Soltanpour, P.; Tabatabai, M.; Johnston, C.; Sumner, M. Total carbon, organic carbon, and organic matter. In *Methods of Soil Analysis: Soil Science Society of America Book Series 5 Part 3—Chemical Methods*; Sparks, D.L., Page, A.L., Helmke, P.A., Loeppert, R.H., Eds.; ASA and SSSA: Madison, WI, USA, 1996; pp. 961–1010.
38. Bray, R.H.; Kurtz, L. Determination of total, organic, and available forms of phosphorus in soils. *Soil Sci.* **1945**, *59*, 39–46. [CrossRef]
39. Mulvaney, R.L. Nitrogen-Inorganic forms. In *Methods of Soil Analysis: Soil Science Society of America Book Series 5 Part 3—Chemical Methods*; Sparks, D.L., Page, A.L., Helmke, P.A., Loeppert, R.H., Eds.; ASA and SSSA: Madison, WI, USA, 1996; pp. 1123–1184.
40. Jien, S.H.; Chen, W.C.; Ok, Y.S.; Awad, Y.M.; Liao, C.S. Short-term biochar application induced variation of C and N mineralization in a compost-added rural soil. *Environ. Sci. Pollut. Res.* **2018**. [CrossRef]
41. Lee, C.H.; Wang, C.C.; Lin, H.H.; Lee, S.S.; Tsang, D.C.W.; Jien, S.H.; Ok, Y.S. In situ biochar application conserves water and nutrients while simultaneously mitigating the erosion of an Fe-oxide-enriched tropical soil. *Sci. Total Environ.* **2018**, *619*, 665–671. [CrossRef] [PubMed]
42. Cantrell, K.B.; Hunt, P.G.; Uchimiya, M.; Novak, J.M.; Ro, K.S. Impact of pyrolysis temperature and manure source on physicochemical characteristics of biochar. *Bioresour. Technol.* **2012**, *107*, 419–428. [CrossRef] [PubMed]
43. Singh, B.; Singh, B.P.; Cowie, A.L. Characterisation and evaluation of biochars for their application as a soil amendment. *Soil Res.* **2010**, *48*, 516–525. [CrossRef]
44. Mukherjee, A.; Zimmerman, A.R.; Harris, W. Surface chemistry variations among a series of laboratory-produced biochars. *Geoderma* **2011**, *163*, 247–255. [CrossRef]
45. Kasozi, G.N.; Zimmerman, A.R.; Kizza, P.N.; Gao, B. Catechol and Humic Acid Sorption onto a Range of LaboratoryProduced Black Carbons (Biochars). *Environ. Sci. Technol.* **2010**, *44*, 6189–6195. [CrossRef] [PubMed]
46. Steiner, C.; Glaser, B.; Teixeira, W.G.; Lehmann, J.; Blum, W.E.H.; Zech, W. Nitrogen retention and plant uptake on a highly weathered central Amazonian Ferralsol amended with compost and charcoal. *J. Plant Nutr. Soil Sci.* **2008**, *171*, 893–899. [CrossRef]
47. Lentz, R.D.; Ippolito, J.A. Biochar and manure affect calcareous soil and corn silage nutrient concentrations and uptake. *J. Environ. Qual.* **2011**, *41*, 1033–1043. [CrossRef]
48. Hagemann, N.; Kammann, C.; Schmidt, H.P.; Kappler, A. Nitrate capture and slow release in biochar amended compost and soil. *PLoS ONE* **2017**, *12*, e171214. [CrossRef]

5

The Application of Artificial Mussels in Conjunction with Transplanted Bivalves to Assess Elemental Exposure in a Platinum Mining Area

Marelize Labuschagne [1,*], Victor Wepener [1], Milen Nachev [2], Sonja Zimmermann [1,2], Bernd Sures [2] and Nico J. Smit [1]

1 Water Research Group, Unit for Environmental Sciences and Management, North-West University, 2520 Potchefstroom, South Africa; victor.wepener@nwu.ac.za (V.W.); sonja.zimmermann@uni-due.de (S.Z.); nico.smit@nwu.ac.za (N.J.S.)

2 Department of Aquatic Ecology and Centre for Water and Environmental Research, University of Duisburg-Essen, 45141 Essen, Germany; milen.nachev@uni-due.de (M.N.); bernd.sures@uni-due.de (B.S.)

* Correspondence: marelizelabuschagne94@gmail.com

Abstract: There is increasing evidence that platinum group elements (PGE) are pollutants of emerging concern worldwide. Limited information exists on levels, particularly in regions where PGEs are mined. A passive sampling device (i.e., the artificial mussel (AM)) and transplanted indicator organisms (i.e., the freshwater clam *Corbicula fluminalis africana*) were deployed along a PGE mining gradient in the Hex River, South Africa, and concentrations of As, Cd, Co, Cr, Ni, Pb, Pt, V, and Zn were determined after six weeks of exposure. Results showed differential uptake patterns for Pt, Cr, and Ni between the AMs and clams indicating availability differences. For monitoring purposes, a combination of AMs and indicator organisms provides a more holistic assessment of element exposure in aquatic environments.

Keywords: platinum; bioaccumulation; passive sample; freshwater clam; *Corbicula fluminalis africana*

---

1. Introduction

In recent decades, the environmental concentrations of platinum group elements (PGE) have increased in different environmental matrices worldwide since these precious metals have been used more often in a number of applications [1]. The biggest contributors to PGEs in the environment are the use of automotive catalytic converters and the mining activities. Mining activities in South Africa greatly contribute to PGE emissions in the environment [1]. It is particularly in the platinum mining regions of the North-West Province of South Africa where PGE and associated metals show increased levels in the aquatic environment [2,3].

Bioaccumulation is an important process where living organisms take up toxicants at a greater rate than the rate at which they eliminate these substances [4]. Many factors can influence the bioaccumulation of metals in the environment such as pH, conductivity, temperature, and salinity, as well as biotic factors such as age, body size, and reproductive status of the bioaccumulating organism [4].

It is this process of bioaccumulation that is applied in monitoring studies where organisms such as bivalves are used as indicators of metal exposure [5]. Traditionally, resident bivalve populations are used as indicator species (passive biomonitoring) [6]. However, in instances where no resident species are available, organisms that were collected from an unstressed or otherwise unpolluted population can be translocated to a selected site to determine its degree of pollution (active biomonitoring) [7].

Suitable transplanted organisms are selected based on biological characteristics, such as tolerance and abundance. Transplanted organisms facilitate the investigation of areas where native specimens are absent or compensatory adaptive mechanisms occur in the native populations from contaminated areas [8]. For this study, the freshwater clam *Corbicula fluminalis africana* was selected as a bioindicator organism. Bivalves are excellent bioindicators because of their sensitivity to contaminants, sedentary lifestyle, wide geographical distribution, high abundance, high bioaccumulation capacity, and their sturdiness for both laboratory and field studies [4].

Despite the widespread use of living organisms as indicators of metal bioaccumulation, they also have disadvantages. The kinetics of metal uptake and elimination by these organisms are sometimes not well understood [9] and are affected by factors such as temperature, life stage of the organism (e.g., larval versus adult), size, depth, and reproduction [7]. Different indicator species have different natural distribution patterns. Therefore, they are not always present in all ecosystems that need to be monitored to prevent the comparison between different sites [7].

To overcome these disadvantages, passive sampling devices such as the artificial mussel (AM) have been utilised. The advantages of using AMs for monitoring metals are the provision of time-integrated estimates of metal concentrations within the aquatic ecosystems, the possibility of monitoring water bodies with unfavourable conditions for living indicator species, and the avoidance of killing organisms [10,11]. These devices are not affected by biotic and abiotic factors and, thus, can be deployed in marine and freshwater ecosystems [12,13].

Several studies included both AMs and transplanted organisms to monitor metal exposure. The results of these studies indicate that AMs are less affected by salinity and temperature than transplanted organisms [10,14,15] and that AMs and bivalves accumulate metals at different levels but in a similar pattern [13,15].

The present study aims to determine if the AM can be used as a tool to determine the biologically available concentrations of Pt and other elements in freshwater ecosystems. The AM was tested in a field study alongside an established bivalve bioaccumulation indicator to determine the degree of similarity in element bioaccumulation. Additionally, the concentrations measured in the AMs and bivalves were related to the environmental concentrations in the water at each of the study sites.

2. Materials and Methods

2.1. Description of the Study Area and Sampling Sites

The Hex River flows through the city of Rustenburg, which is the most populated municipality in the North West Province of South Africa. Rustenburg is located on the western limb of the Bushveld Igneous Complex where the main industrial activity is mining [16]. These mining activities include mining for platinum and chromium. Along the course of the Hex River, there are impoundments, which can be affected by mining activities [17].

Olifantsnek Dam is situated to the south of Rustenburg near the origin of the Hex River and is separated from the mining area by the Magaliesberg Mountain range. It was, therefore, chosen as the reference site [17]. According to Roux [18], the part of the Hex River that joins with Olifantsnek Dam is in a good to fair ecological state. The main impacts on this impoundment include water abstraction and farming activities. The ecological category of this stretch of the river classifies it as largely natural/moderately modified. Olifantsnek Dam is managed by private owners, where the main activity is recreation, e.g., fishing and sailing [18].

Bospoort Dam is a small state-owned impoundment situated along the course of the Hex River in Rustenburg [17]. The land-uses in the catchment include urban developments, intensive mining activities, and agricultural activities [19]. The water of Bospoort Dam is also used for irrigation and, in the past, for domestic water supply. However, due to a decrease in the water quality and an increase in algal blooms that influenced the taste and odor of the water, Bospoort Water Treatment Works (WTW) ceased to operate [19]. The Bospoort Dam is classified as being in a hypereutrophic state, where the possible influences include sewage treatment works, agricultural and urban run-offs, as well as recirculation of nutrients that can be found in the sediments [19]. In 2008, a report was published on the water quality of Bospoort Dam stating that the Cd, As, and Pb concentrations in the water are of concern [19]. The system also shows a high salt content, which is generally indicated by high conductivity values [19]. Urban surface runoff and mining activities are possible sources of ions that contributed to the high conductivity in the Bospoort dam, while the possible sources of elevated metal concentrations include mining activities and runoff from agricultural fields [19]. According to Roux [18], the Hex River upstream of Bospoort Dam is classified as seriously to critically modified, whereas the river below the dam is in a moderate to largely modified category.

Two sampling surveys during the low flow periods (dry seasons—May to June) in 2017 and 2018 were undertaken at the selected sites (Figure 1). In the second survey, a third site was added, which was located in a pollution control dam on the premises of a platinum mine to quantify the release of Pt and other elements by that mine. At each of these sampling sites, water samples were taken and AMs and clams were collected after exposure.

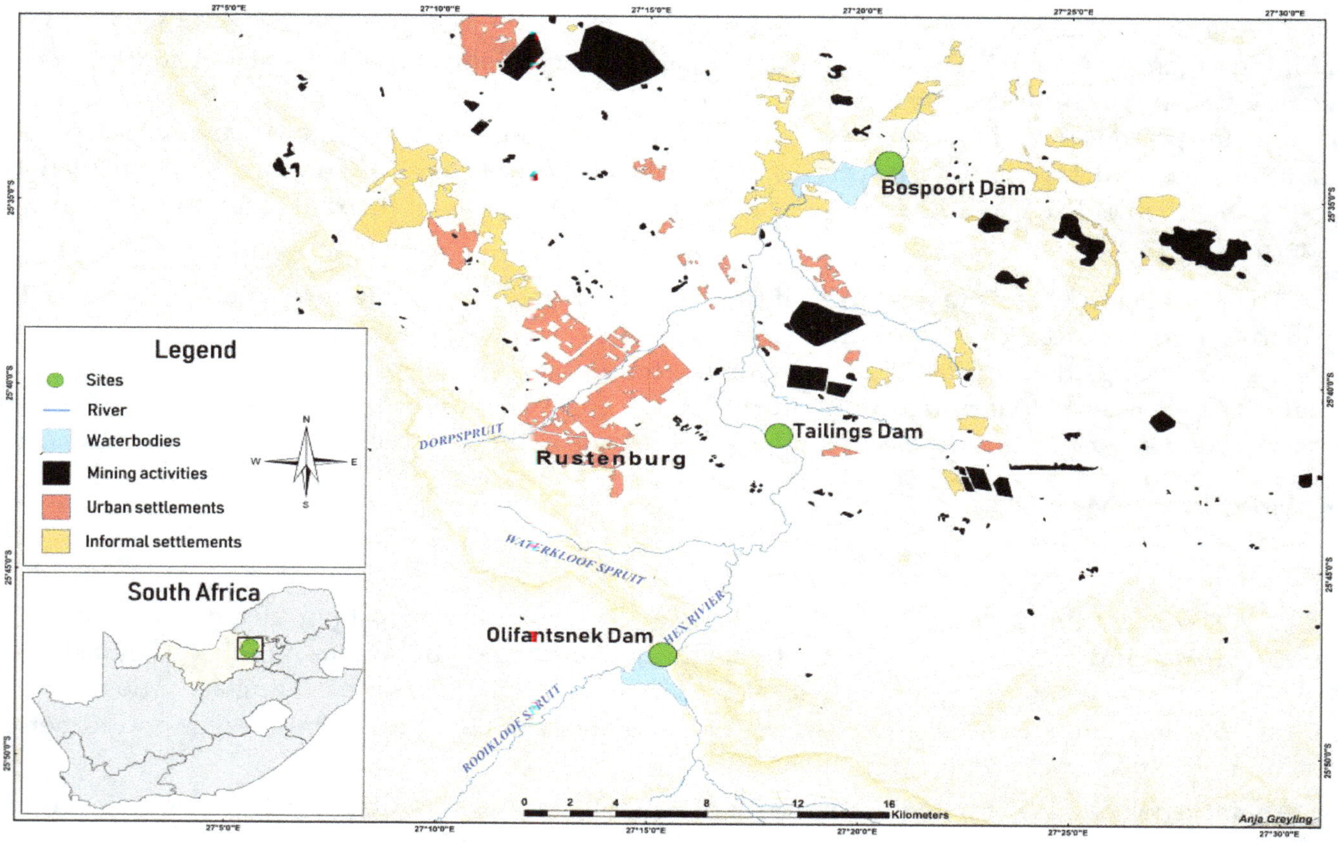

Figure 1. Map of the Hex River System in Rustenburg, North West Province, South Africa indicating the position of the sampling sites.

2.2. *Passive Sampling Device—Artificial Mussel*

The artificial mussel design was adapted for use in freshwater from the original design developed for marine environments by Wu, Lau, Fung, Ko, and Leung [10]. The artificial mussel consists of a Perspex tubing, containing 200 mg Chelex beads suspended in ultrapure water inside a 1 cm Perspex spacer (Figure 2). These components are enclosed between two polyacrylamide gel layers that consist of acrylamide (Acrylamide for electrophoresis, ≥99% (HPLC) powder, Sigma, Germany), N,N methylenebis-acrylamide (BioReagent, suitable for electrophoresis, 99%, Sigma, Germany), ammonium peroxidisulfate (Reagent grade, 98%, Sigma, Germany), and N,N,N′,N′-Tetramethylethylenediamine (BioReagent, for molecular biology, ≥99% (GC), Sigma, Germany) [20].

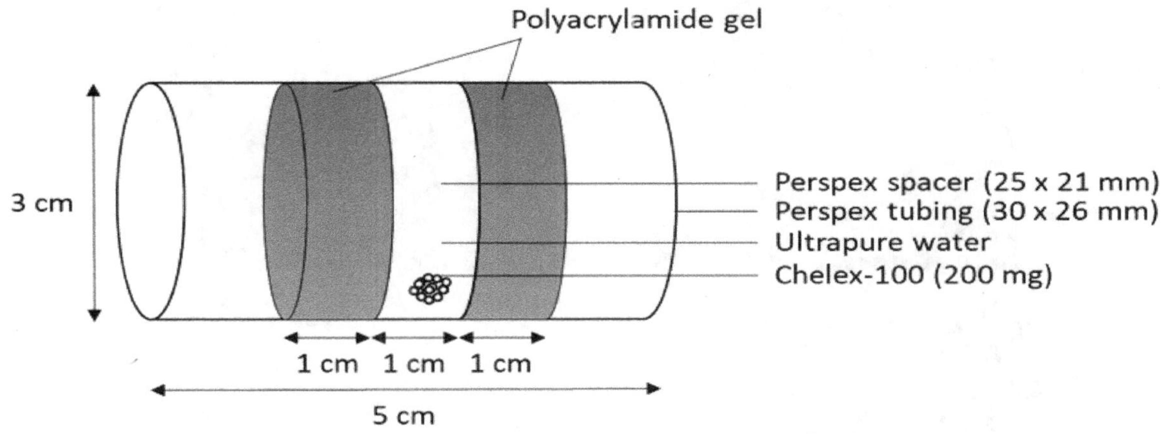

Figure 2. The artificial mussel design (Wu et al., 2007).

Once the AMs were assembled, the devices were placed in containers filled with ultrapure water and stored in the laboratory until being deployed. For the transportation of the AMs to the field, the devices were removed from the water and both ends of the tube were plugged with cotton soaked in ultrapure water. This prevents any damage that may occur during transportation. One set of 10 AMs was kept in the laboratory to serve as a non-exposed control.

2.3. *Corbicula Fluminalis Africana*

In addition to the deployment of the AM, an established bioaccumulation indicator species was concomitantly deployed. Freshwater clams, *Corbicula fluminalis africana,* with an average shell length of 28 ± 5 mm were collected from the Mooi River, Potchefstroom, South Africa. The clams were transported to the laboratory, where they were kept in reconstituted freshwater for mussels [21] for a depuration period of two weeks. During laboratory maintenance, the clams were fed with *Spirulina* cultured in the laboratory at the North-West University. Water was changed regularly throughout the two-week depuration period. For active biomonitoring, the clams were transported to the field in plastic containers filled with reconstituted freshwater and supplied with oxygen using a portable pump. A control group of clams was prepared for metal analyses to determine the initial concentrations before the clams were deployed at the sampling sites.

2.4. *Deployment of the Artificial Mussels and Clams*

The AMs and clams were deployed at each sampling site in two plastic baskets (Figure 3). Each basket was fitted with a mesh bag in the bottom of the basket. In addition, a mesh covered the top of the basket. Once arriving at the site, the mesh bags were filled with sediment from that site. This was done to create a natural habitat for the clams for the exposure period. Thereafter, 10 clams were transferred to each basket. In total, 20 transplanted clams were deployed at each site. After transferring the clams, the mesh bags were closed to prevent the organisms from escaping. The top half of each basket was then used to attach the AMs. Fifteen AMs were deployed at each site by fastening them to the side of the plastic baskets with cable ties. The baskets were covered with mesh to prevent objects from damaging the AMs during the survey. Once the AMs and transplanted clams were secured, the baskets were tied together with rope and cable ties and secured to a weight. The weights ensured the submersion of the baskets within the water body. The baskets were secured to buoys and permanent structures within the impoundments and the pollution control dam on the mine site.

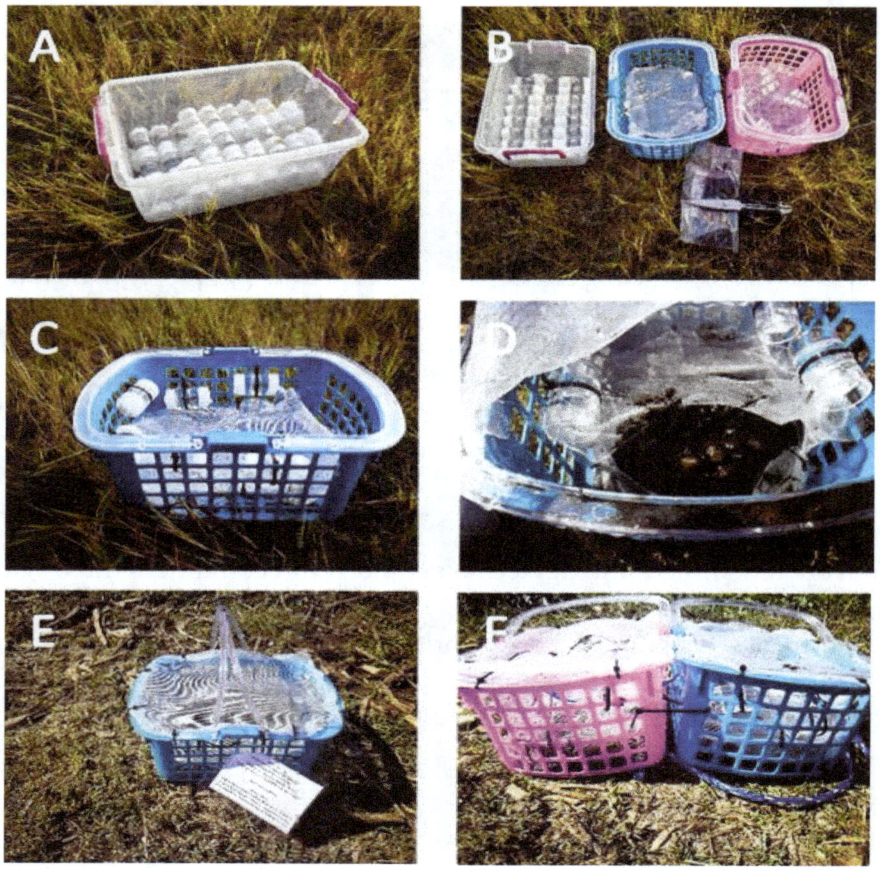

Figure 3. Steps taken to assemble the plastic baskets for the artificial mussels and transplanted clams. (**A**) Storage of artificial mussels before deployment. (**B**) Plastic baskets in which the artificial mussels and transplanted clams were deployed. (**C**) The AMs were secured to the plastic basket with cable ties. (**D**) Sediment from each site was placed in a mesh bag in the bottom of the basket for the transplanted clams. (**E,F**) Two baskets were deployed at each site. These baskets were secured together with rope and cable ties before submerging them in the water.

2.5. Retrieval of Artificial Mussels and Clams

Each batch of AMs and transplanted clams were retrieved after six weeks of exposure. Each AM was then rinsed with water from the site to remove any silt or algae that might have accumulated at the surface during the exposure period [20]. Cotton pads were soaked in water from the site, and placed in both ends of the AM to form a plug. This ensured that the gel would not get damaged during transportation in order to prevent the water and Chelex beads from seeping from the AM. At the laboratory, the containers with the AMs were placed at room temperature until further analysis.

The transplanted clams were retrieved at the same time as the AMs. These organisms were removed from the mesh bags and placed in containers that were fitted with an oxygen pump and water from the site. They were transported back to the laboratory and were killed by deep-freezing after they were measured and weighed. For each sampling site, soft tissues were dissected from the shells, placed in acid pre-washed polypropylene tubes, and frozen at −4 °C until further preparation. The clam tissue samples were freeze-dried before elemental analysis.

2.6. Water Quality Parameters and Sample Collection

In situ water quality parameters, i.e , pH, electrical conductivity (EC), total dissolved solids (TDS), temperature, and dissolved oxygen (DO) concentrations, were determined at each sampling site at the day of deployment as well as at the day of retrieval of the AMs and clams. At the same time, three water samples were collected at each site. The sampling procedure was as follows: 10 mL water was taken and acidified with 10 μL HNO_3 (sub-boiled from 65%; p.a. quality, Merck, Darmstadt, Germany) and stored in polypropylene tubes at room temperature until elemental analysis.

2.7. Element Analysis

For elemental analysis, the content of each control (n = 10) and field-deployed (n = 15) AM was emptied into a 15-mL polypropylene tube. After centrifugation (2 min at 1000× *g*), the supernatant was removed and the beads were rinsed with 5 mL of ultrapure water. The supernatant was removed and the beads were eluted with a mixture of 4.5 mL 6 M HNO_3 (sub-boiled from 65%, Merck, Darmstadt, Germany) and 0.5 mL HCl (37%, suprapure, Merck, Darmstadt, Germany) for approximately 2 hours to ensure that all bound elements were released from the beads. After centrifugation, the supernatant was removed and placed in polypropylene tubes at room temperature for further analysis.

For digestion of the freeze-dried clam soft tissue, five replicate samples of at least 60 mg (dry weight) were weighed into 20-mL TFM® vessels (MarsXpress, CEM, Kamp-Lintfort, Germany). The digestion was carried out in a microwave digestion system (CEM, Mars 6) with a mixture of 2.5 mL H_2O_2 (30%, Suprapur®, Merck, Darmstadt, Germany) and 1.3 mL HNO_3 (sub-boiled from 65%, p.a. quality, Merck, Darmstadt, Germany), according to Zimmermann et al. [22]. The clear digest solutions were then transferred to 5-mL glass flasks and brought to volume with 1% HNO_3. The digested solutions were stored at room temperature in polypropylene tubes until elemental detection.

Concentrations of As, Cd, Co, Ni, Pb, Pt, V, and Zn in the water samples, AMs, and clam tissue samples were determined using a quadrupole Inductively Coupled Plasma Mass Spectrometry (ICP-MS) system (Elan 6000, Perkin Elmer, Rodgau, Germany) equipped with an autosampler system (AS-90, Perkin Elmer, Rodgau, Germany). Cr concentrations were analyzed by atomic absorption spectrometry (AAS) as described by Erasmus et al. [3]. Hafnium oxide interference rates on Pt-194 were below 2%. Therefore, no Hf correction was performed. For the ICP-MS analysis, the wash time was set to 30 s with 2% HNO_3 to avoid contamination. After every 10 samples, a standard solution (10 μg/L), for all elements measured, was used to control the accuracy and stability of the measurements. Before measuring, samples were diluted 1:10 with an internal standard solution, which consisted of 1% HNO_3 and 10 μg/L thulium and yttrium (Certipur®, Merck, Darmstadt, Germany). Calibration of the instrument was performed using a series of 11 dilutions of the standard solution. With this, the concentrations of the sample analytes were calculated using regression lines with a correlation factor of ≥0.999. Detection limits were calculated as three times the standard deviation of the concentrations in 10 analytical blanks (Table 1).

Table 1. Detection limits for metal concentrations analyzed in water, artificial mussels, and transplanted clams.

Metal Analysed	Detection Limit for Water (μg/L)	Detection Limit for Artificial Mussels (μg/g)	Detection Limit for Transplanted Clams (μg/g)
As	0.203	0.0056	0.0169
Cd	0.022	0.0004	0.0018
Co	0.112	0.003	0.0099
Cr	0.151	0.039	0.130
Ni	0.463	0.011	0.0386
Pb	0.0237	0.0003	0.0020
Pt	0.0107	0.00015	0.0009
V	0.96	0.025	0.0798
Zn	17.7	0.053	1.48

The AM data was converted to μg/g units by referring the total eluted analyte mass (in μg) to the mass of Chelex beads (200 mg) in an AM. Thus, the AM data of the present study is comparable with the concentrations measured in the transplanted clams and other studies [20,23].

2.8. Statistical Analysis

Normal distribution of data was checked using the Kolmogorov-Smirnov test with a Dallal-Wilkinson-Lilliefor P-value. For element analyses in water samples of different sites, two-way ANOVAs were performed, which were followed by Tukey's multiple comparison test. For element analyses of both the AMs and transplanted clams, the data were log-transformed to compare all data sets since not all data passed normality. For statistical analyses, all samples that were below the detection limit were set to the detection limit divided by two [24]. A two-way ANOVA was performed to determine significant differences between surveys and sites for the AMs and transplanted clams. Thereafter, Tukey's multi-comparisons test was performed. Statistical significance was set at $p < 0.05$ for all comparisons. Principal component analyses (PCA) were conducted using Canoco 5 showing the grouping of AMs and transplanted clams based on the similarity of element concentrations at the different sites and surveys.

3. Results

The water quality parameters, i.e., pH, electrical conductivity (EC), total dissolved solids (TDS), temperature (Temp), and dissolved oxygen (DO) concentrations showed differences between sampling sites and survey years, respectively (Table 2). The temperature for the first survey was cooler than during the second survey, which influences the DO concentrations. During the second survey, the DO concentration was less than during the first survey. In both surveys, the Bospoort Dam had lower DO concentrations than the Olifantsnek Dam. The EC and TDS values differed significantly between the three sites. Bospoort Dam had significantly higher EC and TDS values than Olifantsnek Dam, whereas the pollution control dam showed the highest values. The pH was relatively stable throughout both surveys.

Table 2. Water quality parameters were measured at each site during the first survey (2017) and the second survey (2018). The results are the average of parameters measured during a survey. For abbreviations, see text.

Site	pH	Temp (°C)	EC (µS/cm)	TDS (mg/L)	DO (mg/L)	DO (%)
			Survey 2017			
Olifantsnek Dam	8.21	19.3	173.8	119.8	10.12	103.3
Bospoort Dam	8.46	19.45	910.5	570	9.69	81.4
			Survey 2018			
Olifantsnek Dam	8.20	24.3	210	147	5.5	62.1
Bospoort Dam	7.65	28.3	1160	819	2.94	34.3
Pollution control dam	8.00	25.6	1775	1244	-	-

The concentrations of the different elements in the water of the three sampling sites showed different patterns (Figures 4–6, left). The element concentrations in the water of the two impoundments, Olifantsnek Dam and Bospoort Dam, revealed no significant differences in a specific survey except for Ni in the first survey, where the water of Bospoort Dam contained significantly higher concentrations. However, in the second survey, the water of the pollution control dam contained significantly higher As, Co, Ni, V, and Zn concentrations than the other impoundments and significantly higher Cr and Pt concentrations when compared to only the Bospoort Dam.

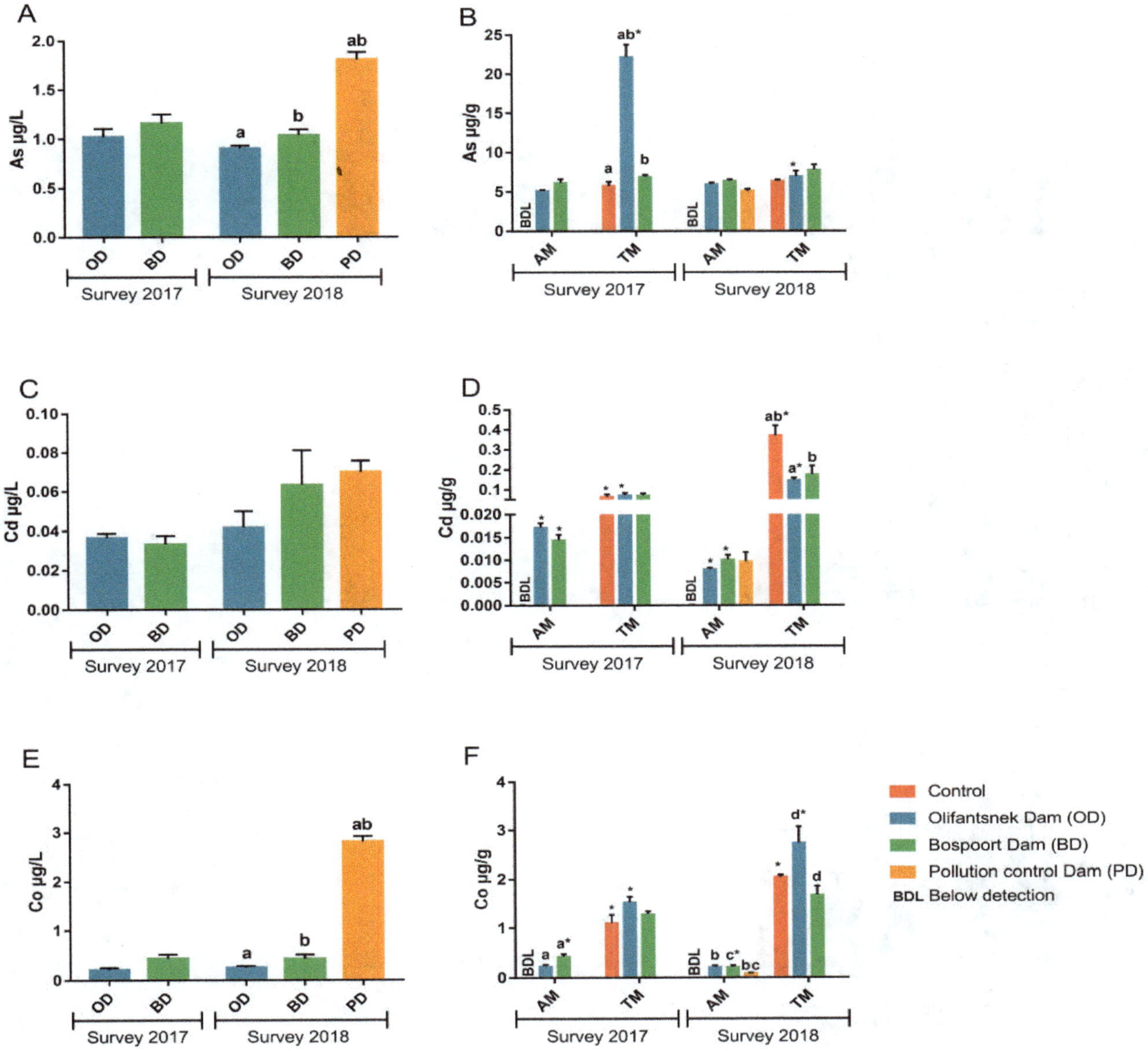

Figure 4. Arsenic (**A**,**B**), cadmium (**C**,**D**), and cobalt (**E**,**F**) concentrations (mean ± SEM) in water (**left**), artificial mussels (AM) and transplanted clams (TM) (**right**) during the two surveys in 2017 and 2018. Concentrations are presented in µg/L for water (n = 3), µg/g Chelex for the AMs (n = 15), and µg/g dry weight for transplanted clams (n = 5). Within a survey period, bars with a common letter superscript represent significant differences between different sites for the particular indicator group. Between surveys, an asterisk (*) indicates a significant temporal difference between the specific indicator at the particular site. Significance was regarded as $p < 0.05$. BDL = below detection limit.

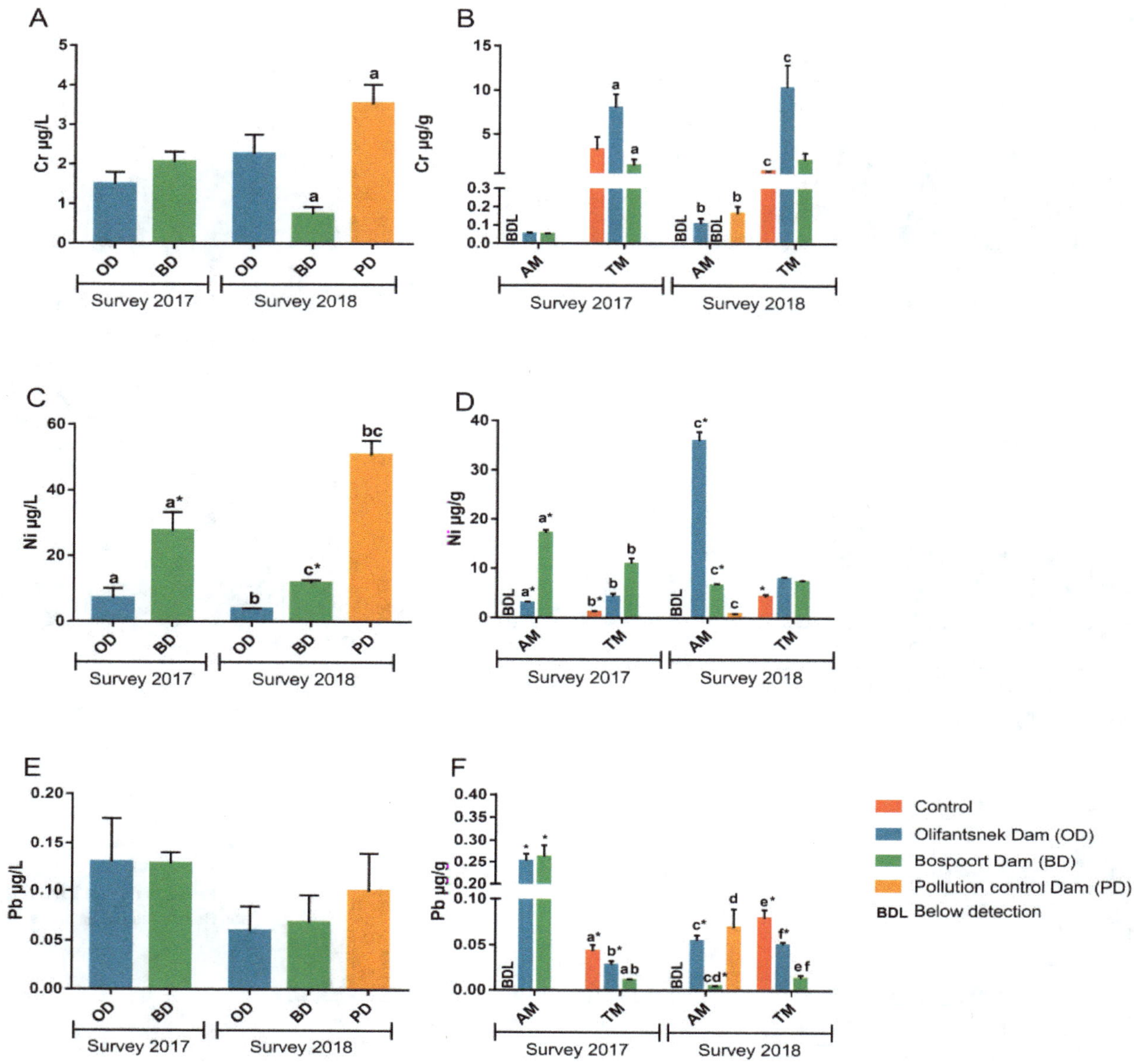

Figure 5. Chromium (**A**,**B**), nickel (**C**,**D**), and lead (**E**,**F**) concentrations (mean ± SEM) in water (**left**), and artificial mussels (AM) and transplanted clams (TM) (**right**) during the two surveys in 2017 and 2018. Concentrations are presented in µg/L for water (n = 3), µg/g Chelex for the AMs (n = 15), and µg/g dry weight for transplanted clams (n = 5). Within a survey, period bars with a common letter superscript represent significant differences between different sites for the particular indicator group. Between surveys, an asterisk (*) indicates a significant temporal difference between the specific indicator at the particular site. Significance was regarded as $p < 0.05$. BDL = below detection limit.

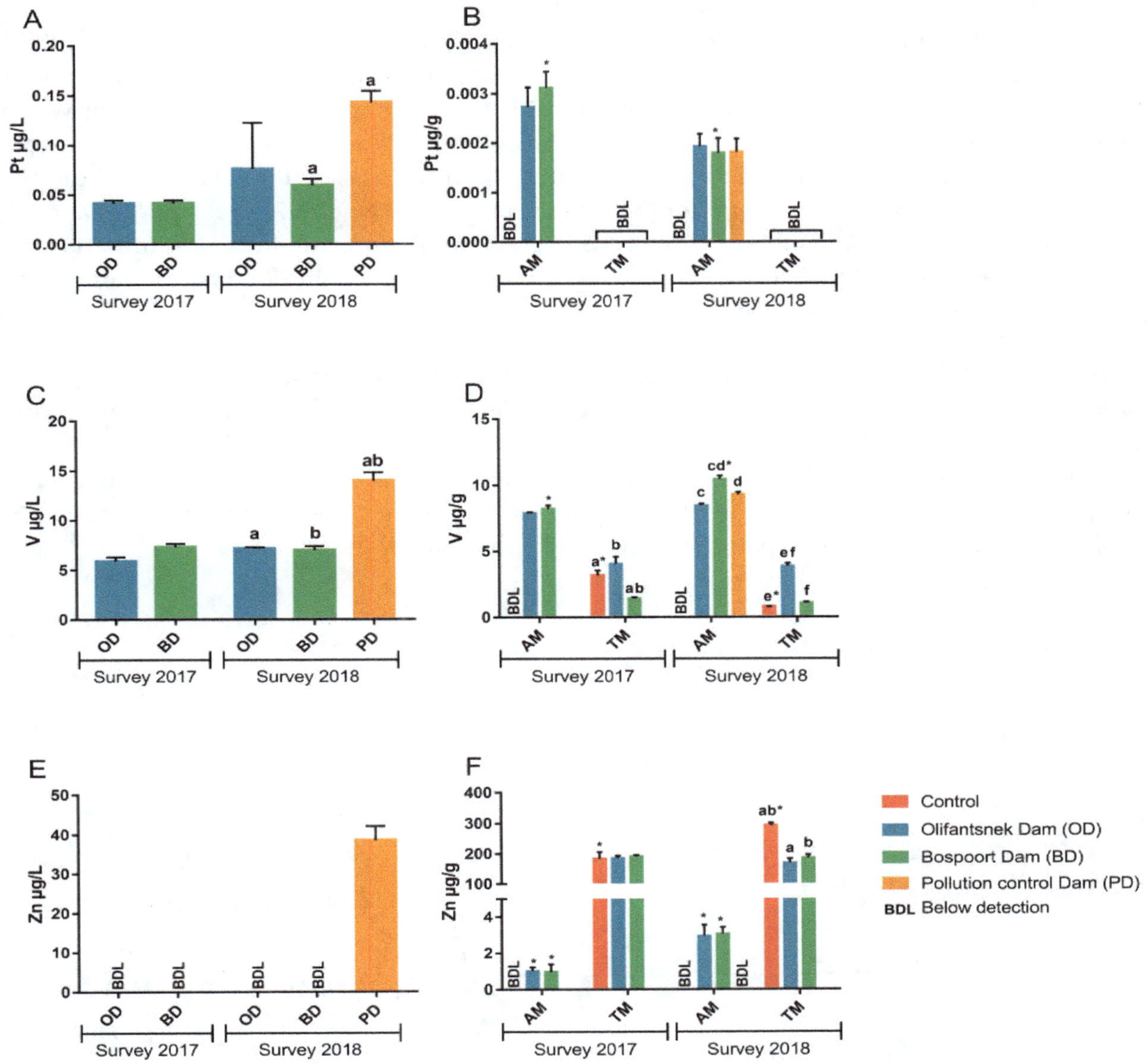

Figure 6. Platinum (**A**,**B**), vanadium (**C**,**D**), and zinc (**E**,**F**) concentrations (mean ± SEM) in water (**left**), and artificial mussels (AM) and transplanted clams (TM) (**right**) during the two surveys in 2017 and 2018. Concentrations are presented in µg/L for water (n = 3), µg/g Chelex for the AMs (n = 15), and µg/g dry weight for transplanted clams (n = 5). Within a survey, period bars with common letter superscript represent significant differences between different sites for the particular indicator group. Between surveys, an asterisk (*) indicates a significant temporal difference between the specific indicator at the particular site. Significance was regarded as $p < 0.05$. BDL = below detection limit.

Concentrations measured in the control AMs were below detection limits (Table 1) for all elements. Therefore, the element concentrations measured in the field-deployed AMs are attributed to accumulation from the ambient environment during the exposure period. In the first survey, the Co and Ni concentrations of the AMs from Bospoort Dam were significantly higher when compared to the Olifantsnek Dam. In the second survey, the AMs of the Olifantsnek Dam contained significantly higher Cr, Ni, and Pb concentrations, but significantly lower V concentrations when compared to Bospoort Dam. The AMs from the pollution control dam in the second survey contained similar As, Cd, and Pt concentrations, but significantly lower Co, Ni, and Zn concentrations when compared to the other impoundments. There were significant differences in the Cr concentrations of the AMs decreasing in the following order: pollution control dam > Olifantsnek Dam > Bospoort Dam.

The element concentrations in the transplanted clams showed a heterogeneous picture (Figures 4–6, right). The concentrations of Cd, Pb, Pt, and Zn in the transplanted clams of all three sampling sites and both surveys did not exceed the respective concentrations of the controls. Significantly higher concentrations occurred only for As and Ni (survey 2017), when compared to the control, as well as for Cr and V (survey 2018) in the clams of Olifantsnek Dam. However, for some elements, the transplanted clams contained lower concentrations as compared to the control, as seen for Cd and Zn (survey 2018) in both impoundments as well as for Pb (both surveys) and V (survey 2017) in Bospoort Dam. The Pt concentrations in all of the clams were below the detection limit. Unfortunately, no comparison can be made between the transplanted clams of the impoundments and the pollution control dam since all of the organisms died during the exposure period.

Thus, the AMs and transplanted clams accumulated different elements from the two impoundments, including Olifantsnek Dam and Bospoort Dam, in different patterns. This can be seen by the clear dissimilarity between the AMs and transplanted clams on the PCA (Figure 7). The AMs correlate strongly with Pb and V, while the transplanted clams correlate with Cd, Cr, Co, Zn, and As. The groupings of element accumulation in the AMs indicated site-specific and survey-specific patterns, while the temporal and spatial groups were less clear for the clams.

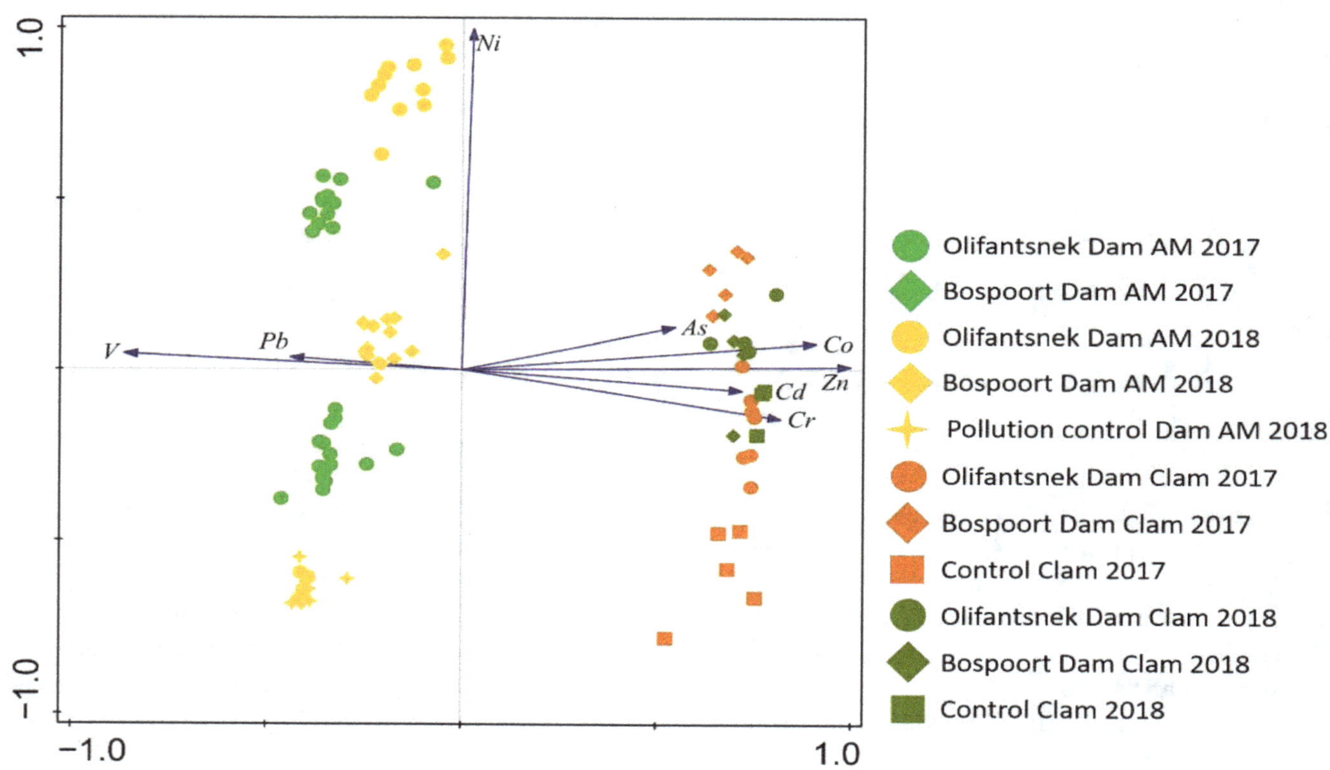

Figure 7. Principal component analysis (PCA) showing the grouping of AMs and transplanted clams based on the similarity of element bioaccumulation at the different sites and surveys. The ordination explains 93.7% of the total variation in the data with 79.5% on axis 1 and 14.2% on axis 2.

The PCA also indicated some groupings and correlations within the AMs (Figure 8). The AMs from the Olifantsnek Dam in 2017 and the pollution control dam in 2018 correlated strongly with

Cd, while there was a strong correlation of the AMs with Pb and Pt in both impoundments in 2017. Olifantsnek Dam correlated with As, Co, and Ni, Bospoort Dam correlated with V, and the pollution control dam correlated with Cd in 2018. The grouping of the AMs of the three impoundments of the survey in 2018 indicated spatial differences.

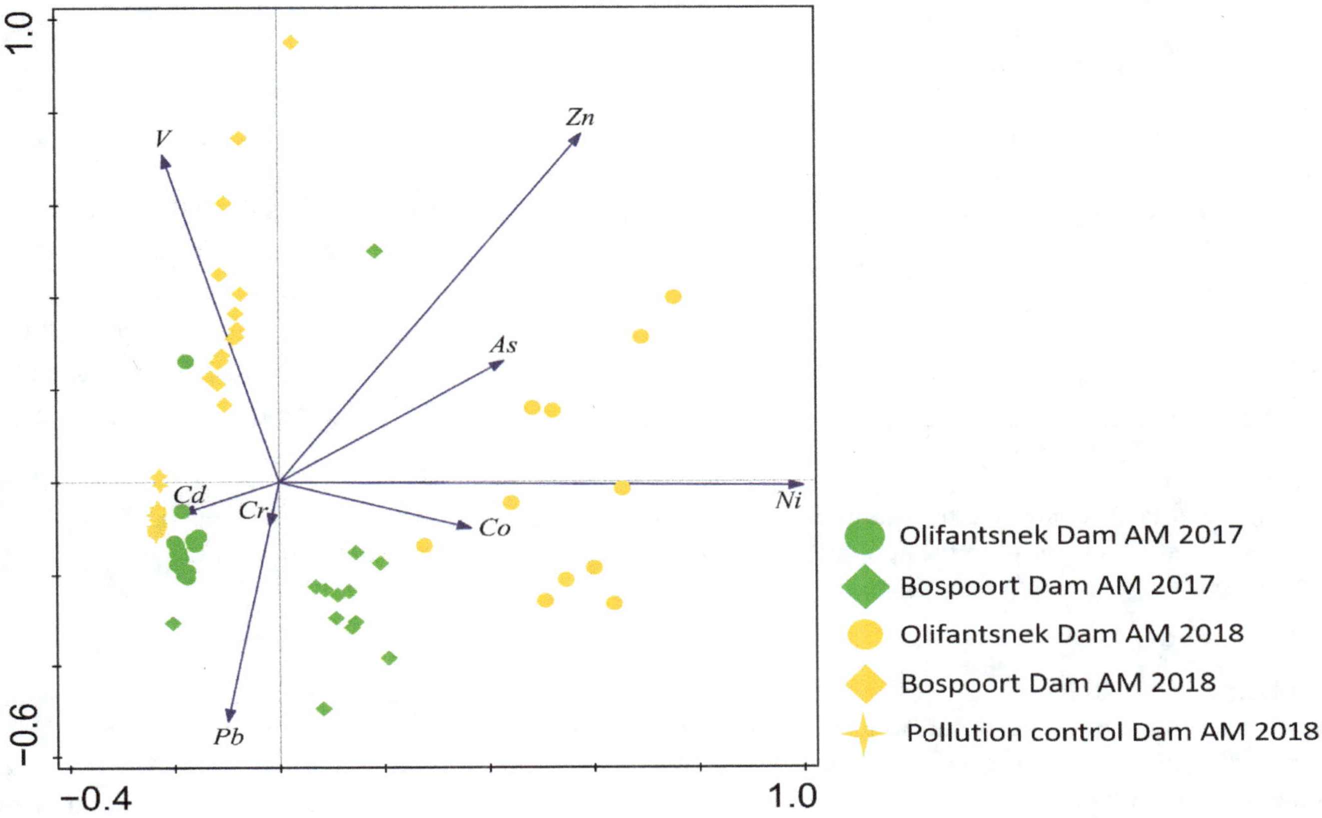

Figure 8. PCA illustrating the grouping of AMs within the three impoundments during the two surveys based on the similarity of metal bioaccumulation at the different sites and surveys. The ordination explains 98.9% of the total variation in the data with 97.3% on axis 1 and 1.6% on axis 2.

Within the transplanted clams, the PCA revealed groupings and correlations, which were different from the AM results (Figure 9). The bioaccumulation patterns of the clams from the Olifantsnek Dam in 2017 were distinctly different from all other clams and correlated strongly with As, Cr, and V. Additionally, the clams from Bospoort Dam correlated strongly with Ni. Furthermore, the control clams in 2018 showed a strong correlation with Cd, Pb, and Zn.

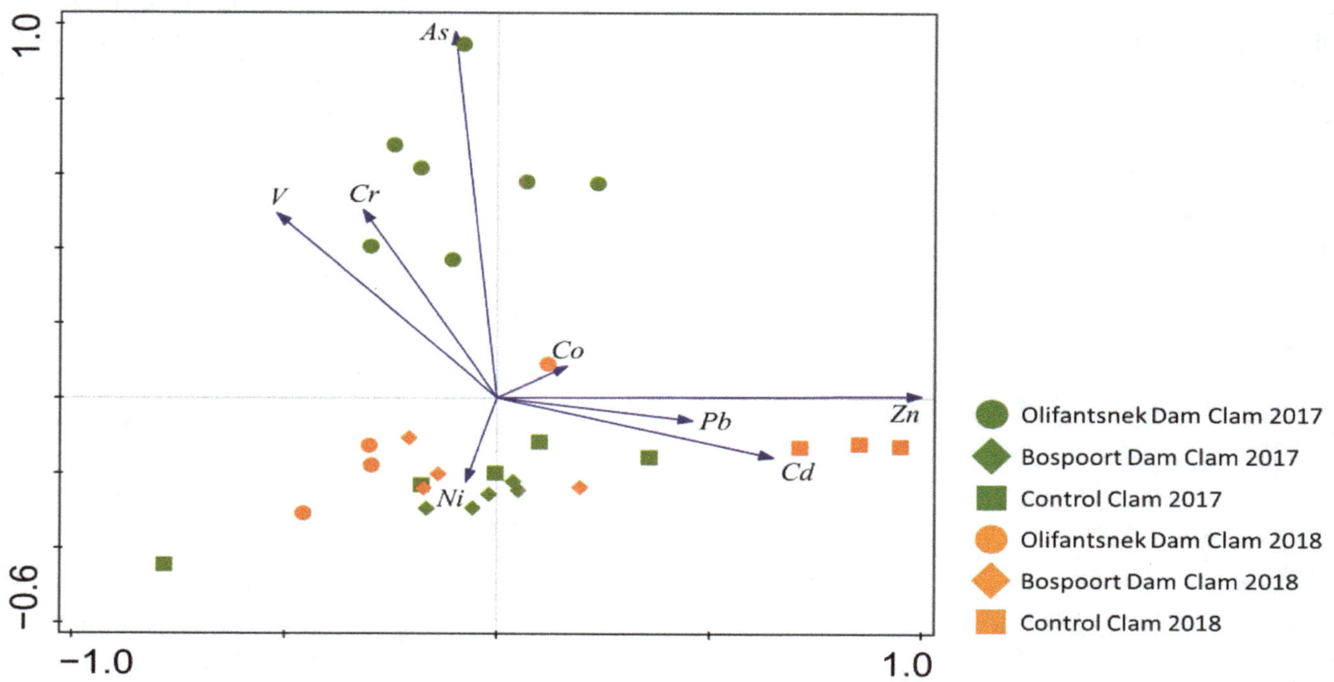

Figure 9. PCA illustrating the grouping of transplanted clams within the two impoundments during the two surveys based on the similarity of metal bioaccumulation at the different sites and surveys. The ordination explains 98.5% of the total variation in the data with 95.7% on axis 1 and 2.8% on axis 2.

4. Discussion

Point and diffuse source pollution from mining and production activities are recognized as an important source of contamination in the environment [16]. Platinum group elements are mined from primary deposits where they are associated with other elements such as Cu and Ni [16]. Elevated concentrations of PGE and other mining-associated elements have been reported in the vicinity of mining areas [2]. Analyses of road dust samples from four highly populated cities in South Africa demonstrated that Rustenburg had the most elevated Pt concentrations associated with mining and production activities [16]. The Hex River, which flows through Rustenburg, drains one of the most important mining areas in the western limb of the Bushveld Igneous Complex [2]. Rustenburg is located in a large basin, which forms part of the Bushveld Igneous Complex, where the ore mainly consists of PGEs. In addition to the PGEs and chromite, the ore contains deposits of gold and sulphides of nickel, copper, and iron [17].

4.1. Metals in Water

Olifantsnek Dam is geographically separated from the mining area in Rustenburg, while the Bospoort Dam is located in the vicinity of many informal settlements as well as many of the mines found in this area [17]. The element concentrations in the water showed that Bospoort Dam was slightly more polluted than the Olifantsnek Dam. The Hex River flows in a northerly direction through all the industrial and mining areas [2]. Consequently, wastes from these activities, which is discharged into the environment, will end up in the Hex River, where Bospoort Dam acts as a sink by accumulating the discharged elements. Since both of these impoundments are located in an area that contains geologically-rich deposits, it is assumed that some elements will occur naturally in high concentrations in both impoundments [2,25].

According to the weather data of the Rustenburg region, there was almost no rain during the first survey [26]. This resulted in very low flow to no flow in the river between the two impoundments during the first survey. The rainfall and ensuing run-off were much greater and more variable during the second survey in 2018. This is also reflected in the far greater variation in physic-chemical parameters. This is supported by Prathumratana et al. [27] who showed that rainfall and associated run-off influences variations in site-specific water quality.

The increase in element concentrations measured in Bospoort Dam compared to Olifantsnek Dam can be linked to a local point source located along the course of the Hex River and not from run-off from the greater catchment. Bospoort Dam contained slightly higher Cd, Co, and Ni concentrations than the Olifantsnek Dam. In general, element concentrations in the water of the Bospoort Dam were higher when compared to the Olifantsnek Dam. The significant difference in metal concentrations from the two impoundments at the end of the first survey can be attributed to the change in season, where there was almost no rainfall during this period. During this period, metal concentrations within the water became more concentrated due to low inflow and outflow of the impoundment and water evaporation.

The water data indicated that almost all of the analysed elements were present in concentrations below South Africa's target water quality range (see supplement data Table S2), for both aquatic ecosystems and domestic use [28,29]. However, the Zn concentrations in the pollution control dam were above the Target Water Quality Range (TWQR) (2 μg/L) for aquatic ecosystems. The concentrations found in the pollution control dam were above the acute effect value (36 μg/L) [29].

The pollution control dam from the mine contained significantly higher As, Co, Cr, Ni, Pt, V, and Zn concentrations than the other two impoundments. Pollution control dams are generally constructed to collect the sludge containing the residues rejected from the ore, to allow solids to settle and to let the liquid phase evaporate [17]. The main environmental risk of such pollution control dams is the possibility that the wall of the dam comes apart and the sludge flows over the mining areas, farmlands, and roads and into aquatic ecosystems [17]. Furthermore, the pollution control dam can be overflown during heavy rain so that over-spilling water containing soluble contaminants and re-mobilized particles can enter the nearby aquatic system. Therefore, it is important to control the element concentrations within these dams. The pollution control dam of the present study is located near the river, which facilitates an entry of contaminants into the Hex River system.

South Africa is a mineral resource-rich region and, consequently, metal and metalloid concentrations are elevated in rivers that drain through mining regions. The Olifants River is generally regarded as the most metal-enriched system in South Africa. Studies by Gerber et al. recorded average concentrations of 0.75 μg/L As, 7.41 μg/L Cd, 5.26 μg/L Co, 4.24 μg/L Cr, 1.43 μg/L Ni, 4.74 μg/L Pb, and 2.68 μg/L Zn [30]. In this case, Cd concentrations were orders of magnitude higher than the values in the present study. Additionally, the highest Cr concentrations measured in the Olifants River were approximately double than those measured in the Olifantsnek Dam and the Bospoort Dam. However, the Ni concentrations were 3, 13, and 35 times higher in the Olifantsnek Dam, the Bospoort Dam, and the pollution control dam, respectively. In addition, the Zn concentrations in the water of the pollution control dam of the present study were 14 times higher than in the study by Gerber et al. [30].

For the water of the Marico River system, which is one of the least impacted systems in a mining region in South Africa [31], levels of 0.858 ± 0.1912 μg/L As, <0.0001 μg/L Cd, 0.693 ± 0.056 μg/L Co, 5.9 ± 0.154 μg/L Cr, 7.158 ± 0.586 μg/L Ni, 0.009 ± 0.007 μg/L Pb, and 2.275 ± 1.33 μg/L Zn were reported. The aqueous As concentrations in the Olifantsnek Dam and the Bospoort Dam were similar. The Co concentration was even lower when compared to the Marico River. The pollution control dam, however, contained 2-times and 4-times the concentrations of As and Co, respectively, which was found within the Marico River. However, the As, Co, Ni, and Zn concentrations in the water of the pollution control dam significantly exceeded the respective concentrations in the two other impoundments as well as in the Marico River system, which indicated that these elements were derived from the nearby mines. In contrast, although the mining activities near Rustenburg include Cr mining, the Cr concentrations within the impoundments and the pollution control dam were lower than those found in the Marico River. A different picture shows the aqueous Cd and Pb concentrations, which were higher at all three sites of the present study. Higher Cd and Pb concentrations observed upstream of the mining and other human activities may be the result of the geogenic input from the natural geology [2].

In recent decades, the demand and consumption of platinum group elements, especially in the automobile catalytic converter production, has continuously increased, which resulted in increasing anthropogenic emissions of these metals worldwide [32]. In general, the water solubility of PGE from road dust as well as their biological availability decreased in the order Pd > Pt > Rh [32,33]. Studies on Pt concentrations in the water of rivers without a PGE mining impact recorded levels between 0.006–2.6 ng/L [34–37]. In the present study, the concentrations found within the two impoundments were considerably higher and ranged between 41.5–76.2 ng/L. The water of the pollution control dam contained Pt concentrations (145 ng/L), which were three times higher than the highest concentration found in the two impoundments. This clearly indicates the entry of Pt from the mining activities into the aquatic ecosystem.

4.2. Transplanted Organisms

In the present study, *C. fluminalis africana* was used for active monitoring of metal and metalloid contaminations. Significantly higher concentrations when compared to the control occurred only for As, Cr, Ni, and V, but not for all sampling sites and surveys, respectively. The pre-exposure of the transplanted organisms from a presumed non-contaminated reference site could have resulted in high background levels so that additional bioaccumulation due to the active biomonitoring is not detectable or depressed. In a study by Ruchter and Sures, *Corbicula* sp. were analysed to determine the influence of road runoff near Karlsruhe, Germany [38]. Reference *Corbicula* sp. were analysed to determine background levels of Cd, Cr, Ni, Pb, Pt, and Zn in the organisms before the inlet [38]. These concentrations were much lower than the concentrations measured in the control group in this study except Pb. They were able to detect Pt concentrations. Furthermore, the effects of upswelling and the change in habitat of the transplanted clams can affect the outcome of active biomonitoring studies [7]. The transplanted clams contained higher As, Cd, Co, Cr, and Zn concentrations. The Olifantsnek Dam in 2017 was the only group that was distinctly different.

In studies that make use of transplanted organisms, it is difficult to compare the results to data found in the literature. Nevertheless, when comparing the bioaccumulation by the clams with aquatic snails that were transplanted to a gold mining area [13], both transplanted organisms show similar Co concentrations. However, the As, Cr, Ni, V, and Zn concentrations of the clams were lower when compared with the snails, while the Cd and Pb concentrations were much higher [13].

The higher water temperatures recorded in Bospoort Dam may have resulted in the metabolism of the transplanted clams being higher than in the Olifantsnek Dam since increased temperature and the subsequent metabolism results in aquatic organisms accumulating more metals [39].

When PGEs from road dust accumulated in the soft tissue of bivalves, e.g., *Dreissena polymorpha*, the bioaccumulation of metals decreased in the following order: Cu > Cd > Ag > Pd > Pb > Sb > Fe > Pt > Rh [33]. Platinum concentrations in bivalves sampled from different freshwater ecosystems ranged between 0.00001–0.0013 μg/g [33,38,40], whereas the Pt levels of the transplanted clams from the present study were all under the detection limit of 0.0009 μg/g.

The bioaccumulation in the transplanted clams did not correlate with the exposure concentration in the water column. Claassens et al. [13] made use of transplanted organisms at different localities within the Koekemoer Spruit, South Africa, and found no correlation between the concentrations in the transplanted organisms and the ambient water. This may be explained by the fact that the indicator organisms integrate the pollutants over the whole exposure period, while water samples only indicate the exposure concentration at the time of sampling. Many factors can influence the bioaccumulation of metals in field studies, i.e., among others' seasonality, the bioavailability of metals, and toxicokinetic processes in the indicator organism [41]. Bivalves are filter feeders. Therefore, both the dissolved metal fraction and the particle-bound metals play a crucial role in the metal bioaccumulation.

4.3. Artificial Mussels

The present study reported the uptake of Pt in AMs for the first time. Although the AMs are only able to take up dissolved metals, the availability of Pt could only be shown by the use of the AMs and not by the transplanted clams, where the Pt levels were below the detection limit.

Artificial mussels have successfully been used as monitoring tools for metal bioaccumulation in the aquatic environment [11,13,20,23,42]. Many of these studies endorse the use of AMs during biomonitoring studies since they are not affected by water quality conditions that would greatly effect a bioindicator. Artificial mussels that were used for monitoring aquatic systems in a gold mining region [13] accumulated higher concentrations of Cr, Co, Cd, Pb, V, and Zn than in the present study. In contrast, AMs exposed to potentially less impacted sites along the Nyl River, Limpopo, South Africa [20] contained either similar or higher element concentrations when compared to this study.

The element concentrations in the water indicated that Bospoort Dam contained higher concentrations than those found in the Olifantsnek Dam. During the first survey, the AMs in the two impoundments had similar concentrations, while, in the second survey, there were some significant differences between the sampling sites. However, these results (AMs) did not indicate a specific site has higher concentrations. Other field studies using AMs also show variations in data.

The AMs contained higher Pb and V concentrations when compared to the transplanted clams. The element accumulation in the AMs indicated site-specific and survey-specific patterns.

4.4. Comparison of Artificial Mussels and Transplanted Clams

For almost all of the elements, in the present study, the transplanted clams contained significantly higher concentrations than the AMs. According to Wu et al. [10], the AM and any bioindicator species bioaccumulate different metal fractions during field exposures. Due to the assembly of the AM, large molecules and particle-bound metals are not able to cross the polyacrylamide gel barrier, which acts as a surrogate on the double lipid layer of a cell membrane. However, metals and metalloids occur in the environment in different complexes or may be attached to particles. Many factors can influence their speciation. In contrast to AMs, the transplanted clams can ingest different metals and metalloids from the food that is available within the environment and living organisms can regulate their concentrations by toxicokinetic processes [10]. The characteristics of the permeability of the AM and the binding capacity of the Chelex beads can also have an influence [10,23]. A combination of all of these factors may explain the differences in the accumulation capacity between AM and TM.

The partitioning (dissolve vs. particulate bound) and speciation of metals are dependent on several chemical and physical properties, such as pH, temperature, and the presence of ligands and particulates [43]. Morquecho and Pitt [43] filtered several stormwater samples to determine the colloidal bound metals, metals bound to ligands, and the ionic/dissolved fraction. After adding Chelex beads to the filtered samples, they found that the ionic form of the metals had bound to the beads, while the particle-bound metals remained in the solution. More than 90%, 80%, 80%, and 30% of the filtered Zn, Cr, Pb, and Cd were in an ionic form, respectively [43]. These percentages reflect the amount of each metal that was bound to the Chelex beads during the filtering process. Therefore, in the present study, the elements being accumulated by the AMs belong to the dissolved fraction, whereas the TMs accumulated the dissolved and particulate bound fraction. This may explain the different accumulation patterns for the AMs and TMs.

Claassens et al. [13] found that the AMs accumulated higher concentrations of Cd, Pb, V, and Zn than the transplanted organisms. This is partly in accordance with the present study showing that the AMs accumulated higher concentrations of Pb, Pt, and V. However, a direct comparison of the concentrations in the AM and TM is problematic since, for the AMs, the concentrations were calculated by dividing the total mass of the element that has bound to the Chelex beads by the mass (200 mg) of the Chelex beads. Assuming that not all binding sites of the Chelex beads were saturated, a lower mass of Chelex beads would have resulted in the same amount of metal uptake but a higher calculated concentration in the AM. Thus, it should be considered in future studies that the total mass is a more reliable measure for the uptake by the AM than the concentration. However, as in all studies on AMs published so far, 200 mg of Chelex beads were used since a comparison between these studies is possible by using the concentrations.

Other studies demonstrated that AMs and transplanted organisms accumulate metals in similar patterns but at different concentrations [13,23]. In contrast, in the present study, the accumulation patterns of the AMs indicated significant accumulation when compared to the initial concentrations, whereas, for the transplanted clams for most elements, no clear bioaccumulation was found during the biomonitoring period. Thus, since the AMs have low initial background levels, they can indicate the bioaccumulation of low element concentrations in biomonitoring studies.

During the second survey, over six weeks, there was a massive water hyacinth (*Eichhornia crassipes*) bloom that occurred in the Bospoort Dam. These water plants covered more than 50% of the impoundment at the end of the exposure period. The water hyacinth can accumulate high levels of metals from solutions such as Cd, Co, Cr, Cu, Ni, Pb, V, and Zn [44,45] and the PGEs Pt, Pd, Os, Ru, Ir, and Rh [42]. When considering the biomass of the water hyacinth in the dam, the plants likely had reduced the metal concentrations in the water. This may explain why the concentrations in the water of Bospoort Dam were not significantly different from those found in the Olifantsnek Dam during the second survey whereas the concentrations during the first survey were much higher in the Bospoort Dam.

During the second survey, transplanted clams and AMs were deployed in the pollution control dam on the mine ground, but none of the transplanted clams survived. It is presumed that the transplanted clams died due to a layer of oil on the water surface and decreasing water levels than due to very high metal concentrations. This shows the practicability of the AMs in ecosystems where adverse environmental conditions affect the survival of living organisms. However, it is possible that the oil layer obstructed the metals from diffusing through the gel layers. The metal and metalloid concentrations within the water from the pollution control dam were significantly higher than the impoundments. However, the AMs did not indicate this difference.

5. Conclusions

Different bioindicator species have different accumulation strategies. They are affected by many factors within the environment and can regulate internal metal concentrations through physiological processes [13,15]. The difference between the results of the AMs and the transplanted clams should not be seen as a shortcoming and was expected based on results from previous studies. The present study demonstrated that the transplanted clams and AMs accumulated different metals at different concentrations, which could be attributed to the accumulation of metals in different forms (dissolved or particle-bound) showing different biological availabilities. The AMs and transplanted clams accumulated similar As concentrations, while the AMs preferentially accumulated Pb, Ni, Pt, and V. The AMs are a promising tool for metals and metalloids occurring at low environmental concentrations close to the analytical detection limit and when organisms for the transplantation during active biomonitoring approaches already show high background levels. Metals can be found in various forms within the aquatic environment. The semi-permeable membrane of the AM permits the bioavailable metal ions to pass through and bind to the Chelex beads. Some metal ions have a stronger binding affinity toward the Chelex beads, where these ions may compete for binding sites within the AMs. For that reason, it is possible that some metals can be more abundant in the environment but the AMs will not be able to indicate these concentrations due to competition by other metals. Nonetheless, in this study, it was shown that the AM successfully accumulated Pt and all the other metals that were studied. Traditional bioindicator bivalves were more successful in indicating exposure to metals such as Cd, Co, Cr, and Zn. Thus, the combination of AMs and living bioaccumulation indicator organisms will provide a holistic assessment of metal and metalloid exposure in aquatic environments.

Author Contributions: Conceptualization, B.S. and N.J.S. Formal analysis, M.L., M.N., S.Z., B.S., and N.J.S. Investigation, M.L. Methodology, M.L. and V.W. Resources, B.S. and N.J.S. Supervision, V.W. and S.Z. Validation, V.W., M.N., and S.Z. Writing-original draft, M.L. Writing-review & editing, V.W., S.Z., B.S., and N.J.S. All authors have read and agreed to the published version of the manuscript.

Acknowledgments: This is the contribution number 360 of the North-West University (NWU) Water Research Group. The assistance of Anja Greyling, Hannes Erasmus, and Rian Pienaar is hereby acknowledged for their help during sampling trips and the map of the study site.

References

1. Rauch, S.; Peucker-Ehrenbrink, B. Sources of Platinum Group Elements in the Environment. In *Platinum Metals in the Environmen*; Zereini, F., Wiseman, C.L.S., Eds.; Springer: Heidelberg, Germany, 2015; Volume 1, pp. 3–17.
2. Almécija, C.; Cobelo-García, A.; Wepener, V.; Prego, R. Platinum group elements in stream sediments of mining zones: The Hex River (Bushveld Igneous Complex, South Africa). *J. Afr. Earth Sci.* **2017**, *129*, 934–943. [CrossRef]
3. Erasmus, J.; Malherbe, W.; Zimmermann, S.; Lorenz, A.; Nachev, M.; Wepener, V.; Sures, B.; Smit, N. Metal accumulation in riverine macroinvertebrates from a platinum mining region. *Sci. Total Environ.* **2019**, *703*, 134738. [CrossRef] [PubMed]
4. Zhou, Q.; Zhang, J.; Fu, J.; Shi, J.; Jiang, G. Biomonitoring: An appealing tool for assessment of metal pollution in the aquatic ecosystem. *Anal. Chim. Acta* **2008**, *606*, 135–150. [CrossRef]
5. Gupta, S.K.; Singh, J. Evaluation of mollusc as sensitive indicator of heavy metal pollution in aquatic system: A review. *IIOAB J.* **2011**, *2*, 49–57.
6. Farris, J.L.; Van Hassel, J.H. *Freshwater Bivalve Ecotoxicology*; CRC Press: Boca Raton, FL, USA, 2006. [CrossRef]
7. Wepener, V. Application of active biomonitoring within an integrated water resources management framework in South Africa. *S. Afr. J. Sci.* **2008**, *104*, 367–373.
8. Giarratano, E.; Duarte, C.A.; Amin, O.A. Biomarkers and heavy metal bioaccumulation in mussels transplanted to coastal waters of the Beagle Channel. *Ecotoxicol. Environ. Saf.* **2010**, *73*, 270–279. [CrossRef]
9. Wu, R.S.S.; Lau, T.C. Polymer-ligands: A novel chemical device for monitoring heavy metals in the aquatic environment? *Mar. Pollut. Bull.* **1996**, *32*, 391–396. [CrossRef]
10. Wu, R.S.S.; Lau, T.C.; Fung, W.K.M.; Ko, P.H.; Leung, K.M.Y. An 'artificial mussel' for monitoring heavy metals in marine environments. *Environ. Pollut.* **2007**, *145*, 104–110. [CrossRef]
11. Kibria, G.; Lau, T.; Wu, R. Innovative 'Artificial Mussels' technology for assessing spatial and temporal distribution of metals in Goulburn–Murray catchments waterways, Victoria, Australia: Effects of climate variability (dry vs. wet years). *Environ. Int.* **2012**, *50*, 38–46. [CrossRef]
12. Hossain, M.M.; Kibria, G.; Nugegoda, D.; Lau, T.; Wu, R. *A Training Manual for Assessing Pollution (trace/heavy metals) in Rivers, Estuaries and Coastal waters-Using Innovative "Artificial Mussel (AM) Technology"—Bangladesh Model. Research collaboration between Scientists of the IMSF, University of Chittagong, Bangladesh, RMIT University, Australia, the City University of Hong Kong, and the University of Hong Kong*; FAO: Rome, Italy, 2015; Volume 1, pp. 1–24. [CrossRef]
13. Claassens, L.; Dahms, S.; Van Vuren, J.; Greenfield, R. Artificial mussels as indicators of metal pollution in freshwater systems: A field evaluation in the Koekemoer Spruit, South Africa. *Ecol. Indic.* **2016**, *60*, 940–946. [CrossRef]
14. Leung, K.M.Y.; Furness, R.W.; Svavarsson, J.; Lau, T.C.; Wu, R.S.S. Field validation, in Scotland and Iceland, of the artificial mussel for monitoring trace metals in temperate seas. *Mar. Pollut. Bull.* **2008**, *57*, 790–800. [CrossRef] [PubMed]
15. Degger, N.; Wepener, V.; Richardson, B.J.; Wu, R.S.S. Application of artificial mussels (AMs) under South African marine conditions: A validation study. *Mar. Pollut. Bull.* **2011**, *63*, 108–118. [CrossRef] [PubMed]
16. Rauch, S.; Fatoki, O.S. Impact of Platinum Group Element Emissions from Mining and Production Activities. In *Platinum Metals in the Environment*; Zereini, F., Wiseman, C.L.S., Eds.; Springer: Heidelberg, Germany, 2015; pp. 19–29. [CrossRef]

17. Wittmann, G.; Förstner, U. Heavy metal enrichment in mine drainage: 1. The Rustenburg platinum mining area. *Science* **1976**, *72*, 242–246.
18. Roux, H. Status of the yellowfish populations & River health Programme in NW Province. In Proceedings of the 15th Yellowfish Working Group Conference, Mpumalanga, South Africa, 18–20 February 2011.
19. Mogakabe, D.; Van Ginkel, C. *The Water Quality of the Bospoort Dam*; Internal Report No. N/0000/00/DEQ/0108; Department of Water Affairs and Forestry: Pretoria, South Africa, 2008.
20. Dahms-Verster, S.; Baker, N.; Greenfield, R. A multivariate examination of 'artificial mussels' in conjunction with spot water tests in freshwater ecosystems. *Environ. Monit. Assess.* **2018**, *190*, 427. [CrossRef] [PubMed
21. Brand, S.J.; Erasmus, J.H.; Labuschagne, M.; Grabner, D.; Nachev, M.; Zimmermann, S.; Wepener, V.; Smit, N.; Sures, B. Bioaccumulation and metal-associated biomarker responses in a freshwater mussel, Dreissena polymorpha, following short-term platinum exposure. *Environ. Pollut.* **2019**, *246*, 69–78. [CrossRef]
22. Zimmermann, S.; Menzel, C.M.; Berner, Z.; Eckhardt, J.-D.; Stüben, D.; Alt, F.; Messerschmidt, J.; Taraschewski, H.; Sures, B. Trace analysis of platinum in biological samples: A comparison between sector field ICP-MS and adsorptive cathodic stripping voltammetry following different digestion procedures. *Anal. Chim. Acta* **2001**, *439*, 203–209. [CrossRef]
23. Degger, N.; Chiu, J.M.Y.; Po, B.H.K.; Tse, A.C.K.; Zheng, G.J.; Zhao, D.-M.; Xu, D.; Cheng, Y.-S.; Wang, X.-H.; Liu, W.-H.; et al. Heavy metal contamination along the China coastline: A comprehensive study using Artificial Mussels and native mussels. *Environ. Manag.* **2016**, *180*, 238–246. [CrossRef]
24. Kushner, E. On determining the statistical parameters for pollution concentration from a truncated data set. *Atmos. Environ.* **1976**, *10*, 975–979. [CrossRef]
25. Cawthorn, R.G. The platinum group element deposits of the Bushveld Complex in South Africa. *Platin. Met. Rev.* **2010**, *54*, 205–215. [CrossRef]
26. Thorsen, S. Past Weather in Rustenburg, South Africa—May 2017. Available online: https://www.timeanddate.com/weather/south-africa/rustenburg/historic?month=5&year=2017 (accessed on 1 October 2018).
27. Prathumratana, L.; Sthiannopkao, S.; Kim, K.W. The relationship of climatic and hydrological parameters to surface water quality in the lower Mekong River. *Environ. Int.* **2008**, *34*, 860–866. [CrossRef]
28. Department of Water Affairs and Forestry (DWAF). *South African Water Quality Guidelines (Second Edition). Volume 1: Domestic Use*; DWAF: Pretoria, South Africa, 1996.
29. Department of Water Affairs and Forestry (DWAF). *South African Water Quality Guidelines. Volume 7: Aquatic Ecosystems*; DWAF: Pretoria, South Africa, 1996.
30. Gerber, R.; Smit, N.J.; van Vuren, J.H.; Wepener, V. Metal concentrations in *Hydrocynus vittatus* (Castelnau 1861) populations from a premier conservation area: Relationships with environmental concentrations. *Ecotoxicol. Environ. Saf.* **2016**, *129*, 91–102. [CrossRef] [PubMed]
31. Kemp, M.; Wepener, V.; de Kock, K.; Wolmarans, C. Metallothionein Induction as Indicator of Low Level Metal Exposure to Aquatic Macroinvertebrates from a Relatively Unimpacted River System in South Africa. *Bull. Environ. Contam. Toxicol.* **2017**, *99*, 662–667. [CrossRef] [PubMed]
32. Hoppstock, K.; Sures, B. Platinum-Group Metals. In *Elements and Their Compounds in the Environment: Occurrence, Analysis and Biological Relevance*, 2nd ed.; Merian, E., Anke, M., Ihnat, M., Stoeppler, M., Eds.; WILEY-VCH Verlag GMBH & Co. KGaA: Weinheim, Germany, 2004; pp. 1047–1086.
33. Zimmermann, S.; Alt, F.; Messerschmidt, J.; von Bohlen, A.; Taraschewski, H.; Sures, B. Biological availability of traffic-related platinum-group elements (palladium, platinum, and rhodium) and other metals to the zebra mussel (*Dreissena polymorpha*) in water containing road dust. *Environ. Toxicol. Chem.* **2002**, *21*, 2713–2718. [CrossRef] [PubMed]
34. Hoppstock, K.; Alt, F. Voltammetric determination of ultratrace platinum and rhodium in biological and environmental samples. In *Anthropogenic Platinum-Group Element Emissions*; Fathi, Z., Friedrich, A., Eds.; Springer: Heidelberg, Germany, 2000; pp. 145–152. [CrossRef]
35. Moldovan, M.; Gomez, M.M.; Palacios, M.A. On-line preconcentration of palladium on alumina microcolumns and determination in urban waters by inductively coupled plasma mass spectrometry. *Anal. Chim. Acta* **2003**, *478*, 209–217. [CrossRef]
36. Monticelli, D.; Carugati, G.; Castelletti, A.; Recchia, S.; Dossi, C. Design and development of a low cost, high performance UV digester prototype: Application to the determination of trace elements by stripping voltammetry. *Microchemical* **2010** *95*, 158–163. [CrossRef]

37. Cobelo-García, A.; López-Sánchez, D.E.; Almécija, C.; Santos-Echeandía, J. Behavior of platinum during estuarine mixing (Pontevedra Ria, NW Iberian Peninsula). *Mar. Chem.* **2013**, *150*, 11–18. [CrossRef]
38. Ruchter, N.; Sures, B. Distribution of platinum and other traffic related metals in sediments and clams (Corbicula sp.). *Water Res.* **2015**, *70*, 313–324. [CrossRef]
39. van Heerden, D.; van Rensburg, P.J.; Nikinmaa, M.; Vosloo, A. Gill damage, metallothionein gene expression and metal accumulation in *Tilapia sparrmanii* from selected field sites at Rustenburg and Potchefstroom, South Africa. *Afr. J. Aquat. Sci.* **2006**, *31*, 89–98. [CrossRef]
40. Ruchter, N.; Zimmermann, S.; Sures, B. Field studies on PGE in aquatic ecosystems. In *Platinum Metals in the Environment*; Zereini, F., Wiseman, C.L.S., Eds.; Springer: Heidelberg, Germany, 2015; pp. 351–360. [CrossRef]
41. Greenfield, R.; Brink, K.; Degger, N.; Wepener, V. The usefulness of transplantation studies in monitoring of metals in the marine environment: South African experience. *Mar. Pollut. Bull.* **2014**, *85*, 566–573. [CrossRef]
42. Genç, T.O.; Po, B.H.; Yılmaz, F.; Lau, T.C.; Wu, R.S.; Chiu, J.M. Differences in metal profiles revealed by native mussels and artificial mussels in Sarıçay Stream, Turkey: Implications for pollution monitoring. *Mar. Freshw. Res.* **2018**, *69*, 1372–1378. [CrossRef]
43. Morquecho, R.; Pitt, R. Stormwater heavy metal particulate associations. *Proc. Water Environ. Fed.* **2003**, *2003*, 774–803. [CrossRef]
44. Nor, Y.M. The absorption of metal ions by Eichhornia crassipes. *Chem. Speciat. Bioavailab.* **1990**, *2*, 85–91. [CrossRef]
45. Shirinpur-Valadi, A.; Hatamzadeh, A.; Sedaghathoor, S. Study of the accumulation of contaminants by Cyperus alternifolius, Lemna minor, Eichhornia crassipes, and Canna generalis in some contaminated aquatic environments. *Environ. Sci. Pollut. Res. Int.* **2019**, *26*, 21340–21350. [CrossRef] [PubMed]

6

Biodegradation of Amoxicillin, Tetracyclines and Sulfonamides in Wastewater Sludge

Chu-Wen Yang, Chien Liu and Bea-Ven Chang *

Department of Microbiology, Soochow University, Taipei 11101, Taiwan; ycw6861@scu.edu.tw (C.-W.Y.); jennifer719123@yahoo.com.tw (C.L.)
* Correspondence: bvchang@scu.edu.tw

Abstract: The removal of antibiotics from the aquatic environment has received great interest. The aim of this study is to examine degradation of oxytetracycline (OTC), tetracycline (TC), chlortetracycline (CTC), amoxicillin (AMO), sulfamethazine (SMZ), sulfamethoxazole (SMX), sulfadimethoxine (SDM) in sludge. Four antibiotic-degrading bacterial strains, SF1 (*Pseudmonas* sp.), A12 (*Pseudmonas* sp.), strains B (*Bacillus* sp.), and SANA (*Clostridium* sp.), were isolated, identified and tested under aerobic and anaerobic conditions in this study. Batch experiments indicated that the addition of SF1 and A12 under aerobic conditions and the addition of B and SANA under anaerobic conditions increased the biodegradation of antibiotics in sludge. Moreover, the results of repeated addition experiments indicated that the efficiency of the biodegradation of antibiotics using the isolated bacterial strains could be maintained for three degradation cycles. Two groups of potential microbial communities associated with the aerobic and anaerobic degradation of SMX, AMO and CTC in sludge were revealed. Twenty-four reported antibiotics-degrading bacterial genera (*Achromobacter, Acidovorax, Acinetobacter, Alcaligenes, Bacillus, Burkholderia, Castellaniella, Comamonas, Corynebacterium, Cupriavidus, Dechloromonas, Geobacter, Gordonia, Klebsiella, Mycobacterium, Novosphingobium, Pandoraea, Pseudomonas, Rhodococcus, Sphingomonas, Thauera, Treponema, Vibrio* and *Xanthobacter*) were found in both the aerobic and anaerobic groups, suggesting that these 24 bacterial genera may be the major antibiotic-degrading bacteria in sludge.

Keywords: sludge; antibiotics; biodegradation

---

1. Introduction

Antibiotics are discharged from animals (and humans) and enter wastewater treatment plants (WWTPs) through the sewage system [1]. In general, antibiotics are not easy for wastewater treatment facilities to remove [2]. The release and persistence of antibiotics in the environments may lead to an increase in antibiotic-resistant bacteria [3]. Tetracycline antibiotics (TCs), such as tetracycline (TC), oxytetracycline (OTC), and chlortetracycline (CTC), are broad-spectrum antibacterial drugs that inhibit bacterial protein synthesis [4]. Sulfonamide antibiotics (SAs) act against most Gram-positive and many Gram-negative bacteria. SAs act as competitive inhibitors of *p*-aminobenzoic acid in the folic acid synthetic pathway to inhibit the growth of bacteria [5]. Three sulfonamides, sulfamethoxazole (SMX), sulfadimethoxine (SDM), and sulfamethazine (SMZ), are found in sludge in many WWTPs and rivers [6–8]. Amoxicillin (AMO) is a β-lactam family antibiotic drug. This class of antibiotics disrupts bacterial cell walls during bacterial growth [9]. AMO in the environment may lead to an increase in antibiotic-resistant bacteria [10].

Several methods for antibiotic removal from wastewater sludge have been proposed. These approaches include chemical oxidation methods such as UV- and solar-based procedures (UV/H_2O_2, $solar/H_2O_2$), ozonation, photocatalysis by Fe^{2+} or Fe^{3+}/H_2O_2 and TiO_2 photocatalysis [11].

Antibiotic degradation by fungal extracellular enzymes extracted from spent mushroom composts has also been reported [12]. Thermal hydrolysis/anaerobic digestion [13] and the use of aerobic granular sludge treated with manganese oxides [14,15] have also been proposed. Nevertheless, many antibiotics continue to be detected in wastewater sludge [16–18]. The use of microorganisms to eliminate antibiotics is a promising strategy [19–21]. The aerobic degradation of antibiotics has been observed in pure bacterial cultures and microbial consortiums. *Rhodococcus rhodochrous* and *Aspergillus niger* degrade pharmaceuticals via cometabolism [22]. *Stenotrophomonas maltophilia* DT1 biotransforms tetracycline [23]. *Klebsiella* sp. SQY5 degrades tetracycline (TEC) [24]. Sulfamethoxazole is biodegraded by individual and mixed bacteria [25,26]. Antibiotic sulfanilamide biodegradation is performed by acclimated microbial populations [27]. Chlortetracycline can be used as the sole carbon and nitrogen source by the acclimated microbiota [28].

The aim of this study was to examine the degradation of antibiotics in sludge by antibiotic-degrading bacteria under aerobic and anaerobic conditions. The microbial communities involved in the aerobic and anaerobic degradation of antibiotics in sludge are revealed.

2. Materials and Methods

2.1. Chemicals

SMX, SDM, SMZ, TC, OTC, CTC, AMO (Table S1) of 99.0% purity and all other chemicals were purchased from Sigma Chemical Co. (St. Louis, MO, USA). Solvents were purchased from Mallinckrodt (Paris, KY, USA).

2.2. Sludge Sample and Medium

The sludge was a semisolid slurry (total solids 0.87 g/L, pH 6.7) that was produced as sewage sludge from the Dihus wastewater treatment plant in Taipei, Taiwan. Fresh sludge was used for antibiotic adaptation. The temperature of the sampling site was 30 °C. The medium used in aerobic experiments contained the following chemicals (mg/L): K_2HPO_4, 65.3; KH_2PO_4, 25.5; $Na_2HPO_4 \cdot 12H_2O$, 133.8; NH_4Cl, 5.1; $CaCl_2$, 82.5; $MgSO_4 \cdot 7H_2O$, 67.5; and $FeCl_3 \cdot 6H_2O$, 0.75. The medium used in anaerobic experiments contained the following chemicals (mg/L): NH_4Cl, 2.7; $MgCl_2 \bullet 6H_2O$, 0.1; $CaCl_2 \bullet 2H_2O$, 0.1; $FeCl_2 \bullet 4H_2O$, 0.02; K_2HPO_4, 0.27; KH_2PO_4, 0.35 and resazurin, 0.001. The pH of the medium was adjusted to 7.0 using potassium hydroxide or nitric acid. Resazurin is an indicator, which exhibits red color under aerobic conditions and is colorless under anaerobic conditions. Titanium citrate of 0.9 mM was used as a reducing reagent. All anaerobic operations were performed in an anaerobic glove box.

2.3. Sludge Adaptation

Aerobic adaptation was performed by adding 1 mg/L CTC, SMX and AMO, simultaneously, to 1000 mL serum bottles containing 450 mL of aerobic medium, 50 mL sludge, and incubated on a rotary shaker (120 rpm) at 30 °C without light for 6 months. Anaerobic adaptation was performed by adding 1 mg/L CTC, SMX and AMO, simultaneously, to 1000 mL serum bottles containing 450 mL of anaerobic medium, 50 mL sludge and capped with butyl rubber stoppers and crimp seals, wrapped in aluminum foil, and then incubated without shaking at 30 °C without light for 6 months. All anaerobic operations were performed in an anaerobic glove box. In this paper, the sludge was referred to as antibiotic-adapted sludge.

2.4. Enrichment, Isolation, and Identification of Antibiotic-Degrading Bacteria

The enrichment procedure was performed using 5 mL of antibiotic-adapted sludge in a 125-mL serum bottle containing 45 mL of aerobic or anaerobic medium, with CTC, SMX and AMO of final concentration of 0.2 mg/L, simultaneously, and incubated at 30 °C. The second to the fourth enrichment transfers were amended with gradually increasing concentrations of CTC, SMX and AMO from 0.2 to

1 mg/L. An elevated concentration is commonly used for enrichment to ensure that the CTC, SMX and AMO degraders are selected. After the fourth enrichment, aerobic or anaerobic medium agar plates containing CTC, SMX and AMO (1 mg/L) were inoculated with 100 μL of the liquid part of the sludge by streaking to isolate pure strains of bacteria. The AnaeroPack system was used for anaerobic cultivation, including AnaeroPack®-Anaero and AnaeroPouch®-Anaero. All anaerobic operations were performed in the anaerobic glove box. To confirm that the bacterial strains were antibiotic degraders, degradation experiments were performed using 5 mL (10^6 CFU/mL) of bacterial culture and 45 mL of medium with 1 mg/L antibiotics on a shaker (120 rpm) at 30 °C in the dark. Samples were taken periodically to analyze residual CTC, SMX and AMO.

The 16S rRNA gene of the isolated bacterial strains were amplified by PCR with the 5′-primer F8 (5′-AGAGTTTGATCCTGGCTCAG-3′) and the 3′-primer R1510 (5′-GGTTACCTTGTTACGACTT-3′). The PCR parameters included initial denaturation at 94 °C for 10 min, followed by 35 cycles of 45 s at 94 °C, 1 min at 60 °C, and 1 min at 72 °C, with a final extension at 72 °C for 10 min. The PCR products were sequenced on an ABI Prism automatic sequencer. The 3′-end sequence was converted into the reverse complementary sequence. The overlapping parts of sequences from the 5′ and 3′-ends were identified with the Align Sequences Nucleotide BLAST tool at the National Center for Biotechnology Information (NCBI) website. Finally, the two sequences were assembled into a single contig sequence based on the overlapping sequences. The 16S rRNA sequences of the four isolates have been submitted to Genbank; accession number: A12: MT678104, B: MT678105, SANA: MT678106, SF1: MT678107. The 16S rRNA gene sequences of the four isolated bacterial strains were used to search the NCBI 16S rRNA database with the Basic Local Alignment Search Tool (BLASTn). The top five sequences (with the highest scores) in the Blast results for each 16S rRNA sequence were retrieved and used to construct the phylogenetic tree. Phylogenetic analysis was performed using Clustal X 2.0 with 1000 bootstrapping repetitions [29]. The neighbor-joining algorithm was used to construct the phylogenetic tree.

2.5. Settings of Batch and Readdition Experiments

Two sets of experiments (batch and continuous addition) were performed under aerobic and anaerobic conditions. In the batch experiments, the aerobic experiments were performed in 125 mL serum bottles with the following contents: 45 mL of medium, 5 g of sludge, 5 mL (10^6 CFU/mL) of bacterial culture, and 2 mg/L SAs (SMX, SDM, SMZ), TCs (TC, OTC, CTC) or AMO, and which were incubated on a rotary shaker (120 rpm) at 30 °C in the dark. The anaerobic experiments were performed in 125 mL serum bottles containing 45 mL of medium, 5 g of sludge, 5 mL (10^6 CFU/mL) of bacterial culture, and 2 mg/L SAs (SMX, SDM, SMZ), TCs (TC, OTC, CTC) or AMO, and were conducted in an anaerobic glove box. The bottles were capped with butyl rubber stoppers and crimp seals, wrapped in aluminum foil, and then incubated without shaking at 30 °C. Inoculated controls containing 45 mL of medium, 5 g of sludge, and 2 mg/L SAs (SMX, SDM, SMZ), TCs (TC, OTC, CTC) or AMO, and were incubated without degrading bacteria at 30 °C. Sterile controls containing 45 mL of medium and 5 g of sludge were autoclaved at 121 °C for 30 min. SAs (SMX, SDM, SMZ), TCs (TC, OTC, CTC) or AMO were added at 2 mg/L after autoclaving. The remaining percentage of the original antibiotic content in the sludge ranged from 98.5 to 96.8%, indicating that the aerobic and anaerobic antibiotic degradation observed in all of the following experiments was due to microbial activity. Each experiment was repeated three times.

For the repeated addition experiments, sludge samples were used by adding 2 mg/L CTC, SMX and AMO three times under aerobic or anaerobic conditions. Aerobic experiments were conducted using 1000 mL bottles containing 450 mL of aerobic medium, and 50 g of sludge, with or without 50 mL (10^6 CFU/mL) of mixed bacterial culture. Anaerobic experiments were filled using 1000 mL bottles containing 450 mL of anaerobic medium, and 50 g of sludge, with or without 50 mL (10^6 CFU/mL) of mixed bacterial culture. CTC, SMX and AMO was readded into each medium when antibiotics were decreased to an undetectable level. The aerobic experiments were aerated by an air diffuser and the mixtures were stirred. The anaerobic experiments were conducted in an anaerobic glove box;

capped with butyl rubber stoppers and crimp seals, wrapped in aluminum foil, and then incubated without shaking. The repeated addition experiments were performed at 30 °C.

2.6. Analytical Methods

SAs were extracted twice from the samples with an extraction solution of 10:3:1 water (containing 0.1% formic acid): acetonitrile:methanol. The extracts were analyzed using an Agilent 1260 HPLC system equipped with a 4.6 × 250 mm column (Zorbax Eclipse Plus C18, Agilent, Santa Clara, CA, USA) with photodiode array detector monitoring at 270 nm. The mobile phase was 30%:70% acetonitrile:water (containing 0.1% formic acid). TCs were extracted twice from the samples with an extraction solution of 4:1 water (containing 0.1% HCl): acetonitrile. The mobile phase consisted of acetonitrile and water (containing 0.1% HCl); AMO was extracted twice from the samples with methanol. TCs, and AMO were quantified with an Agilent 1260 HPLC equipped with a 4.6 × 100 mm column (Poroshell 120 EC-C18, Agilent) with a photodiode array detector monitoring at 270 nm and 229 nm. The recovery percentages for TC, OTC, CTC, SDM, SMX, SMZ, and AMO were 91.2%, 95.5%, 89.3%, 94.7%, 93.4%, 90.3% and 92.6%, respectively. The detection limit was 0.1 mg/L in all cases. The remaining percentage and removal percentage were computed using the following formula: remaining percentage [%] = (residue concentration/initial concentration) × 100; removal percentage [%] = [1 − (residue concentration/initial concentration) × 100]. The degradation data collected in this study fit well with first-order kinetics (i.e., $t = -\ln(C/C_0)/k$, where C_0 is the initial concentration, C is the concentration, t is the time period, $t_{1/2}$ is the half-life, and k is the degradation rate constant). The coefficient of determinations (R^2) ranged from 0.923 to 0.982.

2.7. Microbial Community Analysis

Total DNA was extracted from the sludge at the end of the readdition experiments using a PowerSoil DNA Isolation kit (QIAGEN, Venlo, Netherlands). The purified DNA was used as a template for the amplification of the 16S rRNA gene sequence containing the V5-V8 variable regions using a 5′-primer containing an Illumina adaptor and a 16S rRNA gene-specific sequence (5′-CCTACGGGNBGCASCAG-3′). The sequence of the 3′-primer contained an Illumina adaptor and a 16S rRNA gene-specific sequence (5′-GACTACNVGGGTATCTAATCC-3′). PCR was performed as previously described [30,31]. Next generation sequencing (NGS) was performed at the Genome Center of National Yang-Ming University, Taiwan, using the MiSeq platform (Illumina, Inc, San Diego, CA, USA.). A chimera check was used to analyze the 16S rRNA gene sequences to remove chimeras. The remaining sequences were subsequently analyzed using the classifier software of the Ribosomal Database Project (http://pyro.cme.msu.edu/) for phylogenetic assignment. Similarity (95%) was used for bacterial grouping. Xenobiotic biodegradation-associated bacteria were identified using the Kyoto Encyclopedia of Genes and Genomes (KEGG) database [32]. A cluster analysis of the bacterial community compositions was performed using the Heatmap function in the ComplexHeatmap package of R (www.r-project.org).

2.8. Scanning Electron Microscopy

Sludge samples were fixed in 2.5% glutaraldehyde in phosphate-buffered saline for 2 h at 4 °C and postfixed in 1% osmium tetroxide in the same buffer for 1 h. The samples were dehydrated in a graded ethyl alcohol series, critical point-dried, gold coated, and viewed under a JSM-5410 scanning electron microscope (JEOL Ltd., Tokyo, Japan).

3. Results and Discussion

3.1. Isolation and Identification of Antibiotic-Degrading Bacteria

Twelve and eight bacterial strains with the ability to use CTC, AMO and SMX as carbon sources under aerobic and anaerobic conditions were isolated from the sludge samples. The five aerobic isolates showing the greatest degrading capability were strains SF1, SF2, SF3, SF4 and A12. When the bacterial counts increased, the level of CTC, AMO and SMX decreased. The order of CTC, AMO and SMX degradation was strain SF1 > strain A12 > strain SF2 > strain SF3 > strain SF4. Strains SF1 and A12 exhibited the greatest aerobic degradation capability. The degradation of CTC, AMO and SMX by strain SF1 after 8 days of incubation was equal to 81.6, 89.1, and 95.9%, respectively (Table 1). The degradation of CTC, AMO and SMX by strain A12 after 8 days of incubation was equal to 69.4, 81.4, and 89.6%, respectively. The degradation of CTC, AMO and SMX by strains SF1 and A12 after 8 days of incubation was equal to 89.4, 93.4, and 99.3%, respectively. The order of the efficiency of the aerobic degradation of antibiotics was as follows: strains SF1 and A12 > strain SF1 > strain A12.

Table 1. Remaining percentages (%) of chlortetracycline (CTC), amoxicillin (AMO) and sulfamethoxazole (SMX) after incubation with antibiotic-degrading bacteria.

	Aerobic Condition [a]				Anaerobic Condition [b]			
	Medium	SF1	A12	SF1+A12	Medium	B	SANA	B + SANA
CTC	97.2 ± 3.70	18.4 ± 0.93	30.6 ± 1.48	10.6 ± 0.53	98.3 ± 4.33	18.4 ± 0.94	40.6 ± 2.13	10.4 ± 0.42
AMO	94.5 ± 2.91	10.9 ± 0.49	18.6 ± 0.88	6.6 ± 0.34	96.7 ± 5.24	10.9 ± 0.58	18.6 ± 0.90	6.4 ± 0.29
SMX	98.6 ± 5.12	4.1 ± 0.12	10.4 ± 0.41	0.7 ± 0.02	99.1 ± 6.78	4.9 ± 0.10	10.4 ± 0.44	0.5 ± 0.01

Note(s): [a] aerobic conditions for 8 days. [b] anaerobic conditions for 15 days. Data from three independent experiments are presented as the means ± SE.

Five anaerobic isolates exhibited the greatest antibiotic-degradation capability were strains B, SANA, SANB, SANC, SANC, and SAND. When the bacterial counts increased, the level of CTC, AMO and SMX decreased. The order of CTC, AMO and SMX degradation is strain B > strain SANA > strain SANB > strain SANC > strain SAND. Strains B and SANA exhibited the greatest antibiotic degradation capability under anaerobic conditions among the five tested isolates (B, SANA, SANB, SANC, SANC, and SAND). The degradation of CTC, AMO and SMX by strain B after 15 days of incubation was equal to 81.6, 89.1, and 95.1%, respectively (Table 1). The degradation of CTC, AMO and SMX by strain SANA after 15 days of incubation was equal to 59.4, 81.4, and 89.6%, respectively. The degradation of CTC, AMO and SMX by strain B and SANA after 15 days of incubation was equal to 89.6, 93.4, and 99.5%, respectively. The order of the efficiency of the anaerobic degradation of antibiotics was as follows: strains B and SANA > strain B > strain SANA. Yang et al. (2019) reported that the cocultures of strains M10 and M12 can enhance malachite green degradation in milkfish pond sediments. The addition of both bacterial strains M10 and M12 produced better results than each of the single cultures [31]. The aerobic stains SF1 and A12 and the anaerobic strains B and SANA were used in subsequent studies.

The colony morphology and scanning electron micrographs of the four bacterial strains are shown in Figure 1. All of them were Gram-negative, rod-shaped bacteria.

Phylogenetic analysis of strains A12, SF1, B and SANA, based on the 16S rRNA gene, is shown in Figure 2. The strains A12, SF1, B and SANA are closely related to *Pseudmonas pseudoalcaligenes* (99%), *Pseudmonas taiwanensis* (96%), *Bacillus flexus* (99%) and *Clostridium butyricum* (99%), respectively. *Pseudomonas* bacteria are widespread in various natural environments, such as soil, plants, animals, air and water. *Pseudmonas taiwanensis* is an aerobic Gram-negative, rod-shaped, motile, nonspore-forming bacterial strain isolated from soil [33]. *Pseudomonas pseudoalcaligenes* is an aerobic Gram-negative bacterium. It is able to use cyanide as a nitrogen source, and can be used for bioremediation [34]. *Bacillus flexus* is a Gram-positive, rod-shaped, endospore-forming bacterium. This bacterium may be isolated from feces and soil. *Bacillus flexus* has been shown to exert activity mainly against polyvinyl chloride additives and exhibits a low biodegradation rate of polyvinyl chloride polymers [35]. *Clostridium butyricum is* a strictly anaerobic endospore-forming Gram-positive bacillus. This bacterium has been studied for its efficiency in decolorizing various remazol reactive dyes [36].

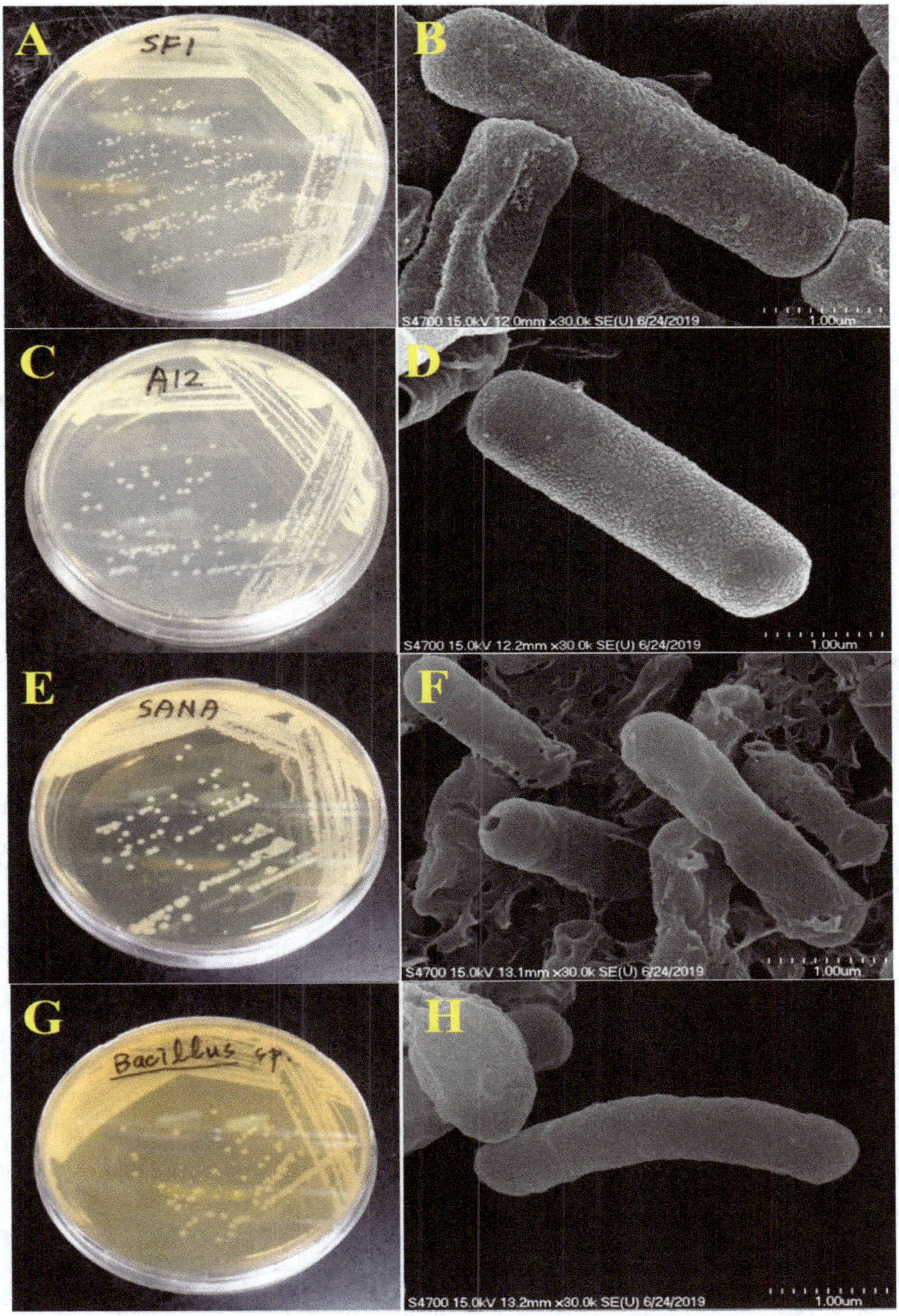

Figure 1. Colony morphology and scanning electron micrographs of isolated bacterial strains SF1 (Pseudmonas sp.) (**A**,**B**), A12 (Pseudmonas sp.) (**C**,**D**), SANA (Clostridium sp.) (**E**,**F**), and B (Bacillus sp.) (**G**,**H**).

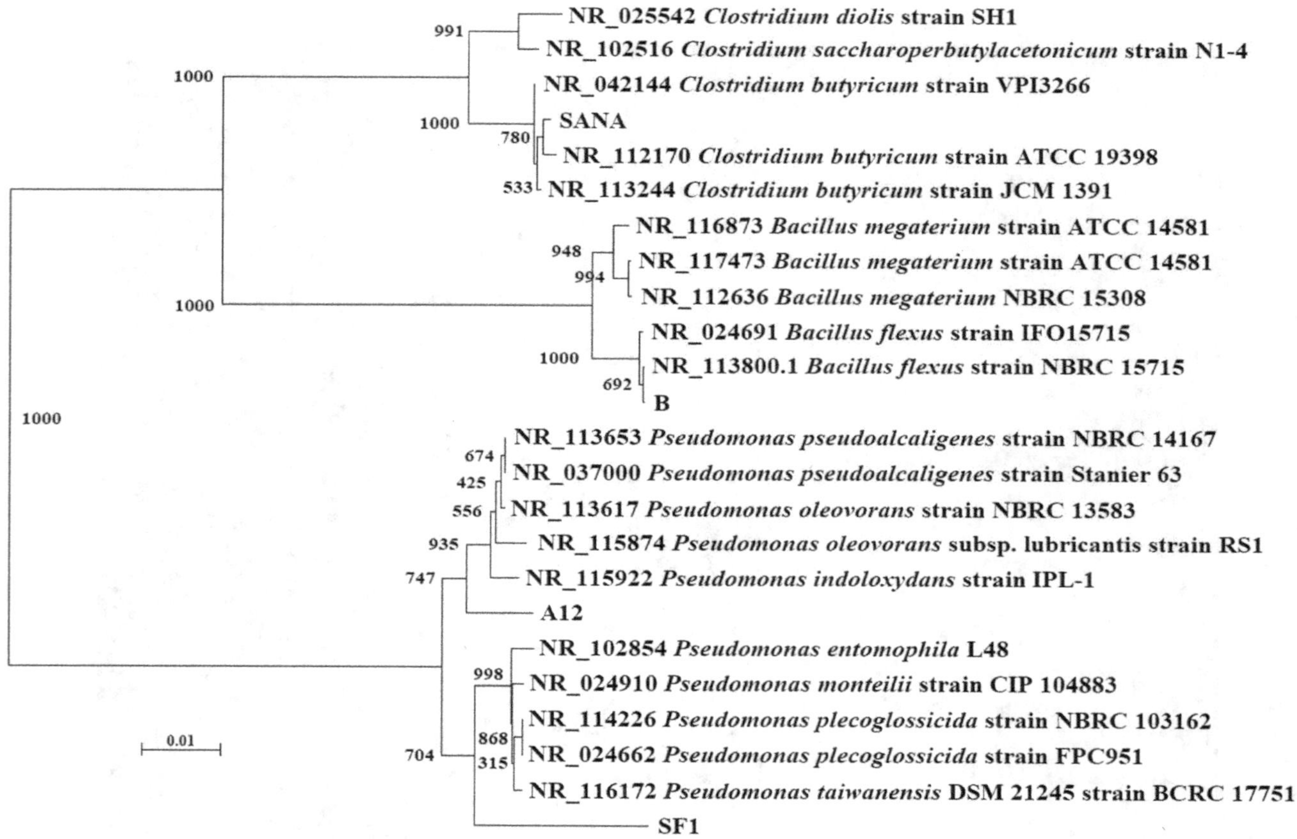

Figure 2. Phylogenetic analysis of the 16S rRNA genes of the four bacterial strains (A12, B, SF1, SANA). Bootstrapping values at branch points indicate the number of times that the same branch was observed out of 1000 repeats of the phylogenetic reconstruction.

3.2. Degradation of Antibiotics in Sludge with Isolated Bacteria

The antibiotic concentrations in the sterile controls were first examined at the end of the 10th or 15th day incubation periods, under aerobic or anaerobic conditions, respectively. The remaining percentages of antibiotics ranged from 91.7 to 95.3%. Therefore, it was concluded that the antibiotic degradation that occurred in all of the following experiments was due to microbial action.

The aerobic and anaerobic degradation of the antibiotics in sludge containing the isolated bacterial strains is shown in Figures 3–5. The degradation of the antibiotics was increased in sludge containing the isolated bacterial strains. As shown in Figures 3A, 4A and 5A, the aerobic degradation half-lives of OTC, TC, CTC, SDM, SMX, SMZ and AMO were 3.0, 4.6, 2.3, 2.5, 3.1, 1.7 and 2.0 days, respectively. The anaerobic degradation half-lives of OTC, TC, CTC, SDM, SMX, SMZ and AMO were 4.4, 7.0, 3.4, 4.1, 6.8, 2.4 and 3.0 days, respectively (Figures 3B, 4B and 5B). The order of the degradation rates of the antibiotics was CTC > OTC > TC> AMO > SMX > SDM > SMZ.

The order of SA degradation rates in sludge was SMX > SDM > SMZ. Similar trends were observed in a study by Yang et al. (2016) in which fungal enzymes were used to increase the degradation of SAs in sludge [37]. The order of the TC degradation rates in sludge was CTC > OTC > TC. Similar results were reported by Suda et al. (2012) from a study in which TC antibiotics were treated with the laccase enzyme in the presence of 1-hydroxybenzotriazole [38]. The order of the antibiotic degradation rates was TCs > AMO > SAs. Antibiotics with lower molecular weights are easier to degrade than those with higher molecular weights. Larger functional groups may hinder degradation by affecting the interactions between target compounds and the bacteria or enzymes [39]. Moreover, the degradation of compounds with complex structures may require more reaction steps.

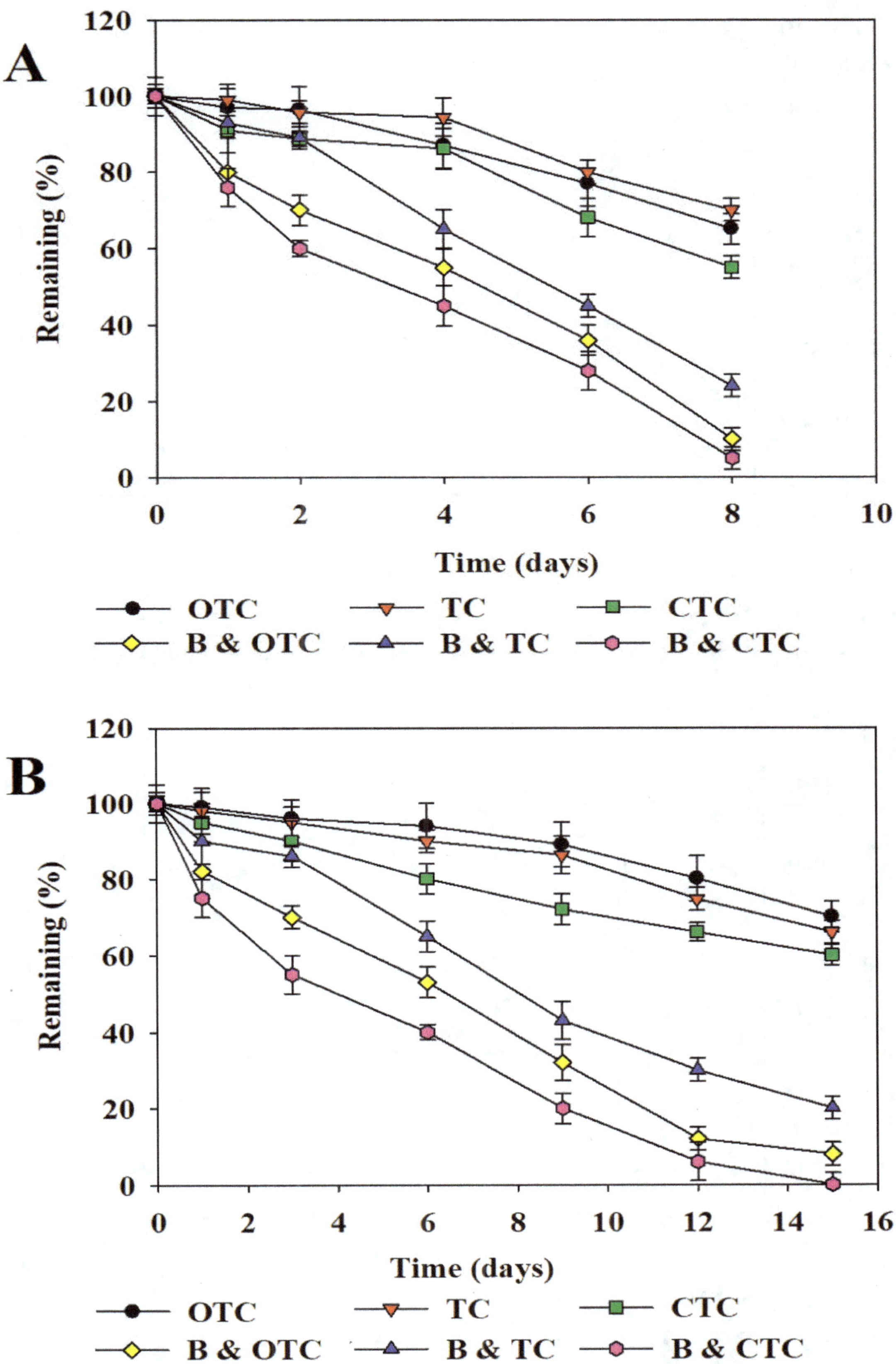

Figure 3. Degradation of tetracyclines with and without degrading bacteria in sludge. Aerobic (**A**) and anaerobic (**B**) degradation of TCs. TC: tetracycline, CTC: chlortetracycline, OTC: oxytetracycline. B: isolated antibiotic-degrading bacteria. Data from three independent experiments are presented as the means ± SE.

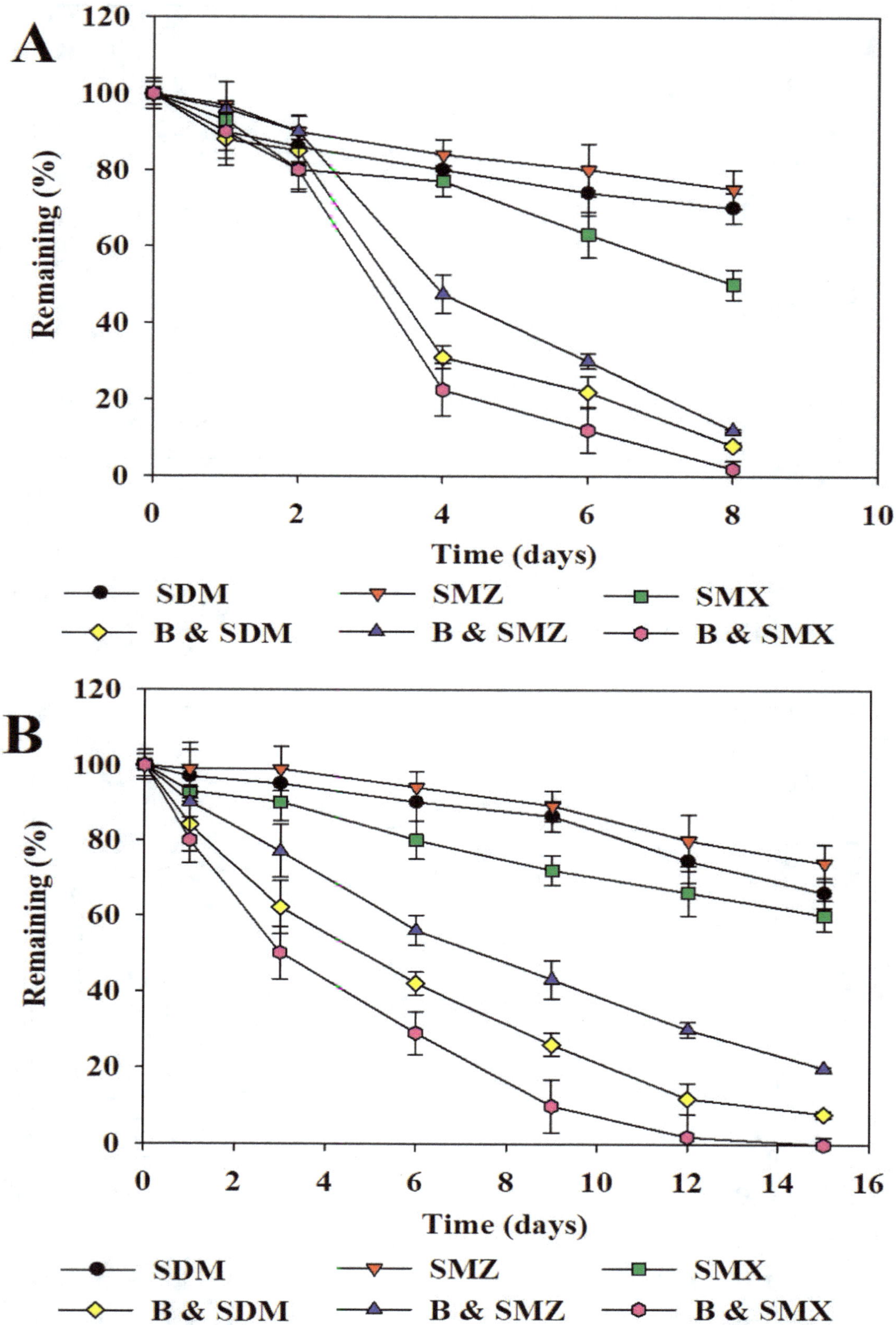

Figure 4. Degradation of sulfonamides with and without degrading bacteria in sludge. Aerobic (**A**) and anaerobic (**B**) degradation of SAs. SMZ: sulfamethazine, SMX: sulfamethoxazole, SDM: sulfadimethoxine. B: isolated antibiotic-degrading bacteria. Data from three independent experiments are presented as the means ± SE.

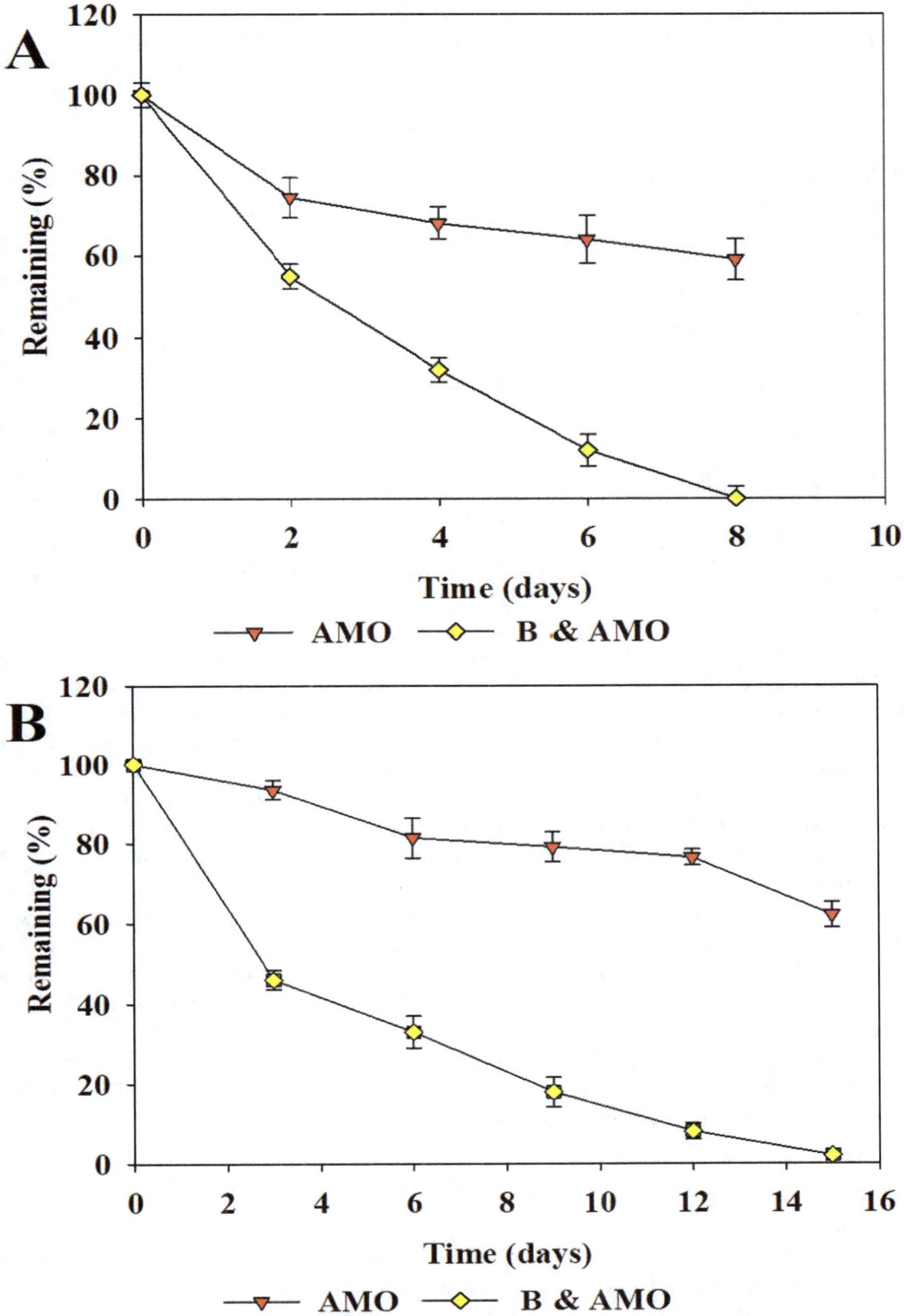

Figure 5. Degradation of amoxicillin with and without degrading bacteria in sludge. Aerobic (**A**) and anaerobic (**B**) degradation of AMO. AMO: amoxicillin. B: isolated antibiotic-degrading bacteria. Data from three independent experiments are presented as the means ± SE.

3.3. Repeated Addition of Antibiotics and Their Degradation in Sludge

To test the long-term degradation ability of aerobic strains SF1 and A12 and anaerobic strains B and SANA in sludge, experiments involving the repeated addition of antibiotics were performed (Table 2).

Table 2. The removal rates (%) after the first (1st), second (2nd), and third (3rd) additions of antibiotics.

		CTC	AMO	SMX
1st	Sludge [a]	35.6 ± 1.6	41.6 ± 1.8	46.3 ± 1.5
	Sludge & SF1 & A12 [a]	82.6 ± 2.5	88.2 ± 2.9	90.6 ± 3.8
	Sludge [b]	33.5 ± 1.2	37.4 ± 1.1	42.2 ± 1.6
	Sludge & B & SANA[b]	85.2 ± 2.4	88.1 ± 2.1	92.3 ± 3.1
2nd	Sludge [a]	43.7 ± 1.5	45.3 ± 1.3	49.1 ± 1.4
	Sludge & SF1 & A12 [a]	88.6 ± 2.5	90.2 ± 2.9	92.5 ± 2.9
	Sludge [b]	41.2 ± 1.5	43.7 ± 1.3	46.4 ± 1.2
	Sludge & B & SANA [b]	90.7 ± 3.3	91.1 ± 3.1	94.3 ± 3.5
3rd	Sludge [a]	46.1 ± 1.3	50.4 ± 1.3	52.5 ± 1.1
	Sludge & SF1 & A12 [a]	90.3 ± 3.2	95.3 ± 4.1	97.5 ± 3.9
	Sludge [b]	42.7 ± 1.1	46.4 ± 1.3	50.4 ± 1.2
	Sludge & B & SANA [b]	92.4 ± 4.2	96.4 ± 4.4	98.1 ± 4.7

Note(s): [a] aerobic conditions for 8 days. [b] anaerobic conditions for 15 days. Data from three independent experiments are presented as the means ± SE.

In the first addition experiments, in the sludge under aerobic conditions, 46.3, 41.6 and 35.6% of SMX, AMO and CTC initially present at 2 mg/L was degraded, respectively. In the sludge under anaerobic conditions, 42.2, 37.4 and 33.5% of SMX, AMO and CTC initially present at 2 mg/L was degraded, respectively. In the sludge under aerobic conditions containing aerobic strains SF1 and A12, 90.6, 88.2 and 82.6% of SMX, AMO and CTC initially present at 2 mg/L was degraded, respectively. In the sludge under anaerobic conditions containing anaerobic strains B and SANA, 92.3, 88.1 and 85.2% of SMX, AMO and CTC initially present at 2 mg/L was degraded, respectively. In the third addition experiments, in the sludge under aerobic conditions, 52.5, 50.4 and 46.1% of SMX, AMO and CTC initially present at 2 mg/L was degraded, respectively. In the sludge under anaerobic conditions, 50.4, 46.4 and 42.7% of SMX, AMO and CTC initially present at 2 mg/L was degraded, respectively. In the sludge under aerobic conditions containing aerobic strains SF1 and A12, 97.5, 95.3 and 90.3% of SMX, AMO and CTC initially present at 2 mg/L was degraded, respectively. In the sludge under anaerobic conditions containing anaerobic strains B and SANA, 98.1, 96.4 and 92.4% of SMX, AMO and CTC initially present at 2 mg/L was degraded, respectively. These results suggest that the degradation rate increased with the repeated addition of SMX, AMO and CTC. The repeated addition of antibiotics might increase the abundance of antibiotic-degrading microbes. These findings are similar to the results of our previous study of SA degradation in sludge [38].

3.4. Microbial Communities in the Repeated Addition Experiments

To obtain deeper insight into the conditions in the repeated addition experiments, the microbial communities in the repeated addition experiments were analyzed. As shown in Figure 6, the five experimental conditions (original sludge, sludge in aerobic conditions after the 3rd addition, sludge and bacteria in aerobic conditions after the 3rd addition, sludge in anaerobic conditions after the 3rd addition, sludge and bacteria in anaerobic conditions after the 3rd addition) exhibited differences in the composition of the major microbial communities (46, 45, 24, 33 and 26 microbial genera). The lists of the five groups of major microbial genera in the five experimental conditions are listed in Table S2.

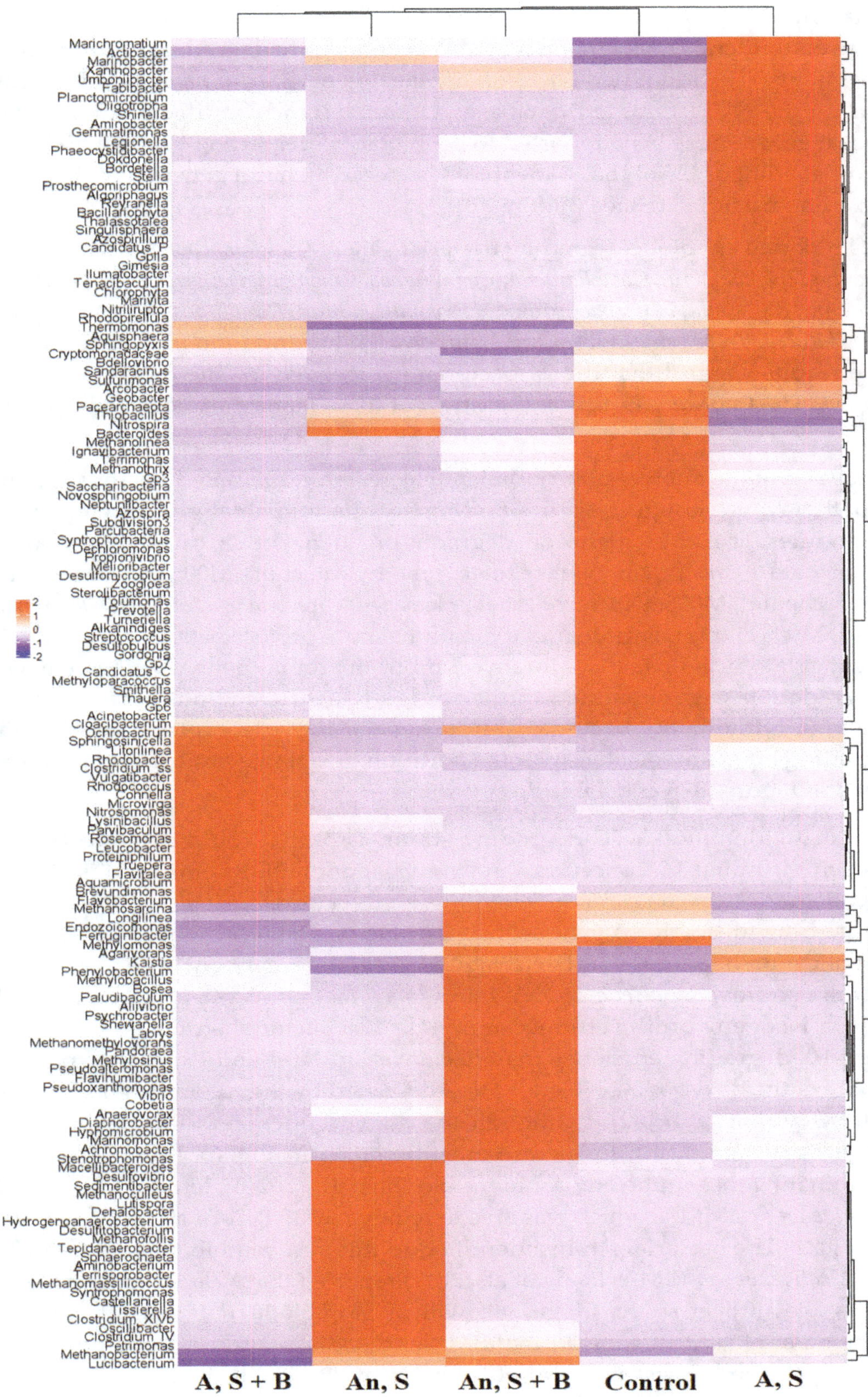

Figure 6. Major microbial communities (genus level) in sludge after the 3rd repeated addition experiment. Control: original sludge. A, S: aerobic conditions, sludge only. A, S + B: aerobic conditions, sludge + bacteria. An, S: anaerobic conditions, sludge only. An, S + B: anaerobic conditions, sludge + bacteria. The major microbial genera (red color in heatmap) under the five experimental conditions are listed in Table S2.

For validation, a list of xenobiotic biodegradation-associated bacteria/archaea in the KEGG database was used to identify bacteria/archaea with xenobiotic degradation ability in the repeated addition experiments. The identified bacterial/archaeal genera were used as key words to perform text mining using the NCBI PubMed database. Under aerobic conditions, 25 of the 63 identified bacterial/archaeal genera have been reported to be antibiotic-degrading bacteria/archaea (Figure 7A). Under anaerobic conditions, 25 of the 63 identified bacterial/archaeal genera have been reported to be antibiotic-degrading bacteria/archaea (Figure 7B).

In total, 24 reported antibiotic-degrading bacterial genera (*Achromobacter, Acidovorax, Acinetobacter, Alcaligenes, Bacillus, Burkholderia, Castellaniella, Comamonas, Corynebacterium, Cupriavidus, Dechloromonas, Geobacter, Gordonia, Klebsiella, Mycobacterium, Novosphingobium, Pandoraea, Pseudomonas, Rhodococcus, Sphingomonas, Thauera, Treponema, Vibrio and Xanthobacter*) are common to the aerobic and anaerobic groups. These results suggest that the 24 bacterial genera may be the major antibiotic-degrading bacteria in sludge. The PubMed ID, title and abstract of these reports are collected and summarized in Table S3.

The distribution of the number of major microbial genera exhibiting different aromatic compound degradation pathways is shown in Figure 8. Most of the microbial genera were associated with six reaction modules: M00548 (benzene degradation, benzene ≥ catechol), M00551 (benzoate degradation, benzoate ≥ catechol/methylbenzoate ≥ methylcatechol), M00568 (catechol ortho-cleavage, catechol ≥ 3-oxoadipate), M00569 (catechol meta-cleavage, catechol ≥ acetyl-CoA/4-methylcatechol ≥ propanoyl-CoA), M00623 (phthalate degradation, phthalate ≥ protocatechuate) and M00638 (salicylate degradation, salicylate ≥ gentisate). Moreover, more microbial genera associated with aromatic compound degradation were identified in the sludge with isolated antibiotic-degrading bacteria than in the sludge without the bacteria. These results provide explanations for the increased antibiotic-degrading effects associated with the addition of isolated antibiotic-degrading bacteria in sludge observed in Figures 3–5 and Table 2.

Yin et al. (2020) identified a TC-degrading strain, TR5, that could degrade TC quickly (~90% within 36 h) when the initial TC concentration was 200 mg/L [40]. Its efficiency is much higher than those of the TC-degrading bacteria identified in this study. However, TR5 was identified as *Klebsiella* pneumoniae according to 16S rRNA gene sequencing and biochemical properties. Great care must be taken in the application of pathogenic bacteria in wastewater treatment. Wu et al. (2020) identified two TC-degrading strains, *Raoultella* sp. XY-1 and *Pandoraea* sp. XY-2, which degraded 81.72% TC within 12 days in lysogeny broth (LB) medium [41]. The bacterial strains identified in this study exhibit a higher TC-degrading efficiency than *Raoultella* sp. XY-1 and *Pandoraea* sp. XY-2. Wen et al. reported that 95% of the doxycycline (50 mg/L) was degraded by the recombinant strain *Escherichia coli* ETD-1, with tetX, within 48 h [42]. Although the doxycycline-degrading ability of recombinant *Escherichia coli* is very significant, great care must be taken in the application of microorganisms carrying recombinant genes in the environment. Sodhi et al. (2020) identified the AMO-degrading bacterium, *Alcaligenes* sp. MMA, which was able to remove up to 84% of amoxicillin in 14 days in M9 minimal media [43]. The bacterial strains identified in this study exhibit a higher AMO-degrading efficiency than *Alcaligenes* sp. MMA. Liang et al. (2019) reported that *Achromobacter* sp. JL9 was able to utilize SMX as its sole nitrogen source for growth, with an SMX biodegradation efficiency of 63.10% [44]. Moreover, Nguyen et al. (2019) showed that SMX and sodium acetate could be cometabolized as carbon sources, with the highest removal efficiencies of 82.44%, 80.2%, and 79.45% for NH_4^+-N, NO_3^--N, and SMX, respectively [45]. The bacterial strains identified in this study exhibit a higher SMX-degrading efficiency than *Achromobacter* sp. Wang et al. (2018) demonstrated that *Acinetobacter* sp. could mineralize 98.8% of SMX, but only 17.5% and 20.5% of sulfadiazine and SMZ, respectively [46]. The bacterial strains identified in this study could degrade 98%, 91% and 88% SMX, SDM and SMZ, respectively. Overall, the antibiotic-degrading efficiency of the bacterial strains identified in this study is very good. Moreover, these bacterial strains exhibit the ability to degrade multiple antibiotics.

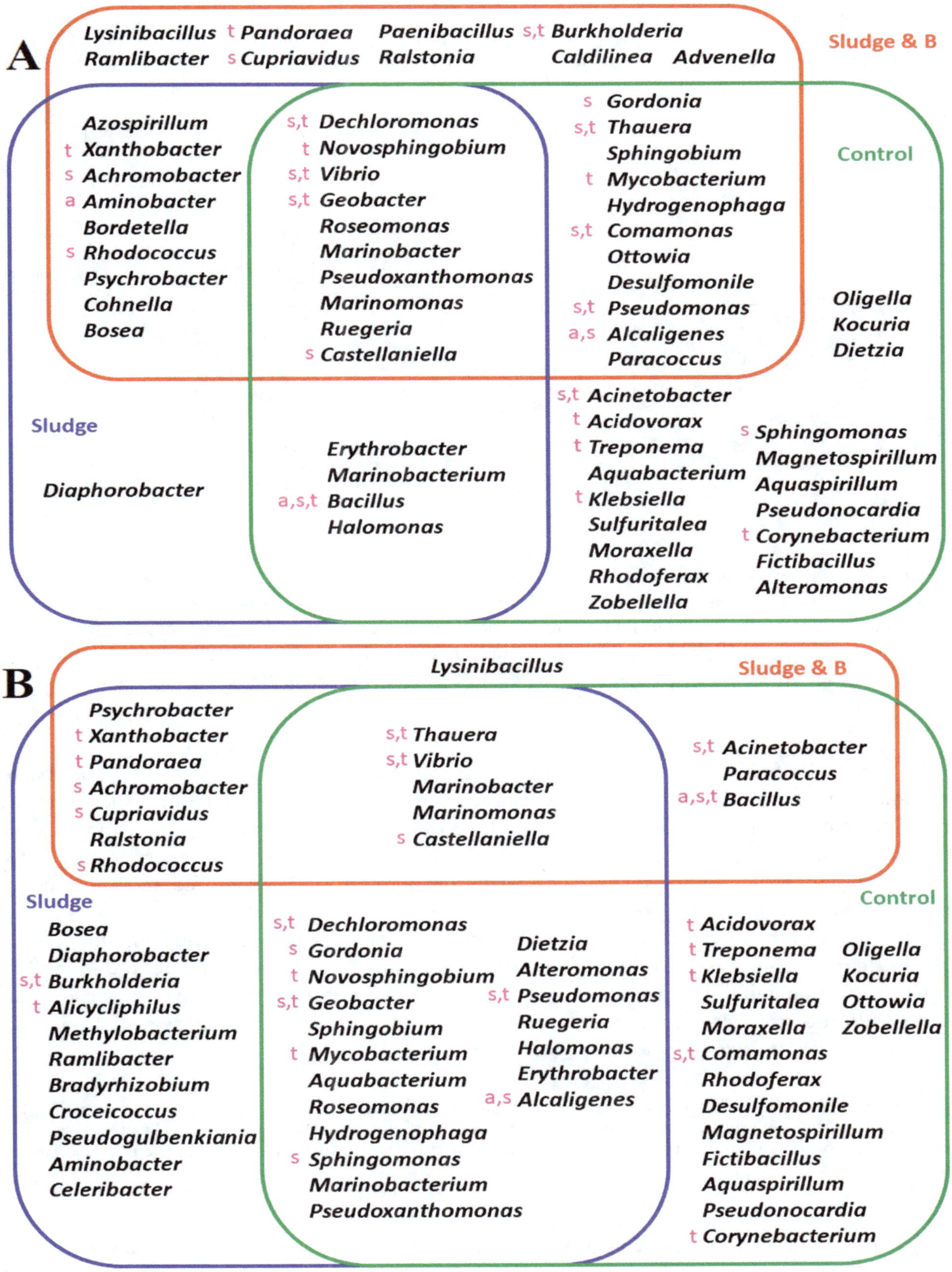

Figure 7. Xenobiotic biodegradation-associated bacteria after the 3rd repeated addition experiment under aerobic (**A**) and anaerobic (**B**) conditions. Control: original sludge. Sludge and B: sludge containing the isolated bacterial strains. a: amoxicillin-degrading bacteria. s: sulfonamide-degrading bacteria. t: tetracycline-degrading bacteria.

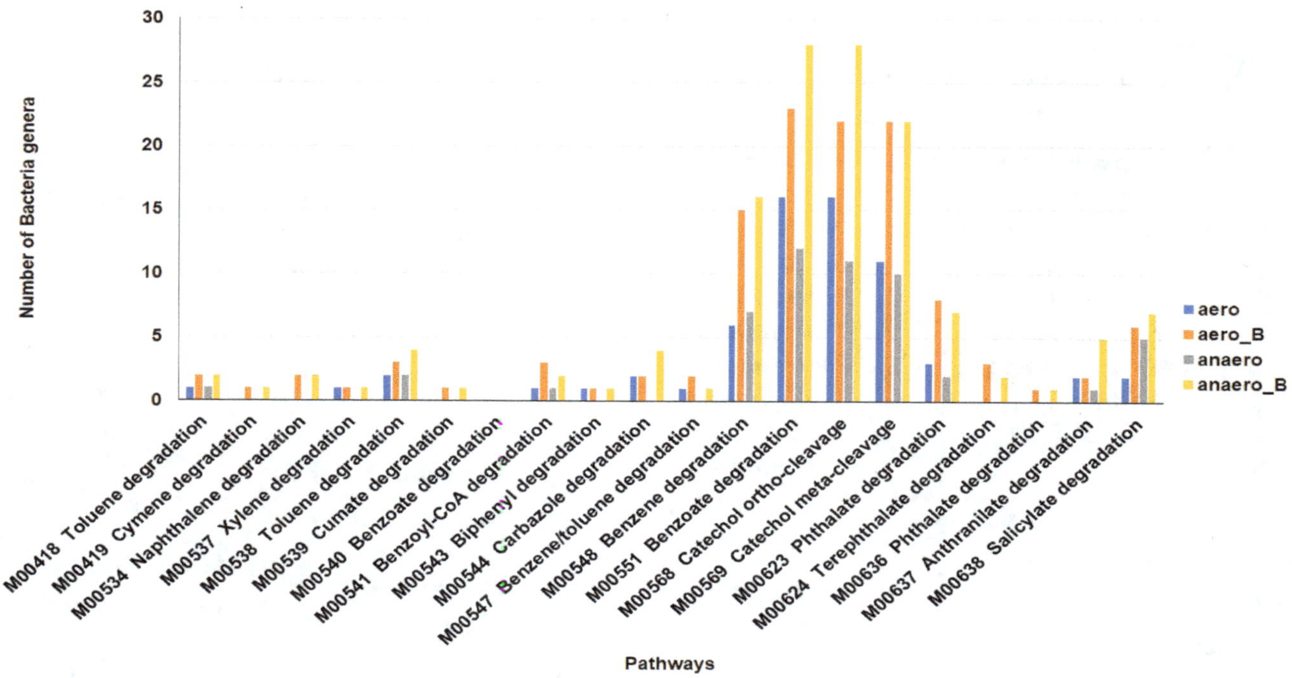

Figure 8. Xenobiotic biodegradation-associated microbial genera after the 3rd repeated addition experiment under aerobic and anaerobic conditions. Mxxxxx are Kyoto Encyclopedia of Genes and Genomes (KEGG) database module IDs. aero: sludge under aerobic conditions. anaero: sludge under anaerobic conditions. aero_B: sludge with antibiotic-degrading bacteria under aerobic conditions. anaero: sludge with antibiotic-degrading bacteria under anaerobic conditions.

4. Conclusion

The removal efficiency of antibiotics in wastewater sludge was enhanced using the bacterial strains isolated in this study. Batch experiments indicated that the combination of two bacterial strains enhanced the biodegradation of antibiotics in sludge to a greater extent than the application of a single bacterial strain under both aerobic and anaerobic conditions. The efficiency was comparable between the aerobic and anaerobic settings. However, anaerobic biodegradation required a longer incubation time (by 2-fold) than aerobic biodegradation. The results of repeated addition experiments indicated that the efficiency of the biodegradation of antibiotics using bacterial strains could be maintained for at least three degradation runs. The added antibiotic-degrading bacteria not only degraded antibiotics directly but also changed the composition of the microbial community in the sludge. As a consequence, the composition of the xenobiotic degradation pathways and the antibiotic degradation capacity were changed.

Author Contributions: Formal analysis, C.L.; Funding acquisition, B.-V.C.; Methodology, C.-W.Y. and C.L.; Project administration, B.-V.C.; Visualization, C.-W.Y.; Writing—original draft, C.-W.Y. and B.-V.C.; Writing—review and editing, C.-W.Y. and B.-V.C. All authors have read and agreed to the published version of the manuscript.

References

1. Salgado, R.; Noronha, J.P.; Oehmen, A.; Carvalho, G.; Reis, M.A. Analysis of 65 pharmaceuticals and personal care products in 5 wastewater treatment plants in Portugal using a simplified analytical methodology. *Water Sci. Technol.* **2010**, *62*, 2862–2871. [CrossRef]
2. Barber, L.B.; Keefe, S.H.; LeBlanc, D.R.; Bradley, P.M.; Chapelle, F.H.; Meyer, M.T.; Loftin, K.A.; Kolpin, D.W.; Rubio, F. Fate of sulfamethoxazole, 4- nonylphenol, and 17ß-estradiol in groundwater contaminated by wastewater treatment plant effluent. *Environ. Sci. Technol.* **2009**, *43*, 4843–4850. [CrossRef]

3. Qiao, M.; Ying, G.G.; Singer, A.C.; Zhu, Y.G. Review of antibiotic resistance in China and its environment. *Environ. Int.* **2018**, *110*, 160–172. [CrossRef]
4. Gzyl, K.E.; Wieden, H.J. Tetracycline does not directly inhibit the function of bacterial elongation factor Tu. *PLoS ONE* **2017**, *12*, e0178523. [CrossRef]
5. Sukul, P.; Spiteller, M. Sulfonamides in the environment as veterinary drugs. In *Reviews of Environmental Contamination and Toxicology*; Springer: New York, NY, USA, 2006; Volume 187, pp. 67–101.
6. Gobel, A.; Thomsen, A.; McArdell, C.S.; Joss, A.; Giger, W. Occurrence and sorption behavior of sulfonamides, macrolides, and trimethoprim in activated sludge treatment. *Environ. Sci. Technol.* **2005**, *39*, 3981–3989. [CrossRef] [PubMed]
7. Nieto, A.; Borrull, F.; Pocurull, E.; Marce, R.M. Occurrence of pharmaceuticals and hormones in sewage sludge. *Environ. Toxicol. Chem.* **2010**, *29*, 1484–1489. [CrossRef]
8. Adamek, E.; Baran, W.; Sobczak, A. Assessment of the biodegradability of selected sulfa drugs in two polluted rivers in Poland: Effects of seasonal variations, accidental contamination, turbidity and salinity. *J. Hazard Mater.* **2016**, *313*, 147–158. [CrossRef] [PubMed]
9. Baghapour, M.A.; Shirdarreh, M.R.; Faramarzian, M. Amoxicillin removal from aqueous solutions using submerged biological aerated filter. *Desalin. Water Treat.* **2014**, *54*, 790–801. [CrossRef]
10. Gavrilescu, M.; Demnerova, K.; Aamand, J.; Agathos, S.; Fava, F. Emerging pollutants in the environment: Present and future challenges in biomonitoring, ecological risks and bioremediation. *New Biotechnol.* **2015**, *32*, 147–156. [CrossRef] [PubMed]
11. Michael-Kordatou, I.; Karaolia, P.; Fatta-Kassinos, D. The role of operating parameters and oxidative damage mechanisms of advanced chemical oxidation processes in the combat against antibiotic-resistant bacteria and resistance genes present in urban wastewater. *Water Res.* **2018**, *129*, 208–230. [PubMed]
12. Chang, B.V.; Fan, S.N.; Tsai, Y.C.; Chung, Y.L.; Tu, P.X.; Yang, C.W. Removal of emerging contaminants using spent mushroom compost. *Sci. Total Environ.* **2018**, *634*, 922–933. [CrossRef] [PubMed]
13. Sun, C.; Li, W.; Chen, Z.; Qin, W.; Wen, X. Responses of antibiotics, antibiotic resistance genes, and mobile genetic elements in sewage sludge to thermal hydrolysis pre-treatment and various anaerobic digestion conditions. *Environ. Int.* **2019**, *133*, 105156. [CrossRef] [PubMed]
14. He, Z.; Wei, Z.; Zhao, Y.; Zhang, D.; Pan, X. Enhanced performance of tetracycline treatment in wastewater using aerobic granular sludge with in-situ generated biogenic manganese oxides. *Sci. Total Environ.* **2020**, *735*, 139533. [CrossRef] [PubMed]
15. Du, B.; Yang, Q.; Li, X.; Yuan, W.; Chen, Y.; Wang, R. Impacts of long-term exposure to tetracycline and sulfamethoxazole on the sludge granules in an anoxic-aerobic wastewater treatment system. *Sci. Total Environ.* **2019**, *684*, 67–77. [CrossRef]
16. Krzeminski, P.; Tomei, M.C.; Karaolia, P.; Langenhoff, A.; Almeida, C.M.; Felis, E.; Gritten, F.; Andersen, H.R.; Fernandes, T.; Manaia, C.M.; et al. Performance of secondary wastewater treatment methods for the removal of contaminants of emerging concern implicated in crop uptake and antibiotic resistance spread: A review. *Sci. Total Environ.* **2019**, *648*, 1052–1081. [CrossRef]
17. Bisognin, R.P.; Wolff, D.B.; Carissimi, E.; Prestes, O.D.; Zanella, R. Occurrence and fate of pharmaceuticals in effluent and sludge from a wastewater treatment plant in Brazil. *Environ. Technol.* **2019**, *2019*, 1–12. [CrossRef]
18. Ezzariai, A.; Hafidi, M.; Khadra, A.; Aemig, Q.; El Fels, L.; Barret, M.; Merlina, G.; Patureau, D.; Pinelli, E. Human and veterinary antibiotics during composting of sludge or manure: Global perspectives on persistence, degradation, and resistance genes. *J. Hazard Mater.* **2018**, *359*, 465–481. [CrossRef]
19. Boonnorat, J.; Kanyatrakul, A.; Prakhongsak, A.; Honda, R.; Panichnumsin, P.; Boonapatcharoen, N. Effect of hydraulic retention time on micropollutant biodegradation in activated sludge system augmented with acclimatized sludge treating low-micropollutants wastewater. *Chemosphere* **2019**, *230*, 606–615. [CrossRef]
20. Nas, B.; Argun, M.E.; Dolu, T.; Ateş, H.; Yel, E.; Koyuncu, S.; Dinç, S.; Kara, M. Occurrence, loadings and removal of EU-priority polycyclic aromatic hydrocarbons (PAHs) in wastewater and sludge by advanced biological treatment, stabilization pond and constructed wetland. *J. Environ. Manag.* **2020**, *268*, 110580. [CrossRef]
21. Kamaz, M.; Wickramasinghe, S.R.; Eswaranandam, S.; Zhang, W.; Jones, S.M.; Watts, M.J.; Qian, X. Investigation into micropollutant removal from wastewaters by a membrane bioreactor. *Int. J. Environ. Res. Public Health* **2019**, *16*, 1363. [CrossRef]
22. Gauthier, H.; Yargeau, V.; Cooper, D.G. Biodegradation of pharmaceuticals by *Rhodococcus rhodochrous* and *Aspergillus niger* by co-metabolism. *Sci. Total Environ.* **2010** *408*, 1701–1706. [CrossRef] [PubMed]

23. Leng, Y.; Bao, J.; Chang, G.; Zheng, H.; Li, X.; Du, J.; Snow, D. Biotransformation of tetracycline by a novel bacterial strain *Stenotrophomonas maltophilia* DT1. *J. Hazard. Mater.* **2016**, *318*, 125–133. [CrossRef] [PubMed]
24. Shao, S.; Hu, Y.; Cheng, J.; Chen, Y. Biodegradation mechanism of tetracycline (TEC) by strain *Klebsiella* sp. SQY5 as revealed through products analysis and genomics. *Ecotoxicol. Environ. Saf.* **2019**, *85*, 109676–109683. [CrossRef]
25. Larcher, S.; Yargeau, V. Biodegradation of sulfamethoxazole by individual and mixed bacteria. *Appl. Microbiol. Biotechnol.* **2011**, *91*, 211–218. [CrossRef] [PubMed]
26. Wang, J.; Wang, S. Microbial degradation of sulfamethoxazole in the environment. *Appl. Microbiol. Biotechnol.* **2018**, *102*, 3573–3582. [CrossRef] [PubMed]
27. Liao, X.; Li, B.; Zou, R.; Xie, S.; Yuan, B. Antibiotic sulfanilamide biodegradation by acclimated microbial populations. *Appl. Microbiol. Biotechnol.* **2016**, *100*, 2439–2447. [CrossRef]
28. Liao, X.; Zou, R.; Li, B.; Tong, T.; Xie, S.; Yuan, B. Biodegradation of chlortetracycline by acclimated microbiota. *Process Saf. Environ.* **2017**, *109*, 11–17. [CrossRef]
29. Larkin, M.A.; Blackshields, G.; Brown, N.P.; Chenna, R.; McGettigan, P.A.; McWilliam, H.; Valentin, F.; Wallace, I.M.; Wilm, A.; Lopez, R.; et al. Clustal W and Clustal X version 2.0. *Bioinformatics* **2007**, *23*, 2947–2948. [CrossRef]
30. Chang, B.V.; Chang, Y.T.; Chao, W.L.; Yeh, S.L.; Kuo, D.L.; Yang, C.W. Effects of sulfamethoxazole and sulfamethoxazole-degrading bacteria on water quality and microbial communities in milkfish ponds. *Environ. Pollut.* **2019**, *252*, 305–316. [CrossRef]
31. Yang, C.W.; Chao, W.L.; Hsieh, C.Y.; Chang, B.V. Biodegradation of malachite green in milkfish pond sediments. *Sustainability* **2019**, *11*, 4179. [CrossRef]
32. Kanehisa, M.; Furumichi, M.; Tanabe, M.; Sato, Y.; Morishima, K. KEGG: New perspectives on genomes, pathways, diseases and drugs. *Nucleic Acids Res.* **2017**, *45*, D353–D361. [CrossRef] [PubMed]
33. Wang, L.T.; Tai, C.J.; Wu, Y.C.; Chen, Y.B.; Lee, F.L.; Wang, S.L. *Pseudomonas taiwanensis* sp. nov., isolated from soil. *Int. J. Syst. Evol. Micr.* **2010**, *60*, 2094–2098. [CrossRef] [PubMed]
34. Huertas, M.J.; Luque-Almagro, V.M.; Martinez-Luque, M.; Blasco, R.; Moreno-Vivian, C.; Castillo, F.; Roldan, M.D. Cyanide metabolism of *Pseudomonas pseudoalcaligenes* CECT5344: Role of siderophores. *Biochem. Soc. Trans.* **2006**, *34*, 152–155. [CrossRef] [PubMed]
35. Giacomucci, L.; Raddadi, N.; Soccio, M.; Lotti, N.; Fava, F. Polyvinyl chloride biodegradation by *Pseudomonas citronellolis* and *Bacillus flexus*. *New Biotechnol.* **2019**, *52*, 35–41. [CrossRef] [PubMed]
36. Ekambaram, S.P.; Perumal, S.S.; Annamalai, U. Decolorization and biodegradation of remazol reactive dyes by *Clostridium* species. *3 Biotech.* **2016**, *6*, 20. [CrossRef]
37. Yang, C.W.; Hsiao, W.C.; Chang, B.V. Biodegradation of sulfonamide antibiotics in sludge. *Chemosphere* **2016**, *150*, 559–565. [CrossRef]
38. Suda, T.; Hata, T.; Kawai, S.; Okamura, H.; Nishida, T. Treatment of tetracycline antibiotics by laccase in the presence of 1-hydroxybenzotriazole. *Bioresour. Technol.* **2012**, *103*, 498–501. [CrossRef]
39. Pepper, L.L.; Gerba, C.P.; Gentry, T.J. Environmental microbiology. In *Microorganisms and Organic Pollutant*; Maier, M.M., Gentry, T.J., Eds.; Elsevier Inc.: Berkeley, CA, USA, 2015; pp. 377–413.
40. Yin, Z.; Xia, D.; Shen, M.; Zhu, D.; Cai, H.; Wu, M.; Zhu, Q.; Kang, Y. Tetracycline degradation by *Klebsiella* sp. strain TR5: Proposed degradation pathway and possible genes involved. *Chemosphere* **2020**, *253*, 126729. [CrossRef]
41. Wu, X.; Gu, Y.; Wu, X.; Zhou, X.; Zhou, H.; Amanze, C.; Shen, L.; Zeng, W. Construction of a Tetracycline Degrading Bacterial Consortium and Its Application Evaluation in Laboratory-Scale Soil Remediation. *Microorganisms* **2020**, *8*, 292. [CrossRef]
42. Wen, X.; Huang, J.; Cao, J.; Xu, J.; Mi, J.; Wang, Y.; Ma, B.; Zou, Y.; Liao, X.; Liang, J.B.; et al. Heterologous expression of the tetracycline resistance gene *tetX* to enhance degradability and safety in doxycycline degradation. *Ecotoxicol. Environ. Saf.* **2020**, *191*, 110214. [CrossRef]
43. Sodhi, K.K.; Kumar, M.; Singh, D.K. Potential application in amoxicillin removal of *Alcaligenes* sp. MMA and enzymatic studies through molecular docking. *Arch. Microbiol.* **2020**. [CrossRef] [PubMed]
44. Liang, D.H.; Hu, Y. Simultaneous sulfamethoxazole biodegradation and nitrogen conversion by *Achromobacter* sp. JL9 using with different carbon and nitrogen sources. *Bioresour. Technol.* **2019**, *293*, 122061. [CrossRef] [PubMed]

45. Nguyen, P.Y.; Silva, A.F.; Reis, A.C.; Nunes, O.C.; Rodrigues, A.M.; Rodrigues, J.E.; Cardoso, V.V.; Benoliel, M.J.; Reis, M.A.; Oehmen, A.; et al. Bioaugmentation of membrane bioreactor with *Achromobacter denitrificans* strain PR1 for enhanced sulfamethoxazole removal in wastewater. *Sci. Total Environ.* **2019**, *648*, 44–55. [CrossRef]
46. Wang, S.; Hu, Y.; Wang, J. Biodegradation of typical pharmaceutical compounds by a novel strain *Acinetobacter* sp. *J. Environ. Manag.* **2018**, *217*, 240–246. [CrossRef] [PubMed]

7

Physiological Profiling and Functional Diversity of Groundwater Microbial Communities in a Municipal Solid Waste Landfill Area

Marco Melita *, Stefano Amalfitano, Elisabetta Preziosi, Stefano Ghergo, Eleonora Frollini, Daniele Parrone and Annamaria Zoppini *

Water Research Institute, National Research Council (IRSA-CNR), 00015 Monterotondo, Rome, Italy; amalfitano@irsa.cnr.it (S.A.); preziosi@irsa.cnr.it (E.P.); ghergo@irsa.cnr.it (S.G.); frollini@irsa.cnr.it (E.F.); parrone@irsa.cnr.it (D.P.)

* Correspondence: melita@irsa.cnr.it (M.M.); zoppini@irsa.cnr.it (A.Z.)

Abstract: The disposal of municipal solid wastes in landfills represents a major threat for aquifer environments at the global scale. The aim of this study was to explore how groundwater geochemical characteristics can influence the microbial community functioning and the potential degradation patterns of selected organic substrates in response to different levels of landfill-induced alterations. Groundwaters collected from a landfill area were monitored by assessing major physical-chemical parameters and the microbiological contamination levels (total coliforms and fecal indicators—Colilert-18). The aquatic microbial community was further characterized by flow cytometry and Biolog EcoPlates™ assay. Three groundwater conditions (i.e., pristine, mixed, and altered) were identified according to their distinct geochemical profiles. The altered groundwaters showed relatively higher values of organic matter concentration and total cell counts, along with the presence of fecal indicator bacteria, in comparison to samples from pristine and mixed conditions. The kinetic profiles of the Biolog substrate degradation showed that the microbial community thriving in altered conditions was relatively more efficient in metabolizing a larger number of organic substrates, including those with complex molecular structures. We concluded that the assessment of physiological profiling and functional diversity at the microbial community level could represent a supportive tool to understand the potential consequences of the organic contamination of impacted aquifers, thus complementing the current strategies for groundwater management.

Keywords: Biolog EcoPlates™; flow cytometry; microbial community; metabolic fingerprint; groundwater quality; hydrogeochemistry

---

1. Introduction

Groundwater is the second largest freshwater deposit representing 14% of all inland waters and one of the most important drinking water resource [1,2]. Several factors may impair groundwater environment including drought, excessive withdrawals, and pollutant contamination derived by anthropogenic activities [3–5]. Awareness of the importance of groundwater resources led the European Commission to define legislative tools for ensuring the groundwater ecosystem quality and fostering the related research efforts [6]. The environmental policy started to consider aquifers as a living ecosystem to be preserved [7]. To date, the anthropogenic impact on groundwater biological communities is largely disregarded, although a wide range of organisms are adapted to live under the limiting conditions imposed by darkness and low trophic resources [8,9]. Among those, prokaryotic microorganisms are the most relevant in terms of biomass [10]. The environmental factors that influence the dynamics of aquatic microbial communities are increasingly attracting the research interest, given the key role played by microorganisms in the ecosystem while processing and cycling organic carbon and nutrients [11,12]. Changes in water physical-chemical characteristics were found to affect the structural and functional properties of the microbial communities [13–15].

Among the global threats for the aquifer environment, the municipal solid waste disposal in landfills is of major concern for either human or environmental health, since pollutants can promptly migrate into groundwater by the leaching of percolate [11,16,17]. To counteract and prevent "significant adverse environmental effects" derived from landfill practices, a detailed characterization of groundwater quality is required and based on representative hydrogeological conceptual models (European Landfill Directive; 1999/31/EC) [18]. However, despite the good chemical status is determined by compliance with water quality standards, the ecological approach to evaluate the anthropogenic impact on groundwaters has been poorly considered [19].

Different approaches are suitable for monitoring and predicting the effects of anthropogenic groundwater alterations, offering an insight into the relationships occurring between geochemical parameters and the microbial functional properties [13,20,21]. The microbial phylogenetic changes, observed so far in many literature studies [22–24], were not likely exhaustive to explain the microbial functional responses to the anthropic impact. A fundamental understanding of the microbially-driven processes and the following ecosystem responses to anthropogenic stress factors is still needed.

Starting from the outcomes of one-year monitoring activities in a municipal solid waste landfill in Central Italy [21], we aimed to explore the functional properties of groundwater microbial communities and their responses to different levels of landfill-induced geochemical alterations. The hypothesis is that changes in groundwater quality will affect the functioning of the resident microbial communities by inducing preferential metabolic pathways. This approach could complement the current geochemical assessments to better elucidate the effects of the anthropic activities on the aquifer environment.

2. Materials and Methods

2.1. Field Sampling and Major Water Characteristics

The study was carried out at a groundwater-monitoring network located within the area of a municipal solid non-hazardous waste landfill facility in Central Italy (Figure 1). The landfill, duly equipped with liners and leachate collection systems, is operating since 2002, following the Landfill Directive 1999/31/EC. Between 2003 and 2018, it received 2.9 million tons of waste in 4 different basins, deriving mainly from the mechanical-biological treatment of municipal waste (65%), and non-hazardous waste of other origin. The yearly amount varied from 170,000 tons to 2,700,000 tons. A water table aquifer (depth 5–15 m) was found only in the northern part of the facility. A confined aquifer (depth 5–27 m) underlies the entire landfill area, with groundwater flowing from North to South through lacustrine sediments formed by clays, silts, and sandy layers [21].

During two sampling campaigns (October 2016 and May 2017), 22 water samples were collected from 12 piezometers (Figure 1), by following specific guidelines for groundwater sampling in industrial areas [25]. The landfill is composed of four different basins, three of which are currently in the "after-care" phase and only the fourth basin was active during sampling. Basin 4 (840,000 m^3) is located between Pz07 and Pz01. The distance of Pz07 from the waste load area is approximately 50 m. Pz01 is at about 50 m from the southern end of the basin 4.

Major physical-chemical water characterization was assessed on-site by measuring electrical conductivity, oxidation-reduction potential (ORP), temperature, pH, dissolved oxygen with a flow cell equipped with AQUAREAD probes. The concentration of NO_3, SO_4, F, Cl (Ionic Chromatography—Dionex DX-120), HCO_3 (HCl titration), Si, Mg, Ca, Na, K (Inductively Coupled Plasma—Optical Emission Spectrometry—Perkin Elmer P400), DOC (Shimadzu TOC-5000A analyser) were successively measured in laboratory [21].

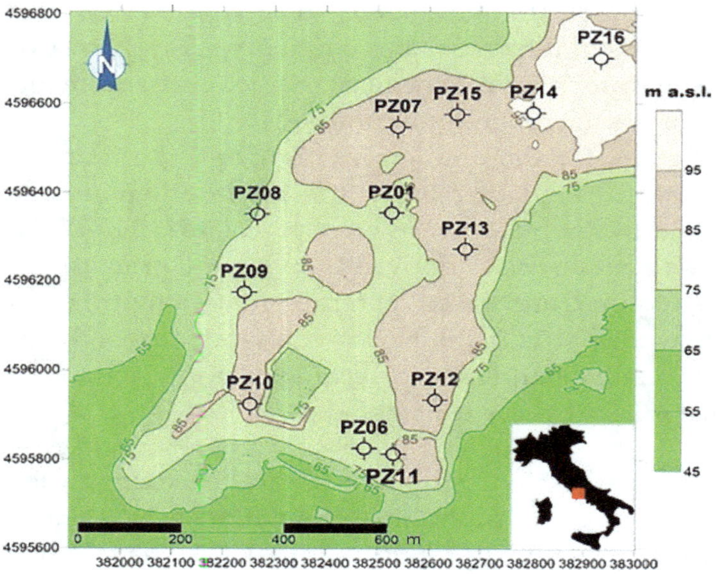

Figure 1. Location of the waste landfill facility area (Central Italy), with the indication of sampling sites.

The occurrence of total coliforms (TC) and fecal indicator bacteria (*E. coli*) was estimated using the Colilert-18 test kit (IDEXX Laboratories Inc., Westbrook, Maine, USA) [26]. Duplicate samples were collected in 100-mL sterile polystyrene bottles and analyzed within 6 h. The positive wells for the activities on ONPG (yellow) and MUG (fluorescent) substrata were counted and referred to the Quanti-Tray/2000 Most Probable Number (MPN) table to quantify the total coliforms and *E. coli*, respectively.

2.2. Microbial Community Characterization by Flow Cytometry

Subsamples for flow cytometric analyses were collected in 2-mL Eppendorf safe-lock tubes, fixed with formaldehyde (2%, final concentration) and analyzed within 24 h. The microbial community characterization was performed by the Flow Cytometer A50-micro (Apogee Flow System, Hertfordshire, UK) equipped with a solid-state laser set at 20 mV and tuned to an excitation wavelength of 488 nm. The total prokaryotic cell counts (TCC) were measured by staining with SYBR Green I (1:10,000 dilution; Molecular Probes, Invitrogen) for 10 min in the dark at room temperature. The light scattering signals (forward and side scatters), the green fluorescence (530/30 nm) and red fluorescence (>610 nm) were acquired with a threshold sets on the green channel. Samples were run at low flow rates to keep the number of events below 1000 event s^{-1}.

The data were analyzed using the Apogee Histogram Software (v2.05). The TCC was determined by their signatures in a plot of the side scatter vs. the green fluorescence. The intensity of green fluorescence emitted by SYBR positive cells allowed for the discrimination among cell groups exhibiting two different nucleic acid content (cells with low NA content—LNA; cells with high NA content—HNA) [27]. This method was successfully applied for prokaryotic cell counting in different groundwater settings [15,21].

2.3. Metabolic Potential, Functional Diversity and Community Level Physiological Profile

The Biolog EcoPlates™ assay (Biolog, Inc., Hayward, California, USA) was used to detect the microbial degradative activity of 31 organic carbon sources, contained in triplicate in 96 micro-well plates (plus three control wells) and labeled with the respiration-sensitive tetrazolium dye. Within 6 h from sampling, 150 µL of sampled water was introduced in each well. The plates were incubated at 20 °C in the dark. The analysis of the color development for each substrate along the incubation time provided information on the degradation rate of each substance. The development of formazan was detected through the measurement of the absorbance at 590 nm (VICTOR™ X3 Multilabel plate reader.

PerkinElmer Inc., Waltham, Maine, USA). Data acquisition was performed at the starting time (time 0) and every 24 h, over a period of 15 days. The optical density (OD 590 nm, expressed as absorbance units) of each substrate was subtracted by the corresponding initial OD value (t = 0) to eliminate the intrinsic absorbance signal [28]. Negative values were set to 0.

The mean degradative activity on all 31 substrates was intended as the microbial metabolic potential, expressed as Average Well Color Development (AWCD), the average of adjusted absorbance of the whole microplate [29]:

$$\mathrm{AWCD} = \frac{1}{31}\sum(\mathrm{OD}_{590}) \tag{1}$$

The AWCD kinetic profiles were calculated over the entire incubation time, based on the density-dependent logistic growth equation:

$$y = \mathrm{AWCD_t} = \frac{\mathrm{K}}{1 + e^{-\mu(\mathrm{t}-\mathrm{s})}} \tag{2}$$

where:

- K = the asymptote/carrying capacity
- μ = exponential rate of AWCD change
- t = time (h)
- s = the time when y = K/2

These parameters were utilized to compare the microbial functional properties in different samples [30]. Data elaboration were performed in the interval between 0 and 168 h of incubation, representing the exponential phase before the stationary phase. In particular, the 24 h endpoint represented the prompt microbial responsiveness in substrate degradation, while the 168-h endpoint represented the maximum potential degradation activity.

Following Weber et al. [31], the functional diversity was related to the rates at which the Biolog substrates were utilized by the aquatic microbial community. Herein, the functional diversity was estimated in terms of equitability (E), taking into account both Shannon index (H) and substrate richness (i.e., number of positive wells in the microplates—S):

$$\mathrm{E} = \frac{\mathrm{H}}{\ln \mathrm{S}} \tag{3}$$

where:

- $\mathrm{H} = -\sum p_i\ (ln\ p_i)$
- p_i = the ratio of the activity of a particular substrate to the sums of activities of all substrates
- S = number of positive wells in the microplates

The standardization of initial density inoculum was carried out by normalization of optical density (OD_{590}) of each substrate by the corresponding Average Well Color Development (AWCD) value [28,32,33]. To simplify the interpretation of the results, the 31 substrates utilized by the Biolog test, were grouped into six categories of chemical compounds as follows: carbohydrates, polymers, carboxylic acids, phenolic compounds, amino acids and amines [34]. The Area Under the Curve (AUC) describing the variation of OD with time was calculated for each well using a trapezoidal approximation [35]. All values, hereafter reported both for classes of compounds and single substrates, were expressed as AUC and utilized for the univariate analysis.

2.4. Statistical Analysis

Geometric means were used to characterize physical, chemical and microbial characteristics of the groundwater groups. In the case of *E. coli* < 1 MPN/100 mL, we assumed 0.5 MPN/100 mL as lowest value.

The multivariate analysis of similarities (ANOSIM) was used to test the difference among the sample groups by considering all major physical, chemical, and microbial parameters. A multi-group similarity percentage test (SIMPER), through the Bray-Curtis similarity measure (multiplied by 100), was run to identify the percentage contribution of selected data to the average dissimilarity.

The non-parametric Kruskal-Wallis univariate test was performed on all parameters in order to identify statistical differences between the groundwater groups.

A non-metric multi-dimensional scaling ordination plot (nMDS), based on the Gower dissimilarity matrix, was used to graphically visualize hydrogeochemical and microbiological variation patterns [36]. The values of the degradation activity performed on the different categories of organic compounds were incorporated in a nMDS analysis with a vector-fitting procedure, in which the length of the arrow is proportional to the correlation between the nMDS-axes and each variable. This method allowed determining which degraded class of compounds were significantly correlated with the nMDS ordination. In particular, we focused on those substrates that contributed for 50% of the overall SIMPER dissimilarities between the groundwater groups.

A non-parametric multivariate analysis of variance (NPMANOVA) was performed on AWCD-normalized Biolog data (i.e., on either substrate classes or single substrates) to highlight differences in the community-level physiological profiles (CLPP) between the identified groundwater groups.

3. Results

3.1. Groundwater Physical, Chemical and Microbial Characteristics

The major groundwater physical, chemical and microbial characterization (Tables S1 and S2) allowed identifying three statistically distinct groundwater groups (ANOSIM, $p < 0.05$), herein defined as "pristine", "mixed", and "altered".

The pristine groundwater group included ten samples from six piezometers (Pz6-2016, Pz6-2017, Pz12-2016, Pz12-2017, Pz13-2017, Pz14-2016, Pz14-2017, Pz15-2016, Pz15-2017, and Pz16-2017) (Figure 1). These samples were anoxic groundwaters characterized by the lowest mean values of ORP and DO, the lowest mean EC along with the lowest mean concentration of SO_4 and Ca (Table 1).

Table 1. Physical and chemical characteristics of different groundwater groups presented as geometric means. The median was used in case of negative ORP values. Data are shown in order of contribution to the variability among groups obtained through the SIMPER test. EC = electrical conductivity; ORP = redox potential; T = temperature; DO = dissolved oxygen; DOC = Dissolved Organic Carbon). a, b, c indicate statistical differences among groups (Kruskal-Wallis, $p < 0.05$).

Parameter	Unit	Pristine	Mixed	Altered
EC	µS/cm	766.4 ^{a}	915.6	1452.7
ORP	mV	−119.3 ^{a}	74.3 ^{b}	−40.8 ^{c}
T	°C	17.8	19.1	17.3
pH		7.2 ^{a}	7.0 ^{b}	6.7 ^{c}
DO	mg/L	0.2 ^{a}	3.1 ^{b}	0.4 ^{a}
NO_3	mg/L	0.9 ^{a}	18.2 ^{b}	0.8 ^{a}
SO_4	mg/L	11.7 ^{a}	45.9 ^{b}	232.5 ^{c}
K	mg/L	10.5 ^{a}	2.1 ^{b}	5.1 ^{ab}
DOC	mg/L	2.2 ^{a}	0.8 ^{b}	2.0 ^{a}
F	mg/L	0.4 ^{a}	0.2 ^{b}	0.2 ^{ab}
Cl	mg/L	7.9 ^{a}	19.5 ^{b}	11.3 ^{a}
Si	mg/L	16.4 ^{a}	7.5 ^{b}	19.7 ^{a}
Mg	mg/L	23.1 ^{b}	22.0 ^{b}	45.7 ^{a}
Ca	mg/L	94.1 ^{b}	135.4 ^{c}	238.3 ^{a}
Na	mg/L	9.9	8.2	8.8
HCO_3	mg/L	419.8 ^{b}	424.5 ^{b}	632.4 ^{a}

The mixed groundwater group included six samples from 3 piezometers, where the water table aquifer is probably mixed with the lower one (Pz7-2016, Pz7-2017, Pz8-2016, Pz8-2017, Pz9-2016, and Pz9-2017). These samples were more oxidized groundwaters and showed the highest mean values of ORP, DO, NO_3 and Cl, along with the lowest mean DOC and Si concentrations.

The altered groundwater group, including six samples from four piezometers (Pz1-2016, Pz10-2016, Pz10-2017, Pz11-2016, Pz11-2017, Pz13-2016), was characterized by anoxic, slightly acidic groundwaters with the highest EC, SO_4, Mg, Ca and HCO_3 concentrations.

Fecal contamination was detected only in the altered groundwater samples in the 2016 campaign. Total prokaryotic cell counts (TCC = $7.93 \times 10^4 - 1.69 \times 10^7$ cells/mL) showed a large variability among the groundwater groups, with relatively high values in altered conditions, along with significant high percentages of LNA cells (Table 2).

Table 2. Microbial community characteristics (geometric means) of the three groundwater groups. In case of negative *E. coli*, half of detection limit was used for mathematical calculation (i.e., 0.5 MPN/100 mL). TCC = Total Cell Counts; HNA cells = cells with high nucleic acid content; LNA cells = cells with low nucleic acid content. a, b, c indicate statistical differences among groups (Kruskal-Wallis, $p < 0.05$).

Parameter	Unit	Pristine	Mixed	Altered
Total Cells Count	10^5 cells/mL	4.4	3.1	6.6
HNA Cells	% of TCC	52.8 ^{b}	52.6 ^{b}	27.1 ^{a}
LNA Cells	% of TCC	41.6 ^{b}	46.4 ^{b}	65.5 ^{a}
Total Coliforms	MPN/100 mL	157.8	103.0	292.6
E. coli	MPN/100 mL	<1	<1	3.3

The different contribution of the macro-chemical elements and microbial characteristics to the identified groundwater groups was graphically summarized in an ordination plot (Figure 2). The multivariate approach allowed for the testing of inter-group dissimilarity, along with highlighting the variation patterns of the groundwater major characteristics.

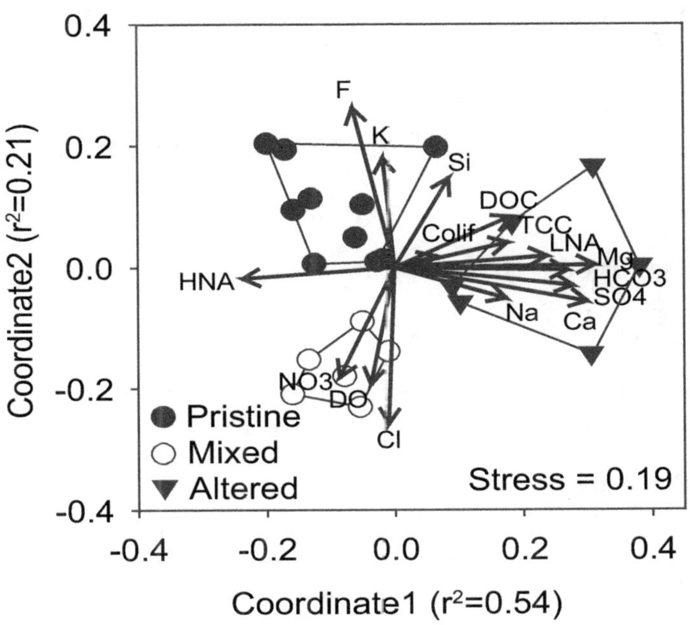

Figure 2. NMDS ordination plot of sampled waters. Vectors represent the major geochemical and microbial community characteristics that concurred to the identification of the three different groundwater groups. Stress value indicates the significant concordance between the distance among samples in the NMDS plot and the actual Gower distance among samples.

3.2. Kinetic Profiles of Metabolic Potential and Functional Diversity

The kinetic parameters, calculated from the sigmoidal curves of the AWCD values over time, highlighted the overall differences in the metabolic responses of microbial communities from the three groundwater groups (Figure 3a and Table 3). The microbial potential activity reached a similar theoretical maximum K value in all groundwater groups. The microbial community from the pristine groundwater differed significantly, showing the highest exponential rate of AWCD change (μ) (Table 3). In altered conditions, the microbial community showed $AWCD_{24h}$ values two-fold higher than that developed under pristine and mixed conditions and the lowest time to reach K/2, hence indicating a higher affinity for the available substrates.

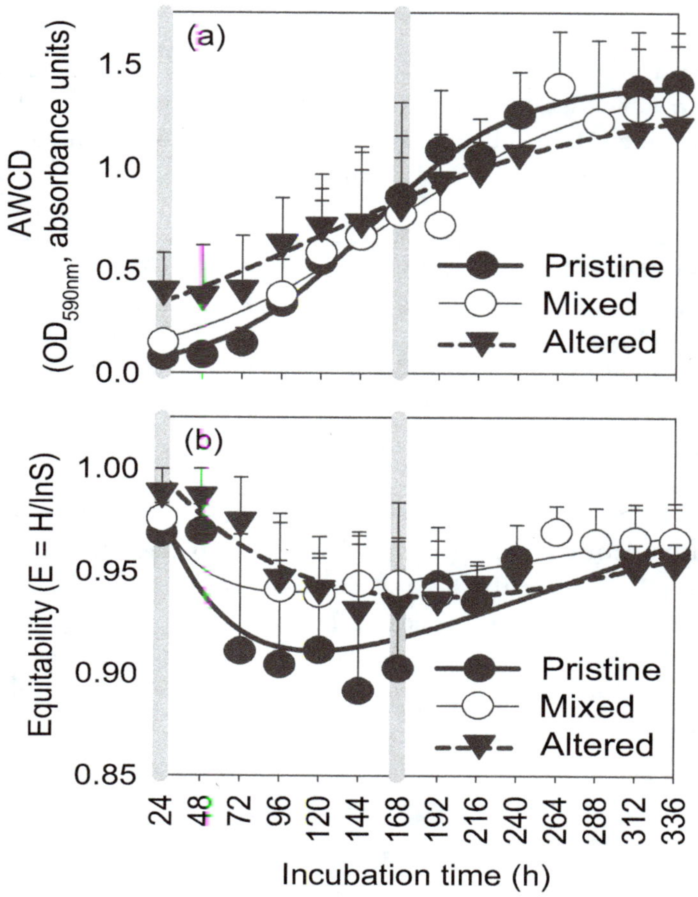

Figure 3. Kinetic profiles of (**a**) metabolic potential (Average Color Well Development—AWCD) and (**b**) functional diversity (substrate utilization equitability—E) (Equation (3)) of the identified groundwater groups. Gray lines indicate 24 h and 168 h incubation time.

Table 3. Kinetic parameters of the fitted logistic growth equations (Equation (2)). K = asymptote, s = time to reach K/2, μ = exponential rate of AWCD change, r^2 = correlation coefficient.

Parameter	Unit	Pristine	Mixed	Altered
K	OD_{590nm}	1.42	1.41	1.33
s	h	157.1	147	117.1
μ	h^{-1}	0.029	0.023	0.021
r^2		0.98	0.99	0.99

After 24 h of incubation, the functional microbial diversity, indicated by the equitability values, of microbial communities from pristine and mixed sample groups showed similar values (E = 0.93 and 0.92, respectively), which were significantly lower than those found in altered conditions (E = 0.96).

The equitability under pristine conditions reached the lowest values with respect to altered and mixed conditions, indicating a heterogeneous distribution in substrate degradation (Figure 3b and Table 4). The equitability showed lower values at 168 h (E = 0.903, 0.944, and 0.934, respectively for pristine, mixed groundwaters) than at 24 h.

Table 4. Utilization patterns of organic substrates in different groundwater groups after 24 h and 168 h of incubation. Values are given as percentage of the total OD_{590}.

24 h

Classes of Substrates	Substrates				Pristine									Mixed						Altered			
Carbohydrates	β-methyl-D-glucoside	4.7	6.4	4.8	4.9	4.6	5.1	5.3	3.8	5.6	4.4	4.8	5.6	3.3	3.4	3.6	4.8	3.3	3.1	4.9	3.5	3.5	3.5
	D-xylose	4.1	1.5	3.2	2.9	4.0	1.6	0.0	5.7	1.5	3.0	3.6	3.6	1.9	3.7	4.5	3.4	3.9	3.1	3.6	3.7	3.3	3.6
	i-erythritol	7.0	4.2	5.2	5.7	6.3	6.6	8.1	5.6	7.0	4.6	5.8	5.3	6.7	4.9	6.0	5.9	4.1	3.7	4.5	4.0	3.7	6.5
	D-mannitol	5.5	7.2	5.1	5.5	5.0	5.2	7.1	4.4	8.6	4.9	5.6	6.8	5.7	3.6	5.3	5.3	3.9	2.9	4.1	3.4	3.6	5.2
	N-acetyl-D-glucosamine	4.9	6.9	5.5	5.4	5.0	4.4	7.3	4.0	6.1	6.2	5.3	4.6	3.8	4.5	4.3	4.1	3.8	2.9	3.3	3.6	3.8	5.0
	D-cellobiose	5.0	2.9	5.1	5.6	6.6	3.7	6.2	4.8	5.9	5.3	5.5	3.2	1.7	4.6	5.0	5.7	3.5	4.7	4.2	3.6	3.7	4.7
	α-D-lactose	6.0	3.3	5.0	6.1	5.2	5.5	1.7	4.7	6.7	5.5	5.2	7.7	1.8	4.1	5.2	6.2	3.2	5.0	4.0	2.9	4.1	4.4
	glucose-1-phosphate	2.7	1.9	1.5	2.7	1.9	0.0	7.3	2.3	0.0	2.0	2.5	1.1	2.9	2.5	1.4	1.8	2.5	2.5	2.0	2.2	3.0	1.6
	D,L-α-glycerol phosphate	2.0	2.8	0.0	0.6	3.1	0.0	0.0	0.9	2.8	1.1	1.5	2.0	2.7	1.8	1.9	1.7	2.5	1.6	2.1	2.5	2.9	0.5
Polymers	tween40	5.7	5.1	5.7	4.1	6.7	3.0	3.3	4.5	4.6	7.6	4.6	4.0	5.3	4.5	7.8	4.2	4.4	5.7	3.6	3.3	4.0	6.4
	tween80	0.0	2.4	3.4	3.6	0.0	2.4	0.1	2.6	0.0	1.6	0.0	0.0	1.7	3.1	2.9	3.4	3.7	3.3	3.0	3.3	3.8	3.3
	α-cyclodextrin	2.3	1.7	2.3	2.3	1.7	1.1	3.1	2.4	2.7	2.4	2.0	2.5	3.7	2.2	1.3	3.0	2.3	2.8	3.1	3.6	2.9	1.7
	glycogen	5.0	6.3	3.8	2.6	2.8	3.8	4.4	3.6	4.2	4.4	3.9	5.5	2.7	2.8	3.3	3.9	3.0	3.4	3.0	3.3	3.7	3.9
Carboxylic Acids	D-galactonic acid γ-lactone	4.5	2.5	4.6	3.0	3.7	3.9	4.3	2.9	2.4	3.4	4.3	2.8	4.4	2.6	4.2	3.5	3.7	3.8	3.4	3.5	2.5	4.3
	D-galacturonic acid	4.6	2.8	0.7	2.4	2.8	2.8	2.2	3.1	2.8	0.0	2.6	3.8	3.5	3.1	3.0	3.5	1.9	2.9	3.2	3.0	3.1	1.9
	γ-hydroxybutyric acid	4.1	1.8	4.3	5.3	3.9	11	1.0	2.7	9.3	7.6	5.0	2.6	2.7	3.8	2.7	3.8	4.1	3.9	2.6	3.7	3.9	5.1
	D-glucosaminic acid	4.3	2.7	3.1	4.2	3.8	3.2	7.0	3.2	5.8	4.4	4.1	4.1	4.5	3.3	3.7	3.6	2.5	2.4	3.1	3.1	3.2	3.0
	itaconic acid	2.6	2.0	2.9	2.6	2.7	2.1	2.7	2.4	3.2	2.2	2.6	3.4	1.9	2.2	2.0	2.2	3.5	3.1	2.6	2.8	2.4	2.4
	α-ketobutyric acid	0.0	2.7	3.7	3.0	0.0	5.9	0.8	3.2	0.0	2.6	4.6	0.0	2.3	2.9	1.5	1.2	2.0	1.5	2.1	2.8	3.1	2.7
	D-malic acid	0.0	2.1	2.8	1.6	3.0	0.0	0.5	2.3	1.7	1.5	0.0	3.4	3.7	3.5	3.3	1.8	2.6	3.8	3.0	3.4	3.6	1.8
	pyruvic acid methyl ester	0.0	1.7	2.9	0.0	0.0	0.0	0.0	0.7	0.0	0.8	0.0	0.0	2.8	1.0	0.5	0.0	2.6	2.2	2.0	3.4	3.6	0.4
Phenolic Compounds	2-hydroxy benzoic acid	0.0	1.9	2.1	0.0	2.2	0.0	0.0	2.2	0.0	0.0	0.0	2.2	3.3	3.5	3.5	5.0	3.2	2.5	4.4	3.3	3.4	1.7
	4-hydroxy benzoic acid	0.0	1.5	3.2	1.0	0.0	1.2	0.2	3.6	0.0	1.4	0.0	0.0	1.2	1.6	0.8	0.0	3.2	1.9	2.2	2.9	2.7	2.8
Aminoacids	L-arginine	4.1	3.6	3.0	4.1	3.8	4.4	6.4	3.7	2.9	5.9	3.6	4.6	4.1	3.2	3.9	3.7	2.8	3.7	3.5	3.5	3.3	3.9
	L-asparagine	3.8	3.8	2.8	3.0	3.4	3.6	3.1	2.7	1.2	2.8	3.5	3.7	4.6	4.1	3.1	2.5	3.8	4.2	3.7	3.1	3.0	3.5
	L-phenylalanine	3.0	2.3	1.6	2.3	2.0	2.4	1.4	2.8	1.9	1.8	2.4	2.0	2.9	2.5	1.5	1.4	3.3	2.1	2.0	2.9	1.3	2.0
	L-serine	3.4	3.8	2.3	3.1	3.5	3.9	5.0	2.7	2.6	2.5	3.6	4.2	3.5	4.2	3.1	3.3	3.4	3.6	3.2	3.3	3.3	2.9
	L-threonine	3.8	2.8	4.7	4.3	4.2	4.0	4.3	3.9	3.7	4.6	4.1	1.5	3.9	4.2	3.7	4.3	4.1	4.1	3.5	3.7	3.6	4.2
	glycyl-L-glutamic acid	3.6	2.6	1.8	2.8	2.2	0.0	1.5	2.3	2.6	2.4	3.0	3.3	3.1	3.4	2.1	2.9	2.3	3.2	3.0	2.7	2.3	2.0
Amines	phenylethyl-amine	0.0	2.7	0.0	2.5	2.3	5.7	1.2	3.9	2.3	0.9	3.3	3.0	2.1	2.9	2.2	1.2	3.8	3.0	4.2	3.1	2.7	2.5
	putrescine	3.2	4.0	2.7	2.8	3.4	3.8	3.6	2.6	2.1	2.3	2.8	3.5	1.5	2.2	2.7	2.7	3.1	3.4	2.9	3.0	3.1	2.6

168 h

Classes of substrates	Substrates				Pristine									Mixed						Altered			
Carbohydrates	β-methyl-D-glucoside	4.8	4.4	8.8	7.6	1.3	2.6	5.8	2.1	11	5.8	3.4	5.7	3.8	2.8	1.5	5.6	4.8	5.5	4.5	5.3	2.6	8.2
	D-xylose	3.7	3.8	0.3	0.5	3.4	0.3	0.0	1.9	0.1	0.2	1.2	0.6	0.9	1.3	0.9	1.5	1.2	1.2	2.7	1.8	2.3	0.7
	i-erythritol	1.7	1.5	1.2	1.7	1.4	1.1	2.9	2.5	0.9	2.4	2.7	2.4	1.7	2.9	2.1	4.2	2.7	1.6	3.4	2.5	2.8	2.9
	D-mannitol	5.0	5.1	8.2	9.2	2.1	3.1	7.6	3.7	5.3	10	6.7	3.4	6.0	5.6	6.1	5.3	5.6	5.3	3.7	4.7	5.9	5.7
	N-acetyl-D-glucosamine	4.0	4.9	6.1	8.1	3.0	6.8	7.4	7.4	1.4	6.0	4.3	7.0	5.1	4.9	4.2	4.0	5.0	5.4	3.0	3.0	7.2	5.3
	D-cellobiose	5.7	4.7	5.5	3.1	4.6	2.0	3.6	3.2	5.0	7.4	4.2	2.2	3.0	4.0	2.3	4.9	4.9	7.5	4.6	6.2	2.6	6.2
	α-D-lactose	4.3	3.8	2.1	2.9	3.0	6.9	2.2	5.0	1.9	4.2	3.8	2.5	1.8	2.4	1.8	4.9	2.3	2.5	3.5	3.0	2.9	2.1
	glucose-1-phosphate	2.2	2.9	2.1	4.6	0.4	3.7	4.0	2.3	0.3	2.0	2.1	0.5	2.8	2.4	0.8	1.6	3.3	2.8	1.6	3.8	4.5	3.8
	D,L-α-glycerol phosphate	1.0	2.0	0.3	1.6	1.7	1.3	1.2	1.7	1.4	0.7	1.0	1.1	1.4	1.2	1.6	1.6	2.4	1.3	2.1	2.2	2.0	1.4
Polymers	tween40	3.2	3.7	3.9	5.8	2.9	6.0	4.0	4.1	7.3	4.0	4.9	8.3	3.0	4.9	7.6	6.1	3.7	5.1	2.7	4.0	4.7	3.3
	tween80	4.4	3.8	4.1	2.4	0.0	8.7	6.7	1.6	7.4	2.5	4.4	5.9	4.1	4.5	8.6	2.8	5.1	3.1	3.4	4.8	5.1	4.6
	α-cyclodextrin	4.5	3.6	2.3	1.5	1.0	3.7	3.2	4.1	5.4	2.6	3.2	3.8	3.9	3.8	3.2	1.8	1.0	2.1	3.2	2.7	3.2	2.5
	glycogen	4.7	4.3	11	3.2	0.4	1.9	2.4	2.8	1.4	7.1	3.0	4.5	4.7	3.0	3.8	3.6	4.8	6.8	2.6	4.2	2.6	10
Carboxylic Acids	D-galactonic acid γ-lactone	1.9	2.9	1.0	1.3	2.3	1.0	2.6	0.8	0.5	6.1	2.8	3.2	2.9	3.7	3.1	5.3	3.5	1.5	3.3	2.0	1.9	2.0
	D-galacturonic acid	3.4	3.0	6.4	2.6	8.2	1.5	3.0	3.4	5.0	0.3	1.6	2.9	3.6	3.2	2.6	2.5	2.8	1.4	2.6	2.0	2.2	3.9
	γ-hydroxybutyric acid	0.8	1.1	0.5	1.5	3.1	1.2	0.4	0.6	1.4	1.1	3.1	0.5	0.6	1.7	1.0	2.9	2.3	1.2	2.2	1.6	2.4	1.9
	D-glucosaminic acid	2.3	2.6	0.7	2.1	3.3	2.9	1.7	1.0	1.4	3.1	3.3	4.0	4.7	4.3	6.3	4.4	1.9	1.7	2.3	2.2	2.2	2.9
	itaconic acid	4.1	3.9	6.5	4.2	5.3	6.3	2.0	6.3	1.9	3.3	3.9	3.2	4.4	3.9	3.8	4.9	4.5	5.1	9.1	3.5	2.0	3.8
	α-ketobutyric acid	1.9	3.0	0.8	1.6	0.1	3.4	0.3	1.4	0.1	0.6	5.8	1.3	1.5	1.7	0.8	0.6	0.9	0.6	1.3	1.7	2.0	0.4
	D-malic acid	1.8	1.5	0.6	0.9	1.9	0.7	0.6	1.0	0.6	2.4	1.0	1.6	5.3	4.5	3.4	1.2	2.3	1.9	2.8	2.7	2.7	0.7
	pyruvic acid methyl ester	4.3	3.0	3.9	2.9	9.7	9.9	3.9	5.5	4.1	3.0	5.1	5.6	3.5	2.9	4.0	2.8	4.5	4.9	5.5	4.3	4.3	4.7
Phenolic Compounds	2-hydroxy benzoic acid	1.0	1.0	0.2	0.0	0.5	0.0	4.0	0.4	0.0	0.1	0.6	0.8	2.0	2.1	0.8	3.6	1.6	0.9	2.9	2.1	2.2	0.5
	4-hydroxy benzoic acid	3.3	4.4	2.1	3.0	6.8	4.4	5.4	6.4	0.9	4.7	3.8	3.9	5.3	3.4	4.4	1.9	4.1	6.1	2.0	3.8	1.7	4.9
Aminoacids	L-arginine	5.1	2.9	1.4	3.4	1.6	1.8	2.1	3.9	0.3	1.4	3.8	4.2	2.6	3.7	3.8	5.5	3.6	3.9	3.4	4.6	2.9	1.9
	L-asparagine	4.2	4.2	5.3	2.4	3.5	1.6	5.4	6.8	3.9	4.4	4.4	6.8	4.6	4.5	3.6	2.3	5.6	3.5	3.7	4.7	3.5	4.2
	L-phenylalanine	3.6	3.2	1.8	4.8	1.6	2.6	1.6	2.3	2.1	1.3	2.1	1.5	2.6	1.6	1.3	1.6	1.9	1.6	1.2	2.7	1.3	1.5
	L-serine	3.6	4.1	5.7	2.0	3.2	1.6	3.0	2.3	0.6	8.0	2.7	2.9	3.6	4.2	1.9	2.9	3.9	4.3	3.7	3.6	6.9	1.1
	L-threonine	1.5	2.0	1.3	2.3	3.7	0.9	2.0	2.9	0.6	0.9	2.3	2.1	2.1	2.6	1.4	2.9	2.6	2.2	3.2	2.4	2.6	2.0
	glycyl-L-glutamic acid	1.6	1.9	1.8	3.9	1.6	1.1	0.7	1.4	1.5	0.5	2.0	1.6	2.1	2.2	2.2	1.6	1.6	3.0	2.4	2.2	1.8	1.1
Amines	phenylethyl-amine	3.8	2.8	0.1	2.2	1.3	7.4	6.1	5.2	0.3	0.2	2.5	2.3	3.8	2.7	4.6	0.4	3.3	1.1	3.6	1.8	1.8	0.7
	putrescine	2.9	4.1	4.1	6.8	0.9	3.6	4.2	5.9	8.2	3.7	4.4	4.2	2.8	3.5	6.4	4.9	2.3	5.2	4.0	3.9	6.9	4.7

<1	1–2	2–3	3–4	4–5	5–6	6–7	7–8	8–9	9–10	>10

3.3. Community Level Physiological Profiles

The CLPP data were expressed as percentage contribution of the degradative activity on each substrate to the total. The data integration by nMDS representations summarizes how changes in the hydro-geochemical conditions corresponded to a shift in the utilization of organic substrates provided (Figure 4).

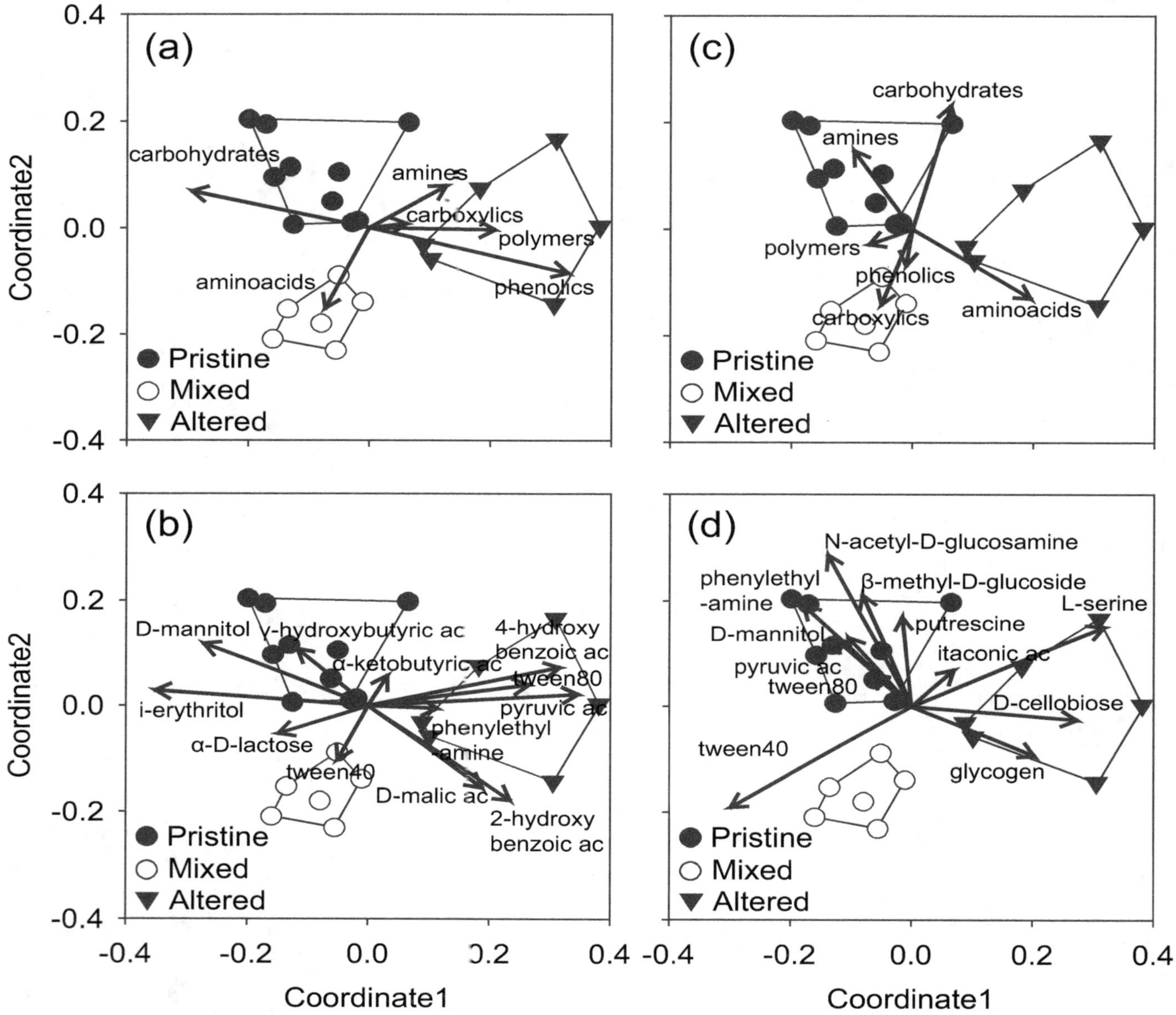

Figure 4. Variation patterns of the six class of compounds and of a selection of 12 substrates (i.e., those explaining 50% of dissimilarity among groups) at 24 h (**a**), (**c**) and 168 h (**b**), (**d**). Data were expressed as AUC values normalized for the AWCD. NMDS ordination plot of sampled waters is based on geochemical and microbial community characteristics (see Figure 2).

The utilization patterns of organic substrates in the early stage of incubation ($CLPP_{24h}$) showed that microbial communities from pristine conditions fed preferentially on carbohydrates (Table 4; Figure 4a), which contributed on average for 38.8% of the total absorbance. Inside this class of compounds, i-erythritol was the most degraded compound (6.0%), followed by D-mannitol and D-lactose, (5.9% and 5.5% respectively) (Figure 4b). Carbohydrates utilization decreased significantly to 33% at 168h incubation (NPMANOVA; $p < 0.01$) and shifted toward the utilization of other compounds such as β-methyl-D-glucoside and N-acetyl-D-glucosamine (Figure 4d). Carboxylic acids contributed to the $CLPP_{24h}$ for 22.0%, although only four out of eight compounds were utilized (Table 4). Most of the contribution was assigned to the γ-hydroxibutyrric acid (5.0%) (Figure 4b). The utilization of pyruvic acid methyl ester increased consistently at 168 h, passing from 0.6% to 5.0%. Aminoacids and polymers contributed for 18.9% and 12.9% of the total absorbance, respectively. The $CLPP_{168h}$ changed significantly due to the significant high utilization (NPMANOVA; $p < 0.05$) of polymers, phenolic compounds and amines (Figure 4c). In particular tween 40 (4.5%), phenylethyl-amine (2.9%) mostly drove this change (Figure 4d).

The $CLPP_{24h}$ in mixed conditions relied preferentially on carbohydrates (36.5%), mainly driven by i-erythritol (5.8%) and D-mannitol (5.4%) (Table 4), that declined after 168 h incubation (NPMANOVA; $p < 0.01$) (27.3%) (Table 4). The carboxylic acids contribution varied significantly in the $CLPP_{24h}$ and $CLPP_{168h}$ (22.1% and 25.1%, respectively) and relied mostly on D-glucosaminic acid (3.9% and 4.5%, respectively) (Table 4). Polymer utilization (13.0%) at 24 h was driven mainly by tween40 degradation (5.1%) (Figure 4a,b). As for pristine conditions, amines and phenolic compounds, contributed at lesser extent to the $CLPP_{24h}$ (5.0% and 3.5% respectively). Polymers and phenolic compounds increased significantly (NPMNOVA; $p < 0.05$) from $CLPP_{24h}$ to $CLPP_{168h}$ (17.9% and 5.4%, respectively) and mainly contributed, together with carboxylic acid degradation, to define the CLPP at 168h (Figure 4c).

In altered conditions, the $CLPP_{24h}$ was characterized by a wide utilization of the provided substrates (Table 4) and no statistical changes were recorded in the degradation patterns over the incubation time (Table 4). Carbohydrates (31.5%) were utilized preferentially, although at lesser extent with respect to the pristine and mixed conditions (NPMANOVA; $p < 0.05$). A relatively higher contribution of carboxylic acids (23.5%), polymers (14.1%), amines (6.2%) and phenolic compounds (5.7%) was also observed (Figure 4a). Compounds that mostly contributed to the $CLPP_{24h}$ were D-malic acid and pyruvic acid methyl ester (3.0% and 2.4%, respectively), tween-80 (3.4%), phenylethyl-amine (3.2%) and 2-hydroxy benzoic acid and 4-hydroxy benzoic acid (3.1% and 2.6% respectively) (Figure 4b). The relative $CLPP_{168h}$ was characterized for enhanced degradative skills toward specific substrata such as: D-cellobiose (5.4%), glycogen (5.2%), itaconic acid (4.7%), and L-serine (3.9%) (Figure 4d).

4. Discussion

In this study, we found that the landfill-induced geochemical alteration determined changes in the microbial functional diversity and metabolic physiological profile with an increased ability in the substrate utilization passing from pristine to altered conditions (i.e., altered > mixed > pristine). Although partly affected by the limitations of all experimental approaches performed in laboratory conditions, the Biolog assay and the kinetic profiles of substrate degradation allowed assessing the potential metabolic capacity of groundwater microbial communities, as a suitable information to better understand the effects of potential inputs of allochthons organic substrates on the aquifer. This methodological approach was consistently applied in several aquatic settings [37–40].

The Biolog assay relies on the different capacity of aquatic microorganisms to adapt to the spatial and temporal variations in quantity and quality of dissolved OM pool available in the aquatic environment [41,42]. Overall, the bioavailable OM fraction decrease by the transit from surface to groundwater, passing through several soil and sediment layers [43–45]. Therefore, groundwater ecosystems can show oligotrophic conditions and scarce productivity [10,46,47].

Microbial communities from the pristine and mixed groundwater showed the typical traits found in subsurface aquatic environments, with a relatively lower affinity for most of the Biolog substrates [10,48,49]. The heterogeneity in substrate utilization and the decreasing values of functional diversity over time suggested a relatively lower metabolic versatility to adapt to newly available substrates with respect to those in altered conditions. The physical-chemical characteristics in pristine conditions were compatible with the preferential utilization of carbohydrates by microbial communities, being their utilization associated to unbalanced nutrient conditions [50] and decreasing redox potential [36,39]. Carbohydrates represent an important energy-rich carbon source, and constitute storage molecules for the aquatic bacterial metabolism [51], being promptly and preferentially utilized through the catabolic heterotrophic pathway that involve oxidation of simple or complex carbohydrates and oxygen as a terminal electron acceptor [52,53]. According to the degradation patterns of C-rich substrates, the microbial communities in pristine and mixed condition were likely to be affected by C-limitation. Only after 168 h, microbial communities in pristine conditions developed the ability to degrade polymers (i.e., tween 80) and amines (phenylethyl-amine and putrescine). Polymers are complex carbon substrates, including compounds of synthetic origin. In particular, tween 40 and tween 80 are nonionic surfactants used as oil-in-water emulsifiers in pharmaceuticals, cosmetics, and cleaning compounds. The delay in utilizing tween 80 may reflect the metabolic plasticity [54] that modulates the microbial community functions, suggesting a higher sensitivity of pristine communities to contingent events of contamination by anthropogenic organic pollutants. The shift toward the utilization of amines in the pristine condition might indicate the presence of opportunistic classes of bacteria and a condition of nitrogen deprivation [37,55]. Unlike C-enriched substrates (i.e., carbohydrates and polymers), amines are enriched in nitrogen and utilized by bacteria to synthetize organic molecules such as amino acids and proteins [56].

In altered conditions, microbial communities were relatively more efficient in metabolizing most of all provided substrates, since the early stage of incubation. Although the degrading activity on Biolog carbon sources did not reflect the presence of these specific compounds in the groundwater, a higher homogeneity in substrates utilization and a relatively higher affinity with organic substrates, suggested a higher number of metabolic pathways, which may allow microbial communities to exploit several carbon sources [20,57,58]. The input of OMs of different origins and quality can induce the development of a well-structured community and rare taxa that contribute to enhance the OM biodegradation [42,59]. In addition, the occurrence of fecal indicator bacteria (i.e., *E. coli*) in altered conditions indicated the input of microorganisms originating from the surface environment, suggesting the co-presence of microorganisms with different functional traits other than those originally available in groundwater. In altered groundwater a higher prokaryotic cell abundance and a structural shift toward LNA cells was observed as similarly found for impacted groundwater [24,57,60]. Notably, LNA cells were found to be specifically composed by selected microbial taxa, which were also considered differently responsive to environmental changes than HNA cells [24,61].

In altered groundwater, the metabolic profile relied mainly on carboxylic acids, phenolic compounds, polymers, and amines. Carboxylic acids can be considered part of the labile pool of the organic matter in the aquatic environment, which is an important carbon source for the aquatic microorganisms [62,63]. These substrates are naturally occurring as products of the bacterial degradation and photochemical degradation of the organic matter in surface waters [64,65]. The prompt ability (24 h) in the utilization of complex polymers (tween 80) might indicate the presence of organic pollutants in altered groundwater, being this functional trait linked to polluted environment [55,63]. In this study, organic pollutants were detected at low concentration in some of the altered groundwater sites [25].

Interestingly, glycogen was one of the polymers utilized by microbial communities in altered conditions in the last stage of incubation. Glycogen, representing a source of readily available glucose for many organisms, is a highly branched polysaccharide that consists of glucose units linked in a linear chain. Glycogen is accumulated in bacterial cells during the stationary phase or during inorganic nutrient limitation, conditions that very likely occurred in the last stage of incubation [66]. Moreover, the presence in these groups with specific functional abilities might be deduced, being the degradation of glycogen performed by a set of extracellular enzymes (i.e., glycogen phosphorylases and glycosidases such as α-amylases) owned only by those bacteria that complete the degradative processes to carbohydrates [67]. The consumption of N-rich substrates (phenylethyl-amine and L-serine) across the incubation time of samples from altered conditions, was compatible with a strong nitrogen depletion of the microbial degradative processes given the relatively high activity occurring with respect to pristine and mixed conditions.

5. Conclusions

Changes in groundwater quality can directly affect the functional properties of the aquatic microbial communities with implications on the pattern of the biogeochemical cycling. The Biolog assay provided valuable information for tracking different levels of groundwater alterations. It is topical to understand the structural and functional diversity of the microbial community implied in the C-cycling and ecosystem services (i.e., nitrogen removal, pollutants degradation, and DOC assimilation). The microbial metabolic potential and the environmental factors shaping microbial community functioning are important to understand the potential drivers of biodegradation, thus helping to develop advanced strategies for groundwater management.

Author Contributions: All authors contributed to conceptualization, writing, and revising the present manuscript.

Acknowledgments: We wish to thank Domenico Mastroianni and Francesca Falconi for major anions, cations and DOC analyses.

References

1. Downing, J.A.; Prairie, Y.T.; Cole, J.J.; Duarte, C.M.; Tranvik, L.J.; Striegl, R.G.; McDowell, W.H.; Kortelainen, P.; Caraco, N.F.; Melack, J.M.; et al. The global abundance and size distribution of lakes, ponds, and impoundments. *Limnol. Oceanogr.* **2006**, *51*, 2388–2397. [CrossRef]
2. Pham, H.V.; Torresan, S.; Critto, A.; Marcomini, A. Alteration of freshwater ecosystem services under global change–A review focusing on the Po River basin Italy and the Red River basin Vietnam. *Sci. Total Environ.* **2019**, *652*, 1347–1365. [CrossRef] [PubMed]
3. Mpelasoka, F.; Hennessy, K.; Jones, R.; Bates, B. Comparison of suitable drought indices for climate change impacts assessment over Australia towards resource management. *Int. J. Climatol.* **2008**, *28*, 1283–1292. [CrossRef]
4. Kløve, B.; Ala-Aho, P.; Bertrand, G.; Gurdak, J.J.; Kupfersberger, H.; Kværner, J.; Muotka, T.; Mykrä, H.; Preda, E.; Rossi, P.; et al. Climate change impacts on groundwater and dependent ecosystems. *J. Hydrol.* **2014**, *518*, 250–266. [CrossRef]
5. Gozlan, R.E.; Karimov, B.K.; Zadereev, E.; Kuznetsova, D.; Brucet, S. Status, trends, and future dynamics of freshwater ecosystems in Europe and Central Asia. *Inland Waters* **2019**, *9*, 78–94. [CrossRef]
6. EU Directive on the protection of groundwater against pollution and deterioration. European Parliament and Council Directive 2006 2006/118/EC of 12 December 2006. *Off. J. Eur. Union* **2006**, *372*, 19–31.
7. Griebler, C.; Stein, H.; Kellermann, C.; Berkhoff, S.; Brielmann, H.; Schmidt, S.; Selesi, D.; Steube, C.; Fuchs, A.; Hahn, H.J. Ecological assessment of groundwater ecosystems-Vision or illusion? *Ecol. Eng.* **2010**, *36*, 1174–1190. [CrossRef]
8. Di Lorenzo, T.; Galassi, D.M.P. Agricultural impact on Mediterranean alluvial aquifers: Do groundwater communities respond? *Fundam. Appl. Limnol. Arch. Hydrobiol.* **2013**, *182*, 271–282. [CrossRef]
9. Griebler, C.; Avramov, M. Groundwater ecosystem services: A review. *Freshw. Sci.* **2015**, *34*, 355–367. [CrossRef]
10. Griebler, C.; Lueders, T. Microbial biodiversity in groundwater ecosystems. *Freshw. Biol.* **2009**, *54*, 649–677. [CrossRef]
11. Moser, D.P.; Fredrickson, J.K.; Geist, D.R.; Arntzen, E.V.; Peacock, A.D.; Li, S.-M.W.; Spadoni, T.; McKinley, J.P. Biogeochemical Processes and Microbial Characteristics across Groundwater−Surface Water Boundaries of the Hanford Reach of the Columbia River. *Environ. Sci. Technol.* **2003**, *37*, 5127–5134. [CrossRef]
12. Yagi, J.M.; Neuhauser, E.F.; Ripp, J.A.; Mauro, D.M.; Madsen, E.L. Subsurface ecosystem resilience: Long-term attenuation of subsurface contaminants supports a dynamic microbial community. *ISME J.* **2010**, *4*, 131–143. [CrossRef] [PubMed]
13. Röling, W.F.M.; van Breukelen, B.M.; Braster, M.; Lin, B.; van Verseveld, H.W. Relationships between Microbial Community Structure and Hydrochemistry in a Landfill Leachate-Polluted Aquifer. *Appl. Environ. Microbiol.* **2001**, *67*, 4619–4629. [CrossRef] [PubMed]

14. D'Angeli, I.M.; Serrazanetti, D.I.; Montanari, C.; Vannini, L.; Gardini, F.; De Waele, J. Geochemistry and microbial diversity of cave waters in the gypsum karst aquifers of Emilia Romagna region, Italy. *Sci. Total Environ.* **2017**, *598*, 538–552. [CrossRef] [PubMed]
15. Amalfitano, S.; Del Bon, A.; Zoppini, A.; Ghergo, S.; Fazi, S.; Parrone, D.; Casella, P.; Stano, F.; Preziosi, E. Groundwater geochemistry and microbial community structure in the aquifer transition from volcanic to alluvial areas. *Water Res.* **2014**, *65*, 384–394. [CrossRef] [PubMed]
16. Röling, W.F.M.; Van Breukelen, B.M.; Braster, M.; Goeltom, M.T.; Groen, J.; Van Verseveld, H.W. Analysis of microbial communities in a landfill leachate polluted aquifer using a new method for anaerobic physiological profiling and 16S rDNA based fingerprinting. *Microb. Ecol.* **2000**, *40*, 177–188. [CrossRef] [PubMed]
17. Castañeda, S.S.; Sucgang, R.J.; Almoneda, R.V.; Mendoza, N.D.S.; David, C.P.C. Environmental isotopes and major ions for tracing leachate contamination from a municipal landfill in Metro Manila, Philippines. *J. Environ. Radioact.* **2012**, *110*, 30–37. [CrossRef] [PubMed]
18. EU Directive on Landfill of Waste. European Union Council Directive 1999/31/EC issued 26 April 1999. *Off. J. Eur. Communities* **1999**, *182*, 228–246.
19. Smith, R.J.; Jeffries, T.C.; Roudnew, B.; Fitch, A.J.; Seymour, J.R.; Delpin, M.W.; Newton, K.; Brown, M.H.; Mitchell, J.G. Metagenomic comparison of microbial communities inhabiting confined and unconfined aquifer ecosystems. *Environ. Microbiol.* **2012**, *14*, 240–253. [CrossRef]
20. Janniche, G.S.; Spliid, H.; Albrechtsen, H.-J. Microbial Community-Level Physiological Profiles (CLPP) and herbicide mineralization potential in groundwater affected by agricultural land use. *J. Contam. Hydrol.* **2012**, *140–141*, 45–55. [CrossRef]
21. Preziosi, E.; Frollini, E.; Zoppini, A.; Ghergo, S.; Melita, M.; Parrone, D.; Rossi, D.; Amalfitano, S. Disentangling natural and anthropogenic impacts on groundwater by hydrogeochemical, isotopic and microbiological data: Hints from a municipal solid waste landfill. *Waste Manag.* **2019**, *84*, 245–255. [CrossRef] [PubMed]
22. Barba, C.; Folch, A.; Gaju, N.; Sanchez-Vila, X.; Carrasquilla, M.; Grau-Martínez, A.; Martínez-Alonso, M. Microbial community changes induced by Managed Aquifer Recharge activities: Linking hydrogeological and biological processes. *Hydrol. Earth Syst. Sci.* **2019**, *23*, 139–154. [CrossRef]
23. Salo, M.J. Anthropogenically Driven Changes to Shallow Groundwater in Southeastern Wisconsin and Its Effects on the Aquifer Microbial Communities. *ProQuest Diss. Theses* **2019**, 1–178.
24. Song, Y.; Mao, G.; Gao, G.; Bartlam, M.; Wang, Y. Structural and Functional Changes of Groundwater Bacterial Community During Temperature and pH Disturbances. *Microb. Ecol.* **2019**, *78*, 428–445. [CrossRef] [PubMed]
25. Frollini, E.; Rossi, D.; Rainaldi, M.; Parrone, D.; Ghergo, S.; Preziosi, E. A proposal for groundwater sampling guidelines: Application to a case study in southern Latium. *Rend. Online Soc. Geol. Ital.* **2019**, *47*, 46–51. [CrossRef]
26. Eaton, A.D.; American Public Health Association; Water Environment Federation; American Water Works Associations. *Standard Methods for the Examination of Water and Wastewater*, 21st ed.; APHA-AWWA-WEF: Washington, DC, USA, 2005.
27. Amalfitano, S.; Fazi, S.; Ejarque, E.; Freixa, A.; Romaní, A.M.; Butturini, A. Deconvolution model to resolve cytometric microbial community patterns in flowing waters. *Cytom. Part A* **2018**, *93*, 194–200. [CrossRef] [PubMed]
28. Insam, H.; Goberna, M. *Molecular Microbial Ecology Manual*; Kowalchuk, G.A., de Bruijn, F.J., Head, I.M., Akkermans, A.D., van Elsas, J.D., Eds.; Springer: Dordrecht, The Netherlands, 2004; ISBN 978-1-4020-2176-3.
29. Garland, J.L.; Mills, A.L. Classification and Characterization of Heterotrophic Microbial Communities on the Basis of Patterns of Community-Level Sole-Carbon-Source Utilization. *Appl. Environ. Microbiol.* **1991**, *57*, 2351–2359.
30. Salomo, S.; Münch, C.; Röske, I. Evaluation of the metabolic diversity of microbial communities in four different filter layers of a constructed wetland with vertical flow by Biolog^{TM} analysis. *Water Res.* **2009**, *43*, 4569–4578. [CrossRef] [PubMed]
31. Weber, K.P.; Gehder, M.; Legge, R.L. Assessment of changes in the microbial community of constructed wetland mesocosms in response to acid mine drainage exposure. *Water Res.* **2008**, *42*, 180–188. [CrossRef]
32. Garland, J.L. Analytical approaches to the characterization of samples of microbial communities using patterns of potential C source utilization. *Soil Biol. Biochem.* **1996**, *28*, 213–221. [CrossRef]

33. Konopka, A.; Oliver, L.; Turco, R.F., Jr. The Use of Carbon Substrate Utilization Patterns in Environmental and Ecological Microbiology. *Microb. Ecol.* **1998**, *35*, 103–115. [CrossRef] [PubMed]
34. Insam, H. A New Set of Substrates Proposed for Community Characterization in Environmental Samples. In *Microbial Communities*; Insam, H., Rangger, A., Eds.; Springer: Berlin/Heidelberg, Germany, 1997; pp. 259–260. [CrossRef]
35. Guckert, J.B.; Carr, G.J.; Johnson, T.D.; Hamm, B.G.; Davidson, D.H.; Kumagai, Y. Community analysis by Biolog: Curve integration for statistical analysis of activated sludge microbial habitats. *J. Microbiol. Methods* **1996**, *27*, 183–197. [CrossRef]
36. Clarke, K.R.; Green, R.H. Statistical design and analysis for a biological effects study. *Mar. Ecol. Prog. Ser.* **1988**, *46*, 213–226. [CrossRef]
37. Tiquia, S.M. Metabolic diversity of the heterotrophic microorganisms and potential link to pollution of the Rouge River. *Environ. Pollut.* **2010**, *158*, 1435–1443. [CrossRef] [PubMed]
38. Taş, N.; Brandt, B.W.; Braster, M.; van Breukelen, B.M.; Röling, W.F.M. Subsurface landfill leachate contamination affects microbial metabolic potential and gene expression in the Banisveld aquifer. *FEMS Microbiol. Ecol.* **2018**, *94*, 1–12. [CrossRef] [PubMed]
39. Preston-Mafham, J.; Boddy, L.; Randerson, P.F. Analysis of microbial community functional diversity using sole-carbon-source utilisation profiles-A critique. *FEMS Microbiol. Ecol.* **2002**, *42*, 1–14.
40. Christian, B.W.; Lind, O.T. Multiple carbon substrate utilization by bacteria at the sediment–water interface: Seasonal patterns in a stratified eutrophic reservoir. *Hydrobiologia* **2007**, *586*, 43–56. [CrossRef]
41. Kothawala, D.N.; Stedmon, C.A.; Müller, R.A.; Weyhenmeyer, G.A.; Köhler, S.J.; Tranvik, L.J. Controls of dissolved organic matter quality: Evidence from a large-scale boreal lake survey. *Glob. Chang. Biol.* **2014**, *20*, 1101–1114. [CrossRef]
42. Logue, J.B.; Stedmon, C.A.; Kellerman, A.M.; Nielsen, N.J.; Andersson, A.F.; Laudon, H.; Lindström, E.S.; Kritzberg, E.S. Experimental insights into the importance of aquatic bacterial community composition to the degradation of dissolved organic matter. *ISME J.* **2016**, *10*, 533–545. [CrossRef]
43. Griebler, C.; Malard, F.; Lefébure, T. Current developments in groundwater ecology—from biodiversity to ecosystem function and services. *Curr. Opin. Biotechnol.* **2014**, *27*, 159–167. [CrossRef]
44. Shen, Y.; Chapelle, F.H.; Strom, E.W.; Benner, R. Origins and bioavailability of dissolved organic matter in groundwater. *Biogeochemistry* **2015**, *122*, 61–78. [CrossRef]
45. Meckenstock, R.U.; Elsner, M.; Griebler, C.; Lueders, T.; Stumpp, C.; Aamand, J.; Agathos, S.N.; Albrechtsen, H.-J.; Bastiaens, L.; Bjerg, P.L.; et al. Biodegradation: Updating the Concepts of Control for Microbial Cleanup in Contaminated Aquifers. *Environ. Sci. Technol.* **2015**, *49*, 7073–7081. [CrossRef] [PubMed]
46. Pretty, J.L.; Hildrew, A.G.; Trimmer, M. Nutrient dynamics in relation to surface–subsurface hydrological exchange in a groundwater fed chalk stream. *J. Hydrol.* **2006**, *330*, 84–100. [CrossRef]
47. Hofmann, R.; Griebler, C. DOM and bacterial growth efficiency in oligotrophic groundwater: Absence of priming and co-limitation by organic carbon and phosphorus. *Aquat. Microb. Ecol.* **2018**, *81*, 55–71. [CrossRef]
48. Gounot, A.-M. Microbial oxidation and reduction of manganese: Consequences in groundwater and applications. *FEMS Microbiol. Rev.* **1994**, *14*, 339–349. [CrossRef]
49. Hemme, C.L.; Tu, Q.; Shi, Z.; Qin, Y.; Gao, W.; Deng, Y.; Nostrand, J.D.; van Wu, L.; He, Z.; Chain, P.S.G.; et al. Comparative metagenomics reveals impact of contaminants on groundwater microbiomes. *Front. Microbiol.* **2015**, *6*, 1205. [CrossRef]
50. Zoppini, A.; Puddu, A.; Fazi, S.; Rosati, M.; Sist, P. Extracellular enzyme activity and dynamics of bacterial community in mucilaginous aggregates of the northern Adriatic Sea. *Sci. Total Environ.* **2005**, *353*, 270–286. [CrossRef]
51. Arnosti, C.; Bell, C.; Moorhead, D.L.; Sinsabaugh, R.L.; Steen, A.D.; Stromberger, M.; Wallenstein, M.; Weintraub, M.N. Extracellular enzymes in terrestrial, freshwater, and marine environments: Perspectives on system variability and common research needs. *Biogeochemistry* **2014**, *117*, 5–21. [CrossRef]
52. Keith, S.C.; Arnosti, C. Extracellular enzyme activity in a river-bay-shelf transect: Variations in polysaccharide hydrolysis rates with substrate and size class. *Aquat. Microb. Ecol.* **2001**, *24*, 243–253. [CrossRef]
53. Rosenstock, B.; Simon, M. Consumption of Dissolved Amino Acids and Carbohydrates by Limnetic

Bacterioplankton According to Molecular Weight Fractions and Proportions Bound to Humic Matter. *Microb. Ecol.* **2003**, *45*, 433–443. [CrossRef]

54. Arnosti, C. Patterns of Microbially Driven Carbon Cycling in the Ocean: Links between Extracellular Enzymes and Microbial Communities. *Adv. Oceanogr.* **2014**, *2014*, 1–12. [CrossRef]
55. Oest, A.; Alsaffar, A.; Fenner, M.; Azzopardi, D.; Tiquia-Arashiro, S.M. Patterns of change in metabolic capabilities of sediment microbial communities in river and lake ecosystems. *Int. J. Microbiol.* **2018**, *2018*, 5–7. [CrossRef] [PubMed]
56. Hollibaugh, J.T.; Azam, F. Microbial degradation of dissolved proteins in seawater1. *Limnol. Oceanogr.* **1983**, *28*, 1104–1116. [CrossRef]
57. Cho, J.-C.; Kim, S.-J. Increase in Bacterial Community Diversity in Subsurface Aquifers Receiving Livestock Wastewater Input. *Appl. Environ. Microbiol.* **2000**, *66*, 956–965. [CrossRef]
58. Feris, K.P.; Hristova, K.; Gebreyesus, B.; Mackay, D.; Scow, K.M. A Shallow BTEX and MTBE Contaminated Aquifer Supports a Diverse Microbial Community. *Microb. Ecol.* **2004**, *48*, 589–600. [CrossRef]
59. Anantharaman, K.; Brown, C.T.; Hug, L.A.; Sharon, I.; Castelle, C.J.; Probst, A.J.; Thomas, B.C.; Singh, A.; Wilkins, M.J.; Karaoz, U.; et al. Thousands of microbial genomes shed light on interconnected biogeochemical processes in an aquifer system. *Nat. Commun.* **2016**, *7*, 13219. [CrossRef]
60. Gözdereliler, E. Groundwater Bacteria: Diversity, Activity and Physiology of Pesticide Degradation at Low Concentrations. PhD thesis, Technical University of Denmark, Lyngby, Denmark, 2012.
61. Proctor, C.R.; Besmer, M.D.; Langenegger, T.; Beck, K.; Walser, J.-C.; Ackermann, M.; Bürgmann, H.; Hammes, F. Phylogenetic clustering of small low nucleic acid-content bacteria across diverse freshwater ecosystems. *ISME J.* **2018**, *12*, 1344–1359. [CrossRef]
62. Obernosterer, I.; Benner, R. Competition between biological and photochemical processes in the mineralization of dissolved organic carbon. *Limnol. Oceanogr.* **2004**, *49*, 117–124. [CrossRef]
63. Sala, M.M.; Estrada, M. Seasonal changes in the functional diversity of bacterioplankton in contrasting coastal environments of the NW Mediterranean. *Aquat. Microb. Ecol.* **2006**, *44*, 1–9. [CrossRef]
64. Ding, H.; Sun, M.-Y. Biochemical degradation of algal fatty acids in oxic and anoxic sediment–seawater interface systems: Effects of structural association and relative roles of aerobic and anaerobic bacteria. *Mar. Chem.* **2005**, *93*, 1–19. [CrossRef]
65. Bertilsson, S.; Tranvik, L.J. Photochemical transformation of dissolved organic matter in lakes. *Limnol. Oceanogr.* **2000**, *45*, 753–762. [CrossRef]
66. Preiss, J.; Romeo, T. *Molecular Biology and Regulatory Aspects of Glycogen Biosynthesis in Bacteria*; Cohn, W.E., Moldave, K.B.T.-P., Eds.; Academic Press: New York, NY, USA, 1994; Volume 47, pp. 299–329.
67. Henrissat, B.; Deleury, E.; Coutinho, P.M. Glycogen metabolism loss: A common marker of parasitic behaviour in bacteria? *Trends Genet.* **2002**, *18*, 437–440. [CrossRef]

8

Aqueous Mercury Removal with Carbonaceous and Iron Sulfide Sorbents and their Applicability as Thin-Layer Caps in Mercury-Contaminated Estuary Sediment

Boon-Lek Ch'ng [1], Che-Jung Hsu [1], Yu Ting [1], Ying-Lin Wang [1], Chi Chen [1], Tien-Chin Chang [2] and Hsing-Cheng Hsi [1,*]

1 Graduate Institute of Environmental Engineering, National Taiwan University, No. 1, Sec. 4, Roosevelt Rd., Taipei 106, Taiwan; r06541135@ntu.edu.tw (B.-L.C.); zacharyhsu01@gmail.com (C.-J.H.); yuting821216@gmail.com (Y.T.); lynn12783@gmail.com (Y.-L.W.); q2461015@gmail.com (C.C.)

2 Institute of Environmental Engineering and Management, National Taipei University of Technology, No. 1, Sec. 3, Chung-Hsiao E. Rd., Taipei 106, Taiwan; tcchang@ntut.edu.tw

* Correspondence: hchsi@ntu.edu.tw

Abstract: This study aimed to investigate the Hg removal efficiency of iron sulfide (FeS), sulfurized activated carbon (SAC), and raw activated carbon (AC) sorbents influenced by salinity and dissolved organic matter (DOM), and the effectiveness of these sorbents as thin layer caps on Hg-contaminated sediment remediation via microcosm experiments to decrease the risk of release. In the batch adsorption experiments, FeS showed the greatest Hg^{2+} removal efficiencies, followed by SAC and AC. The effect of salinity levels on FeS was insignificant. In contrast, the Hg^{2+} removal efficiency of AC and SAC increased as increasing the salinity levels. The presence of DOM tended to decrease Hg removal efficiency of sorbents. Microcosm studies also showed that FeS had the greatest Hg sorption in both freshwater and estuary water; furthermore, the methylmercury (MeHg) removal ability of sorbents was greater in the freshwater than that in the estuary water. Notably, for the microcosms without capping, the overlying water MeHg in the estuary microcosm (0.14–1.01 ng/L) was far lesser than that in the freshwater microcosms (2.26–11.35 ng/L). Therefore, Hg compounds in the freshwater may be more bioavailable to microorganisms in methylated phase as compared to those in the estuary water. Overall, FeS showed the best Hg removal efficiency, resistance to salinity, and only slightly affected by DOM in aqueous adsorption experiments. Additionally, in the microcosms, AC showed as the best MeHg adsorber that help inhibiting the release of MeHg into overlying and decreasing the risk to the aqueous system.

Keywords: mercury; methylmercury; salinity; sediment; remediation

---

1. Introduction

Mercury (Hg) has been known as a global contaminant due to the characteristics of long-range transport in the atmosphere, persistence in the environment, bioaccumulation in the food chain, and adverse effects in human health and ecosystem [1]. The increased accumulation of Hg compounds in sediment may cause the high possibility of Hg being transformed to methylmercury (MeHg) by organisms, which is a neurotoxic compound occurred under anoxic conditions [2–5]. The bioaccumulation and biomagnification of MeHg via food chain transfer may pose a high risk to human through fish consumption [6,7]. Therefore, the strategies for sediment remediation are needed to decrease the Hg contaminant release and the possibility of direct or indirect contact with benthic organisms and water surface.

In-situ capping is a feasible approach to remediate contaminated sediment. The main approach is to allow the sediment left in place but decreasing the chance of further contamination from resuspension of contaminants by the capping layer. This technique could decrease the need for handing sediment and decrease the potential of exposure and consequential spills of sediment. The cost is relatively low as compared to traditional dredging and excavation. Thin layer capping may refer as active capping, which involves a chemically reactive material placing in the subaqueous to sequestrate the emission of contaminants from sediment and decrease the bioavailability, mobility, and toxicity of contaminants [8]. Owing to the physical and chemical sorption properties of reactive materials, the amount of sorbents needed and the thickness of the capping layer to achieve the considerable results are lesser as compared to traditional capping [8]. Besides, remediation technology not only decreases the cost, but also minimizes the exposure of benthic organisms to contaminated sediment and decreases the ecological risk associated with contaminated sediments [9].

The active materials may carry a series of reactions to remove contaminants, adsorption, absorption, and precipitation of contaminants and shift them from the aqueous phase to a solid. It works by increasing the contaminated-solid partition coefficient and extend the isolation time before contaminant breaking through the capping layer. A wide variety of active materials are applied and preferred according to the specific conditions of the remediation site. Carbonaceous materials such as activated carbon (AC), biochars, and surface-modified black carbon could effectively decrease organic contaminants and immobilize Hg. AC has several functional groups such as carboxyl, lactone, and phenolic groups. With its high specific surface area, AC has the potential to be an option for remediation of organic pollutant and Hg-contaminated sites [10–12].

Sulfurized activated carbon (SAC) is generally formed by heating the carbon in the presence of elemental sulfur [13,14] or sulfurous gases [15,16]. SAC provides sulfur-containing functional groups, which show high affinity for Hg compounds to form mercuric sulfide at the surface. Hence, SAC has been verified to further enhance the adsorption capacity of Hg as compared to untreated AC in aqueous adsorption tests [17].

Iron sulfide (FeS) minerals have been widely applied for Hg immobilization on account of the high affinity to Hg ions [18–22]. The mercuric sulfide (HgS) is a stable compound and hardly soluble with a low solubility product constant (Ksp) of 2×10^{-54} for red cinnabar [23] and 4×10^{-54} for black metacinnabar [24,25]. Additionally, FeS can effectively immobilize other divalent metals such as Cd^{2+}, Co^{2+}, Zn^{2+}, and Ni^{2+} through adsorption and coprecipitation [26,27].

The objective of this research attempts to decrease the release of both ionic Hg (Hg^{2+}) and MeHg from sediment to surface water and minimize the negative impacts of Hg contamination on the ecological environment with in-situ thin-layer capping practice, which has been shown to have potentials to decrease Hg contamination in sediment. However, previous application of in-situ capping was primarily focused on systems like contaminated river and lake sediment, there is few studies correlated with wetland and estuary, which are complex systems and effected by tidal flow. Besides, the knowledge of capping material's stability is still limited and easily affected by environmental factors. Therefore, it is necessary to investigate the influence of these environmental factors, such as salinity and dissolved organic material (DOM), on the Hg sorption effectiveness of active capping materials. AC, SAC, and FeS were examined as the capping materials because of their potentially suitable physicochemical properties for Hg sorption. In this study, both aqueous batch sorption experiments and lab-scale vertical up-flow microcosms were conducted to comprehend the impact of environmental factors on the stability of materials and applicability to Hg-contaminated sediment.

2. Materials and Methods

The test sediment was collected from a Hg-contaminated seawater pond in China Petrochemical Development Corporation, An-Shun, Tainan city, Taiwan, designated as An-Shun site. The sediment within 0-15 cm depth was collected using a stainless crab bucket. For the sediment pretreatment procedures, the sediment was air dried in a hood, and the branches and benthic biotas were removed.

The sediment was grounded and sieved through a 20-mesh screen to obtain homogenized sediment. After pretreatment, the sediment was stored at room temperature and covered with a black plastic bag.

Three kinds of sorbents were tested in the experiments. Commercial coconut-shell AC was obtained from Li Jing Viscarb Co. Ltd., Taiwan and sieved to obtain a size range from 18 to 30 mesh. Sieved AC was dried in an oven at 105 °C for 24 h. To obtain SAC, the commercial AC was pretreated with elemental sulfur following the protocol described in Hsi et al. [28]. The prepared SAC had the size range and pretreatment conditions the same as the AC and has been examined in our previous study [29]. FeS was purchased from Sigma-Aldrich.

2.1. *Physicochemical Properties of Sorbents and Sediment*

The physical properties of AC, SAC, and FeS were determined by using a physisorption analyzer (Micromeritics ASAP 2420, Norcross, GA, USA) based on the N_2 adsorption-desorption isotherm at 77K. The Brunauer-Emmett-Teller (BET) equation was used to determine the specific surface area based on ASTM D6556-10 [30], and the micropore surface area and volume were calculated by using the *t*-plot method by using the Jura–Harkins equation: $t = [13.99/(0.034 - \log(p/p_0))]^{0.5}$ [31].

Elemental analyses were conducted to measure the contents of elements including N, C, S, H (Elementar Vario EL cube, Langenselbold, Germany), and O (Thermo Fisher Flash 2000, Waltham, MA, USA) for AC and SAC. The water content of air-dried sediment was measured by the weight method based on the Taiwan Environmental Protection Administration (TEPA) standard method (NIEA S280.62C). Sediment pH value with 1:1 sediment to H_2O ratio was measured by pH meter (Suntex SP-2300, New Taipei City, Taiwan) based on TEPA standard method (NIEA S410.62C). Sediment texture was measured by using the bouyoucos hydrometer method [32]. Sediment organic carbon (OC) content was measured by Walkley-Black wet oxidation [33]. Sediment cation exchange capacity (CEC) was measured by the ammonium acetate method based on the TEPA standard method (NIEA S201.61C). Detailed descriptions pertaining to characterizing the physical and chemical properties of sampled sediment can be found in Supplementary Material.

2.2. *Aqueous Batch Sorption Experiment*

The following steps were conducted in each aqueous batch sorption experiment:

2.2.1. Preparation of Hg Stock Solution

A serial dilution of Hg standard solution was made to the intended concentrations using analyzed reagent-grade Hg standard (1000 μg/mL dilute in nitric acid; Ultra Scientific), and adjust the pH value of the solution to 7.0 for various dosage and Hg concentration experiments. In contrast, the pH value of artificial waters was measured to be 7.6, 8.2, 8.3 for freshwater, estuary water, and seawater conditions, respectively, determined by the ions present (Table S1) to evaluate the effect of salinity and DOM on Hg adsorption. 0.1 M HNO_3 or 0.1 M NaOH were used to adjust the pH of solutions.

2.2.2. Sample Preparation

(1) The Hg stock solution of 50 mL was injected into the glass bottles.
(2) The intended dosage of sorbents was added into the glass bottles and sealed with a rubber plug and aluminum cap.
(3) The bottles were collected, including the samples of triplicate plus the blank.

2.2.3. Adsorption Process

The samples were put into the reciprocating water bath shaker and shaken at 130 rpm at 30 °C for 24 h to achieve an adsorption equilibrium.

In order to determine the suitable dosage of sorbents in the following experiments, the dosages 5, 10, 15, 20, 25, 30, 40 and 50 mg were tested with the Hg concentration of 74 μg/L in 50 mL solution.

To determine the adsorption isotherms of sorbents, 4.3, 25, 56, and 135 μg/L Hg^{2+} were tested with 20 mg of sorbents in 50 mL solution.

In order to study the influence of salinity on Hg removal, the Hg^{2+} solutions (Hg concentration = 197.1 ± 10 μg/L) with three different salinity levels of artificial waters were tested in this study, including freshwater (pH = 7.6 ± 0.1), estuary water (pH = 8.2 ± 0.1), and seawater (pH = 8.3 ± 0.1). The compositions of freshwater and seawater are listed in Table S1. The freshwater was prepared according to Lewis et al. [34]; the seawater was prepared according to Kester et al. [35]. The estuary water was prepared by mixing fresh water and seawater at a 1:1 volume ratio. The concentrations of ion species listed in Table S1 were confirmed by ion chromatography (Metrohm 792 Basic IC, Herisau, Switzerland). Hg(II) species at various salinity levels were simulated by using MINTEQ 3.1 and listed in Table S2.

To evaluate the effect of DOM on sorbents' Hg removal efficiency, the DOM solution was prepared by humic acid (HA, Sigma Aldrich, Saint Louis, MO, USA). HA of 10 mg was dissolved in ultrapure water and pre-adjusted to pH 7 by 0.1 M NaOH. The solution was stirred for 1 h to promote fast dissolution, and then filtered through a 0.2 μm mixed cellulose ester filter (DISMIC-25AS, Toyo Roshi Kaisha, Tokyo, Japan). The filtrate was collected and stored at 4 °C refrigerator before using. The actual concentration of DOM was verified by a Total OC analyzer (OI Analytical Aurora 1030W, College Station, TX, USA). The DOM was controlled at a concentration of approximately 2.5 mg-C/L and the Hg concentration is 196.2 ± 5 μg/L.

2.2.4. Preservation of Sample and Analysis

(1) Each sample was preserved with 0.5% $BrCl_2$ and estimated by cold vapor atomic fluorescence spectroscopy (CVAFS; Brooks Rand Automated Total Mercury System, Seattle, WA, USA).
(2) The aqueous Hg removal efficiency by sorbents was calculated by the following Equation (1):

$$R = \frac{C_0 - C_t}{C_0} \times 100\%, \tag{1}$$

where R (%) is the Hg removal efficiency of sorbents, C_0 (ng/L) is the initial Hg concentrations detected in blank solution, and C_t (ng/L) is the concentration of remaining Hg at any time.

Equation (2) was used to determine the partitioning coefficient (K_D) for Hg adsorption by the adsorbent:

$$K_D = \frac{q_e}{C_e} \tag{2}$$

where q_e is the equilibrium Hg adsorption capacity (mg/g) and C_e is the equilibrium Hg concentration (μM). The calculated values of K_D were listed in Table S3.

2.3. Laboratory Microcosm Experiments

The microcosm was established according to Ting et al. [17] with modifications. Vertical up-flow columns were used to stimulate the release of Hg compounds and examine the efficiency of sorbent cappings on Hg sequestration (Figure 1). The dimensions of the column with an internal height of 15 cm and an internal diameter of 6 cm with glass fiber at the bottom. To investigate how the Hg-contaminated estuary sediment affected by salinity and to understand the stabilizing efficiency of AC, SAC, and FeS cappings, two systems, the freshwater and estuary systems, were set up. Each of the four columns contained 300 g of dried Hg-contaminated sediment from the An-Shun site. The total number of column is 8. Column A was capped with AC (9 g; i.e., 3 wt% AC added); column B was capped with SAC (9 g; i.e., 3 wt% SAC added); column C was capped with FeS (9 g; i.e., 3 wt% FeS added); column D was without capping as the control unit.

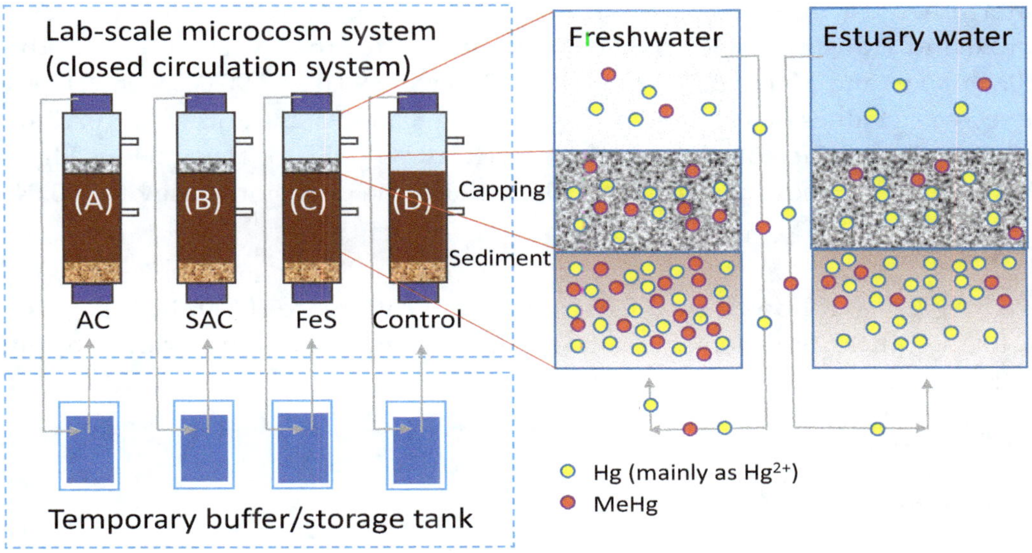

Figure 1. The diagram of vertical up-flow microcosms construction.

To start up the microcosms, dried Hg-contaminated sediment was firstly added into the column, filled with waters and waited for 24 h to settle. The microcosms were activated to circulate the waters in the column for sediment acclimation and stabilization, and counted as the operation day 1. The capping materials were applied on day 25. The total volume of water used for each experiment was 1.5 L. The water first entering into the bottom of the column was referred as inflow, then vertically moved upwards to fill the column, and then discharged through the outlet. The effluent water were stored in a temporary buffer tank and recirculated into the bottom of column in a closed system by using a peristaltic pump (Lead Fluid BT100S, Hebei, China) to maintain a flow rate of 1 mL/min. The reason that the effluent water was recirculated back to the microcosm column is to establish the mass balance of THg partitioning in various phases (i.e., sediment, capping material, overlying water) in a closed system and evaluate the THg accumulation ability of the various capping materials.

Periodic sampling was in progress while the microcosms were in operation. Each time, 100 mL of water sample was collected from the tube of outflow. After that, the temperature, pH (SunTex SR-2300, New Taipei City, Taiwan), dissolved oxygen (DO; Extech EXStik DO600, Nashua, NH, USA), electrical conductivity (EC; Taina EZDO 6021, Taichung, Taiwan), oxidation reduction potential (ORP; SunTex SR-2300, New Taipei City, Taiwan), trace metals concentration, total Hg (THg), MeHg, DOM, and anions were analyzed. The temporary buffer tank was then refilled with the artificial waters to maintain a constant water amount.

To measure the trace compound concentration, including metals, THg, MeHg, DOM, and anions, the water samples were filtered with 0.45 μm mixed cellulose ester filter (DISMIC-25AS, Toyo Roshi Kaisha, Tokyo, Japan). Trace metals were determined by Inductively couple plasma optical emission spectrometry (Agilent 700, Santa Clara, CA, USA) after the sample was acidified with 0.15% HNO_3. Water samples for THg analysis were preserved by adding 0.5% $BrCl_2$ solution and stored in 20 mL glass bottles. THg in sediment and water was analyzed following the USEPA Method 1631 and NIEA W331.50B protocols by using CVAFS. The water samples were preserved by adding 0.2% HCl and stored in 20 mL amber glass bottles in MeHg analysis. MeHg in sediment and water was analyzed following the USEPA Method 1630, NIEA W341.60B and NIEA W540.50B procedures by using gas chromatography/CVAFS (Brooks Rand MERX Integrated Automated MeHg Analyzer, Seattle, WA, USA).

Recovery of Hg in the microcosm system was evaluated based on the Hg concentration in overlying water, cap materials, and sediment. To determine the Hg concentration in cap materials and sediment, digestion was first conducted in a microwave system (Ethos 1600, Milestone, Shelton, CT, USA) with a power setting of 800 W (USEPA method 3051a). After digestion, the THg (μg/mg) in the solid phase was determined using CVAFS (USEPA method 245.7). QA/QC of data were confirmed based on Hsu et al. [29] and Wang et al. [36]. The recovery values for the QC samples of sediment of THg (NIST 2709a, 0.9 mg/kg) and MeHg (SQC-1238, 10 μg/kg) were 98.6 and 90.0%, respectively. For spiked sediment, the recovery values of THg and MeHg were 94.8 and 95.2%, and were 107.1 and 101.4% for water, respectively. The coefficient of determination (R^2) of CVAFS for the aqueous Hg was regularly kept larger than 0.998, the recovery was within 96.2–120%, the precision was within 0.04–5.60%.

2.4. Statistical Analysis

A one-way ANOVA, followed by a least significant difference (LSD) test ($p < 0.05$), was used to determine the significance differences among microcosm tests with various capping materials (IBM SPSS statistics).

3. Results and Discussion

3.1. Physicochemical Properties of Sorbents

The physical and chemical properties of the three test sorbents, AC, SAC, and FeS, were summarized in Table 1. Because FeS is a non-porous mineral, its surface area and pore volume were much smaller than those of AC and SAC. Elemental analyses of AC and SAC showed that the oxygen and sulfur contents in SAC were increased after impregnation of sulfur on AC, hence SAC should be more favored for Hg^{2+} uptake because sulfur-containing groups on SAC had high affinity towards mercuric sulfide formation. The oxygenated groups are also known beneficial for Hg adsorption. The increase in the oxygen content after sulfur impregnation could be due to the reaction of S with C to form vaporized CS_2 evolved from the SAC surface.

Table 1. The physicochemical properties of sorbents.

	S_{BET} (m²/g) *	S_{Micro} (m²/g) *	V_{Total} (cm³/g) *		V_{micro} (cm³/g) *
AC	1024.1	634.1	0.540		0.284
SAC	903.3	528.9	0.502		0.267
FeS	2.811	0.356	0.04		-
	C (%)	**H (%)**	**O (%)**	**N (%)**	**S (%)**
AC	78.3	1.61	7.72	0.791	0.672
SAC	74.9	1.80	13.8	0.36	5.75
FeS	-	-	-	-	36.4

* S_{BET}: specific BET surface area; S_{Micro}: micropore area; V_{Total}: total pore volume; V_{micro}: micropore volume.

3.2. Aqueous Batch Adsorption Experiment

3.2.1. The Effects of the Sorbent Dosage

The influence of sorbent dosage on Hg removal has been investigated to find out the appropriate dosage of sorbents as a basis for subsequently study. The sorbent dosage determines the sorption capacity of sorbents for a given concentration of Hg^{2+} because it controls the sorbent-sorbate equilibrium of a system [37]. The effects of sorbents dosage on Hg sorption were studied in the dosage range of 5–60 mg in 50 mL Hg^{2+} solution. The results of AC, SAC, and FeS dosage are presented in Figure 2, and all experimental data are performed in triplicate. The Hg sorption capacity of AC first increased and then decreased as the Hg removal efficiency increased, due to the increment of AC dosage would provide more sorption active sites to take up Hg in a fixed Hg^{2+} initial concentration. The SAC and FeS showed a similar tendency as AC. Based on the result, the suggested optimum dosages for AC, SAC and FeS at a fixed initial concentration of Hg^{2+} were 30, 10, and 10 mg, respectively. Owning to Hg removal efficiency may be affected by both the properties and dosage of sorbents, the normalization and comparison of different types sorbents in the same mass benchmark were needed. Therefore, the appropriate dosage of sorbents was determined as 20 mg for the subsequent experiments in this research.

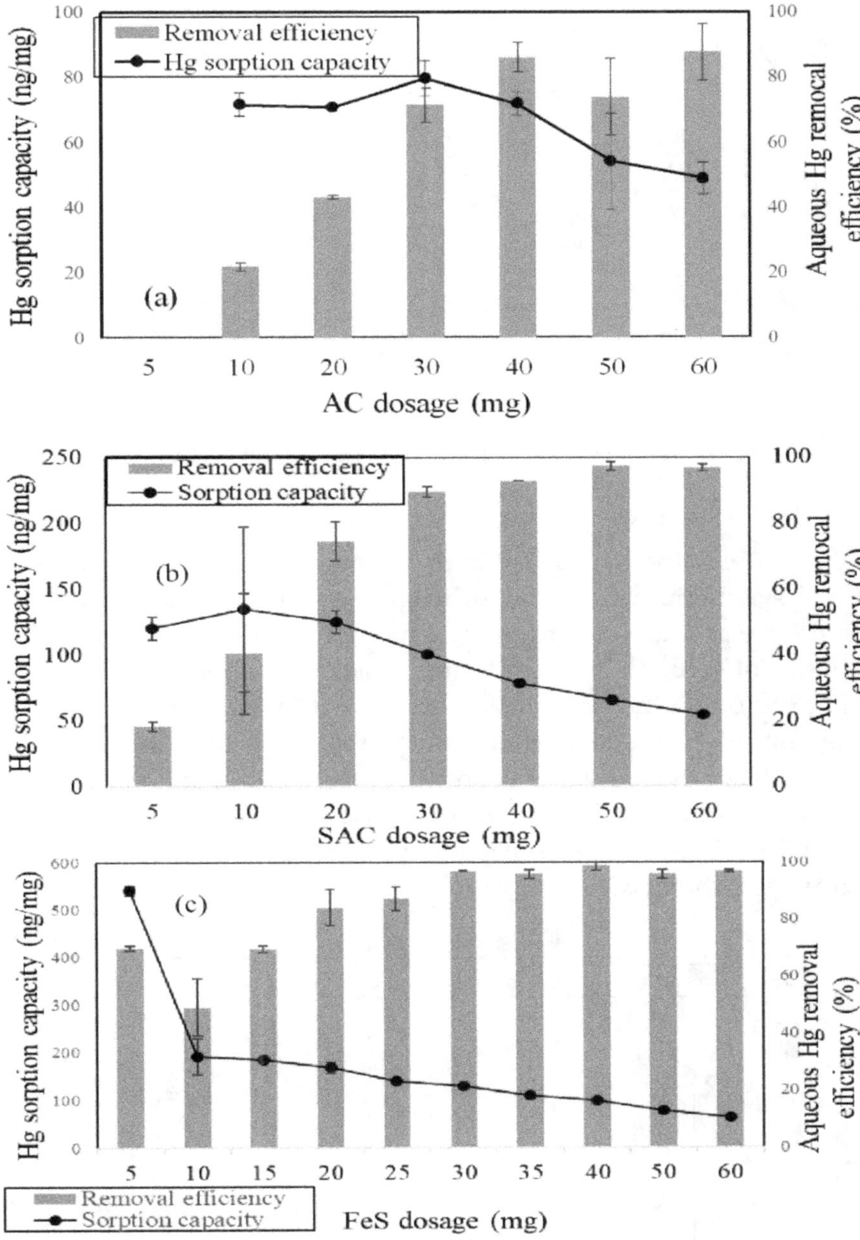

Figure 2. Effect of (**a**) AC; (**b**) SAC; (**c**) FeS dosage on Hg removal efficiency.

3.2.2. Effect of Initial Concentration of Hg^{2+}

The batch experiments were tested within the initial concentration of Hg^{2+} from 5 to 135 μg/L using a diluted standard Hg^{2+} solution. The Hg sorption capacity of sorbents and the initial Hg^{2+} concentration is shown in Figure 3. The experimental results showed that the Hg sorption capacity of sorbents increased with the increment of the initial concentration of Hg^{2+}, with linear adsorption behaviors within the test concentration range. The FeS could maintain a high Hg removal efficiency of up to 90% in the range of the initial concentration of Hg^{2+} given, which illustrates that FeS is an excellent sorbent for Hg removal followed by SAC and AC.

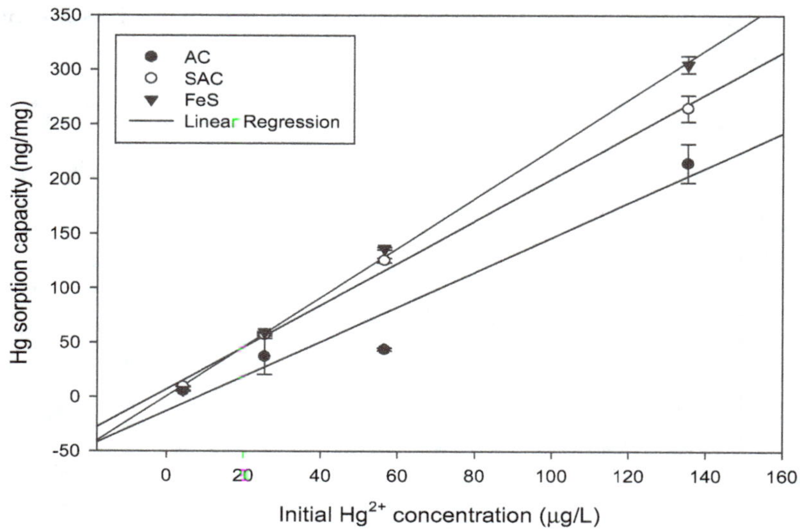

Figure 3. Hg sorption at different Hg^{2+} concentrations.

3.2.3. Effect of Salinity

The salinity level of water in the Hg-contaminated site may affect the Hg removal ability of sorbents. Thereby, three artificial synthetic waters, including freshwater, estuary water, and seawater with different salinity levels, were prepared to study the effects of salinity on Hg removal ability of sorbents. The Hg removal efficiency of sorbents for each water system are presented in Figure 4, indicating that FeS had the largest Hg removal efficiency, followed by SAC and then AC. From the lowest salinity (freshwater) to the highest salinity level (seawater), the Hg removal efficiency increased for both AC and SAC. Although the effect of salinity levels on FeS was insignificant, the Hg removal efficiency of FeS was still the highest as compared to AC and SAC. The calculated K_D values are listed in Table S3, which shows that K_D values increased as the salinity increased. FeS also performed the largest K_D values, followed by SAC and then AC.

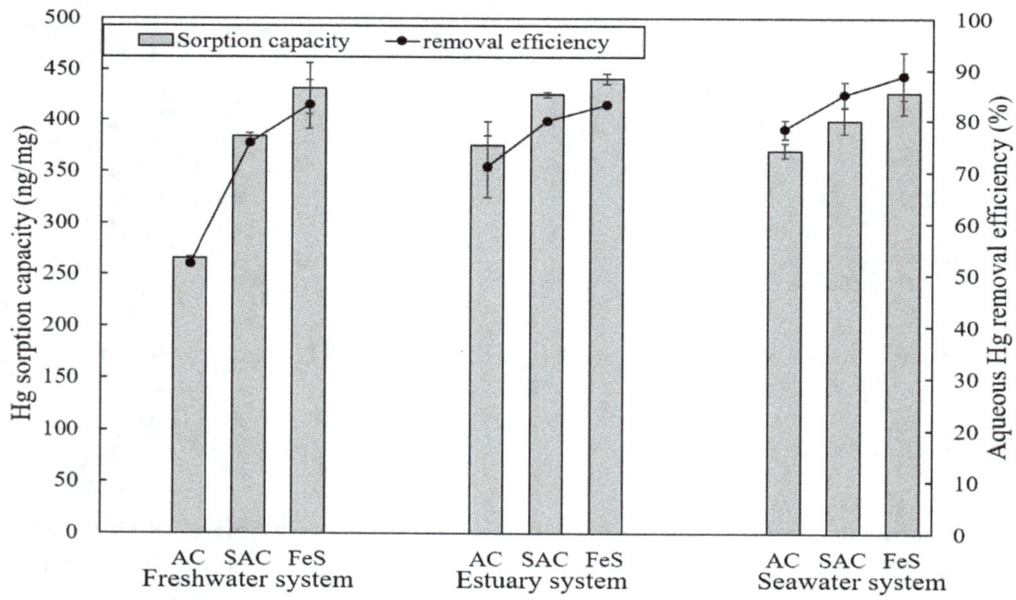

Figure 4. Comparison of various salinity levels affecting the sorbents' Hg removal.

Notably, our results were contrary to those in previous studies, which have shown that a high level of salinity may decrease the removal efficiency of Hg by sorbents [38,39]. The previous studies have demonstrated that an increase in NaCl concentration would decrease the sorption efficiency of

AC and kaolin. However in our research, the different salinity levels were prepared by adding various salts to form the artificial waters. Hence, the complexation and species of Hg in this study may be more complicated than the cases with only presence of NaCl. Table S2 displays the simulated speciation of Hg(II) compounds, indicating that the fraction of $Hg(OH)_2$ in the freshwater system (i.e., 95.97%) was significantly higher than that in both systems of estuary water (i.e., 0.094%) and seawater (i.e., 0.014%). $Hg(OH)_2$ has been reported that it could easily decompose to the elemental form as Hg^0, Hg^0 is more difficult to be captured from the water solution than other oxidized form because of its extremely low solubility (i.e., 5.6×10^{-5} g/L) [29]. Additionally, Thiem et al. [40] showed that the addition of calcium ion would enhance Hg removal of AC. They speculated that the calcium ion may react with the surface group on AC to form a new adsorption site, leading to an increment of Hg removal capacity in the solution. Besides, according to the K_D values, the increase of salinity has a positive effect on the partition behavior of aqueous Hg to the adsorbent; furthermore, FeS has a fabulous capability for converting Hg from the liquid phase to the solid phase.

3.2.4. Effect of Dissolved Organic Matter on Hg Sorption at Various Salinity Levels

DOM is widely spread in environments and may control a number of essential processes relevant for Hg cycling. In this study, artificial waters with a DOM concentration of 2.6 mg-C/L was prepared with three salinity levels, and the results are shown in Figure 5. For the addition of DOM to the salinity test, the Hg removal by sorbents was relatively decreased as compared with those in the salinity test (Figure 4). In freshwater system, the sorption capacity of AC, SAC, and FeS was decreased from 286 to 137 ng/mg, 384 to 270 ng/mg, and 431 to 287 ng/mg, respectively. As for the estuary system, the sorption capacity of AC, SAC and FeS was decreased from 401 to 286 ng/mg, 408 to 366 ng/mg, and 441 to 403 ng/mg, respectively. While the sorption capacity of AC, SAC, and FeS were decreased from 355 to 302 ng/mg, 400 to 322 ng/mg, and 427 to 363 ng/mg in the seawater system, respectively (Figure 6). In Table S3, K_D values also decreased significantly with the presence of DOM. Therefore, DOM may inhibit the Hg adsorption by complexation mechanism because it can compete with sorbents and complex with Hg^{2+} [41–44]. Hg^{2+} may form complex with organic thiol groups in DOM [45,46]. The phenolic hydroxyl groups in DOM may also complex with Hg easily to form a stable chelate that restrained the Hg adsorption [47].

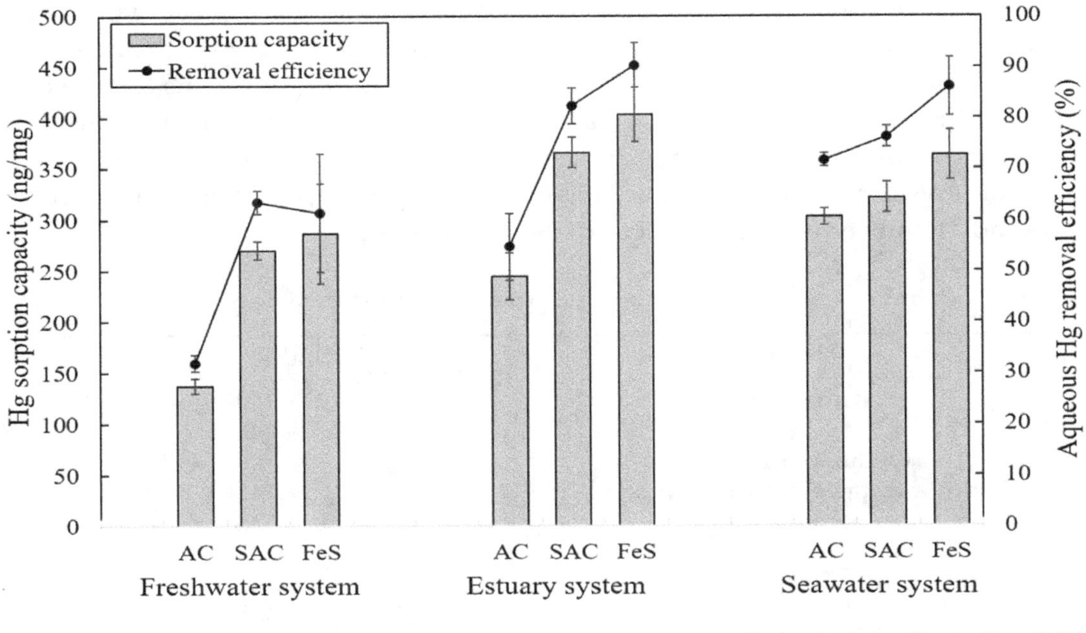

Figure 5. Comparison of the Hg sorption capacity at various salinity levels affected by DOM.

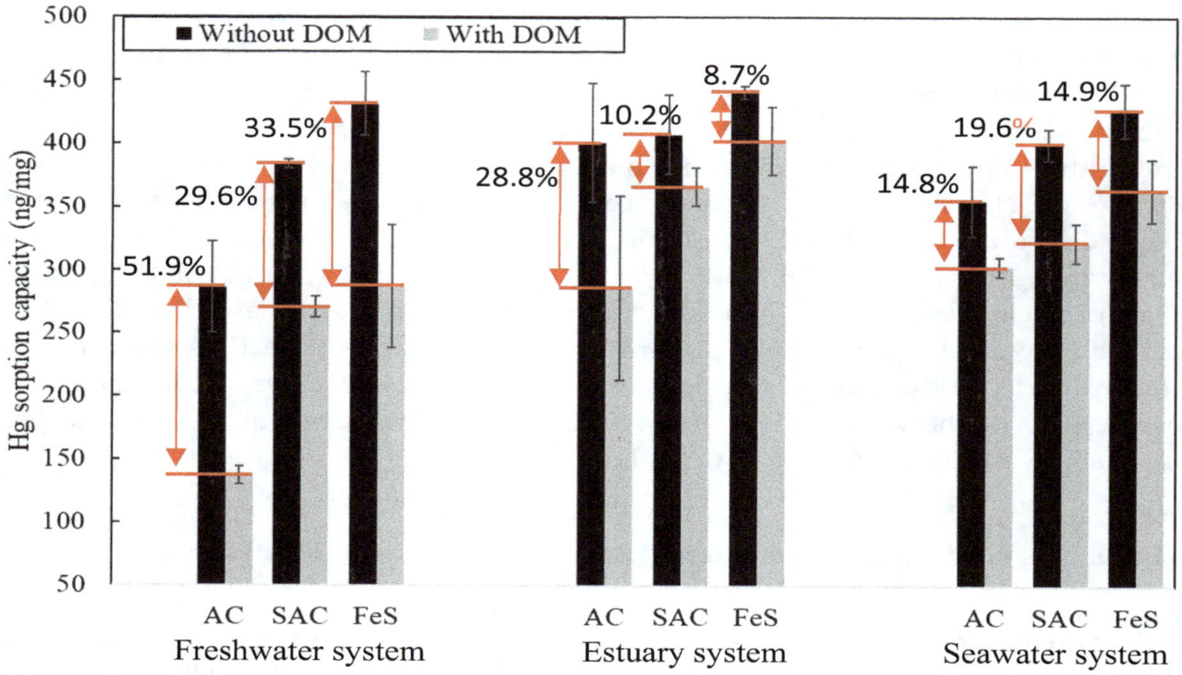

Figure 6. Comparison of the influence of various salinity levels and DOM on Hg sorption capacity in the aqueous batch experiments.

3.3. Laboratory Microcosm Experiments

3.3.1. Physicochemical Properties of Sediments Used in Microcosm Experiments

The physicochemical properties of the An-Shun site sediment are summarized in Table 2. The pH value of the sediment was about 7.5, slightly alkaline due to the estuary water system of that region. The slightly alkaline condition may favor to immobilize and decrease the toxicity of most metals as they may precipitate in the alkaline environment. The OC content in the An-Shun site sediment was about 0.8 wt%, which was relatively low when compared to other sediments. The organic matter in the sediment may potentially affect the interaction between pollutants and sediment. The sand, silt and clay contents of the sediment were 71.4, 14.3, and 14.3 wt%, respectively. The An-Shun site sediment could thus be classified as a sandy loam according to the soil texture classification of the United States Department of Agriculture (USDA) (Figure S2). In general, the texture of sediment is tightly correlated with OC content and CEC. The fine-textured sediment may contain high OC and CEC due to its large surface areas, whereas the OC and CEC of sandy-textured sediment was relatively low. The results of physicochemical properties analysis of An-Shun site sediment were consistent with the fact that a sandy loam showed low OC content and low CEC.

Table 2. The physicochemical properties of An-Shun site sediment.

Parameters	Value
Water content (wt%)	2.33 ± 0.01
pH	7.5 ± 0.03
Organic carbon (wt%)	0.80 ± 0.65
Cation exchange capacity (cmol(+)/kg)	3.3 ± 0.2
Sand (wt%)	71.4 ± 2.9
Silt (wt%)	14.3 ± 0.0
Clay (wt%)	14.3 ± 2.9
THg (mg/kg)	76.0 ± 2.59
MeHg (μg/kg)	1.17 ± 0.31

3.3.2. Environmental Factor Variation during Microcosms Operation

The vertical up-flow microcosms with a closed circulation system are set up as shown in Figure 1 and Figure S1. Two kinds of waters were applied in the microcosms, including freshwater and estuary water. The microcosms were operated for 65 days. The operation time from days 1 to 25 was the process of sediment acclimation. After that, the capping materials were delivered to the microcosms and the water was collected periodically from the outflow as the overlying water sample. The appearance of the microcosms, which was capped with the AC, SAC, or no cap were seen to be clear, while the column capped with FeS was found to gradually appear in brown inside the microcosms. It has been speculated that the addition of FeS may promote a chemical reaction to form iron hydroxide or iron oxide, which was found attached on the wall of the column and the buffer tank.

The recorded temperature in the microcosms is shown in Figure S3. The range of temperature was between 20 and 27 °C during the test period between winter and spring. The DO results are shown in Figure S4. The DO value at the beginning of the experiment was approximately 4 mg/L, started to decrease and remained stable at 3 ± 0.5 mg/L as the operation time extended. This result could be explained by the microbial activity in the microcosms. The microbes would consume DO to carry out respiration process, thus its concentration decreased at the beginning. Owning to the operational defects, the oxygen outside the microcosm fluxed in and caused the increment of DO. The pH value of overlying water is shown in Figure S5. It was between 7.6 and 7.9 in all microcosms.

The results of EC in the freshwater system and estuary system are shown in Figure S6. In the freshwater system, the EC of overlying waters of microcosms was around 2300–3270 µS/cm. The EC values decreased as the operation time increased because the microcosms would be refilled with the artificial freshwater and get diluted after sampling periodically (Figure S6a). Moreover, the An-Shun site sediment was estuary sediment, thus it contained a portion of salt and contributed to the background levels of salinity. In the estuary water system of microcosms, the results of EC are shown in Figure S6b. The EC ranged from 37,000 to 38,000 µS/cm.

The ORP results of the overlying water from freshwater system and estuary system are shown in Figure S7. In the freshwater system, the ORP of overlying water of microcosms was around 40–160 mV. As for the estuary system, the ORP of overlying water of microcosms was about −22–145 mV.

The results of DOM in the overlying water are shown in Figure S8. The DOM content in the overlying water was low, mainly related to the low DOM content in An-Shun site sediment. The An-Shun site sediment contained less OC content according to the previous results shown; consequently, less DOM would be released from the sediment to the overlying water.

The total Fe in overlying water is shown in Figure S9. In the freshwater system, the microcosm capped with FeS would contribute significant dissolved Fe to the overlying water while the concentration of Fe in the microcosm with no capped was also high. The Fe content of those treatment were higher if compared to microcosms with AC and SAC. The similar trend was also observed in the estuary system.

3.3.3. Sequestration of Aqueous THg and MeHg by Thin Layer Capping

The effectiveness of capping materials to immobilize Hg is shown in Figure 7 and Table S4. For the freshwater system, the THg immobilization abilities of three capping materials greatly varied as compared to that of the no-capped control unit. The THg reduction efficiencies of the AC, SAC, and FeS, which were calculated based on the initial THg concentration of the microcosm started and the THg concentration in a given sampling date, fluctuated between −22–82%, −30–78% and −87–62%, respectively, in comparison to the control unit. As for the estuary system, THg reduction efficiencies of the AC, SAC and FeS reached −33–49%, 28–60% and −24–44%, respectively, in comparison to that of the control unit. Therefore, the fluctuation and uncertainty for Hg removal of capping sorbents were observed in both freshwater and estuary systems. According to our previous studies [17,48], the recommended amendment dosage of sorbents for Hg-sediment remediation was approximately 1–5%. The dosage of sorbents being used in this study was 3%, which was supposedly to show significant effect on Hg removal, but the results seemed not be similar to our previous studies. The horizontal flow microcosm set up was mostly used to stimulate the emission of Hg from the sediment according to the previous studied [48]. However, a different design, vertical up-flow microcosm was adapted in this study, in which the water moved in an up-flow direction and then recirculated back into the columns in a closed system. Therefore, a dynamic equilibrium among the sediment, capping material, and overlying water achieved in the closed circulation system. The Hg in the sediments will be released to the water continuously until an equilibrium was reached. The same process happened in the interface between water and capping material. These processes will lead to the depletion of Hg on the sediments because it was diffused and transferred to the capping materials. Owning to these process, the reduction of Hg in the overlying water of the microcosm was less evident, which was also demonstrated by the ANOVA analysis (Table S4); instant and marked releases of Hg from the caps were occasionally found. The dosage of sorbents applied in this microcosms may need to be reconsidered as it might not be capable to sorb Hg efficiently due to the characteristic of this microcosms system.

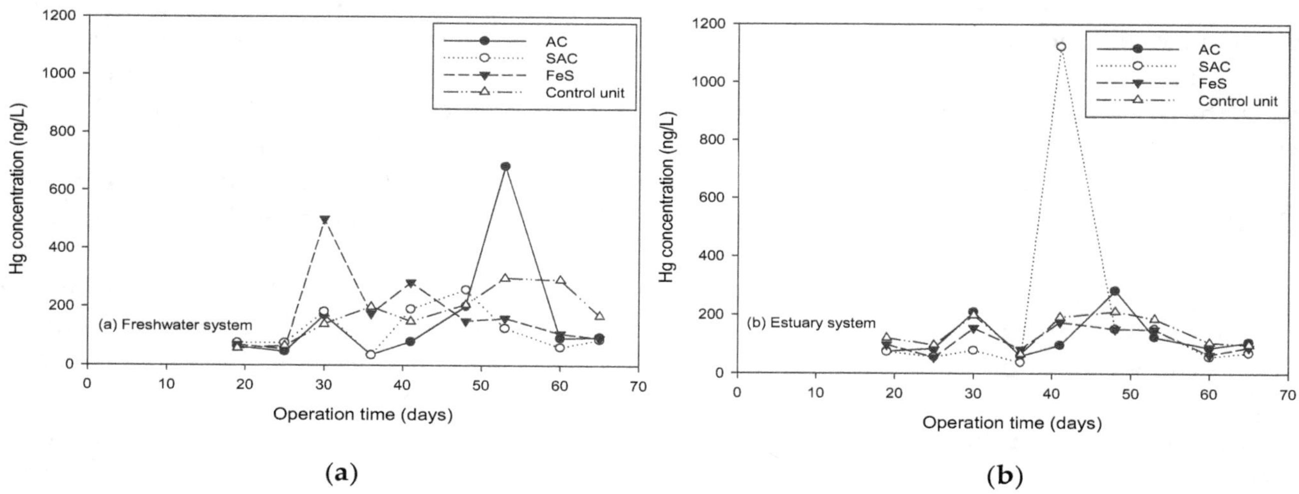

Figure 7. THg concentration in overlying water of (**a**) freshwater system and (**b**) estuary system.

The Hg concentrations in sediment, sorbents and overlying water after 65 days in the microcosms are shown in Table 3. The Hg concentrations in the freshwater sediments at the end of experiment for AC, SAC, FeS and control unit were 65.35 ± 0.65, 68.87 ± 0.46, 50.38 ± 1.03 and 68.77 ± 2.03 mg/kg, respectively. In the estuary system, the Hg concentrations of sediment for AC, SAC, FeS and control unit were 56.16 ± 2.87, 57.55 ± 0.39, 54.80 ± 2.25 and 69.20 ± 4.02 mg/kg, respectively. The initial Hg content in sediment of the An-Shun site was 76.00 ± 2.59 mg/kg, thus, there were varying losses of Hg from sediment with different capping amendment in both freshwater and estuary systems. The sediment can be a reservoir of Hg, which would provide Hg to the water, capping materials, and the microcosm's auxiliary equipment until an equilibrium reached. Hence, the decrease of Hg in the sediment of the study is reasonable. In the freshwater system, the loss of Hg in microcosms capped with AC and SAC was similar to that in the control unit while the loss of Hg in FeS capping was slightly higher than that in the control unit, AC, and SAC capping. As for the estuary system, the Hg loss from sediment with capping materials was decreased significantly as compared to that in the control unit. On the other hand, the contents of Hg in the sorbents of AC, SAC, and FeS were 4.84 ± 0.81, 6.86 ± 3.52, and 37.66 ± 8.34 mg/kg, respectively, in freshwater system. In the estuary system, the contents of Hg in the sorbents of AC, SAC, and FeS were 3.88 ± 1.82, 2.89 ± 1.49, and 43.83 ± 13.04 mg/kg, respectively. The Hg content in FeS was significantly greater as compared to those in AC and SAC in both freshwater and estuary systems. Notably, the Hg sorption capacity of sorbents in the batch experiments was relatively high when compared to those for the sorbents applied in the microcosms. Although the Hg concentration of overlying water, approximately 200 ng/L was relatively low as compared to the batch experiment test, the mechanism of Hg sorption might be different at the low Hg concentration. The recovery of Hg in the microcosms was approximately 67–90% in both freshwater and estuary systems. The low recovery in some cases can be explained by the phenomenon of bacteria respiration, which could produce a large portion of batting biomass and colloids that cause Hg to attach onto. Moreover, the Hg might be adhered to the surface of the microcosm's auxiliary equipment, such as column surface, buffer tank, pipelines, and others that may cause the decreasing recovery of Hg (Figure S1). Therefore, although FeS showed great Hg sorption capacity during the batch and caused Hg partitioning from sediment to the cap during microcosm test, Hg may be released from cap again, through redissolution or release of small-scale FeS particles containing Hg, leading to not only the low recovery but also the risk of unexpected release of Hg for sediment with vertical up-flow.

Table 3. Recovery calculation based on the Hg content in sediment, sorbents and overlying water after 65 days in the microcosms.

	Sediment [1]		Sorbents		Overlying Water	Recovery [2]
	mg/kg	mg	mg/kg	mg	mg	%
F-AC	65.35 ± 0.65	19.61	4.84 ± 0.81	0.04	0.002	86.19
F-SAC	68.87 ± 0.46	20.66	6.86 ± 3.52	0.06	0.02	90.89
F-FeS	50.38 ± 1.03	15.11	37.66 ± 8.34	0.34	0.02	67.78
F-Control	68.77 ± 2.03	20.63	-	-	0.02	90.50
E-AC	56.16 ± 2.87	16.85	3.88 ± 1.82	0.03	0.002	74.05
E-SAC	57.55 ± 0.39	17.27	2.89 ± 1.49	0.03	0.003	75.85
E-FeS	54.80 ± 2.25	16.44	43.83 ± 13.04	0.39	0.002	73.84
E-Control	69.20 ± 4.02	20.70	-	-	0.002	90.80

[1] The Hg content in sediment for the day 65. [2] Recovery was obtained by the sum of Hg content in sediment, sorbents and overlying water divided by the total Hg content in initial sediment (i.e., 76.0 ± 2.59 mg/kg and 22.8 mg) then multiplying by 100.

The results of MeHg content in the overlying water of freshwater system and estuary system are shown in Figure 8. In the freshwater system, the amounts of MeHg in the overlying water from microcosms capped with AC, SAC, and FeS were significantly decreased when compared to that of the control unit; the MeHg concentrations of AC, SAC, FeS, and control unit were 0.04–0.70, 0.14–0.94, 0.13–1.64, and 2.26–11.35 ng/L, respectively. In contrast, the MeHg concentrations of AC, SAC, FeS, and control unit were 0.10–0.14, 0.15–0.87, 0.04–2.77, and 0.14–1.01 ng/L in the estuary system, respectively. Therefore, the production of MeHg under the treatment of AC and SAC capping was decreased tremendously, while the reduction efficiency of MeHg in the microcosms capped with FeS was not significant.

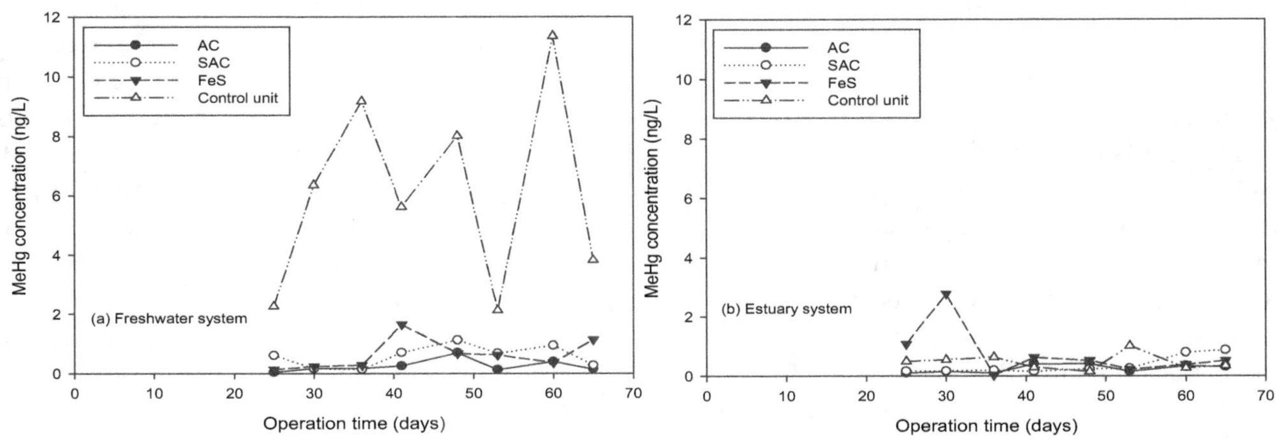

(a) (b)

Figure 8. MeHg concentration in overlying water of (a) freshwater system and (b) estuary system.

The production of MeHg of the control unit in the freshwater system was significant as compared with that in the estuary system. A previous study showed that the salinity level of the environment was negatively correlated with MeHg production [49]. In the high salinity condition, Hg has tendency to form complex with chlorine salt and sulfurous compounds, reducing the trends of Hg methylation [50]. The researchers suggested that a high-salinity condition may inhibit Hg methylation because sulfate with high concentration would reduce to toxic sulfides by microorganisms, and poisons the Hg methylating bacteria, reducing the formation of MeHg as well. In the freshwater system, the MeHg reduction efficiency in microcosms capped with AC was the best, SAC followed, and FeS the least, suggesting that the SAC and FeS provided a portion of the iron and sulfur elements to dissolve into water and enhanced the methylation ability of sulfate-reducing bacterial and Fe(III)-reducing bacteria. As a result, the MeHg reduction efficiency in microcosms capped with SAC and FeS was poorer as compared to that with AC.

The MeHg contents in the sediment are shown in Table 4. The formation of MeHg in the control unit of freshwater system was higher than that in the estuary system. The formation of MeHg capped with AC and FeS was low when compared to that in the control unit in freshwater system, except for the microcosm capped with SAC, which was slightly higher. As a results, the MeHg released to overlying water in freshwater system was inhibited by both limiting the MeHg formation and the effectiveness of caps. In the estuary system, the MeHg contents in both capped with sorbents and control unit were similar and smaller than those in freshwater system.

Table 4. The MeHg content in sediment on day 65.

Freshwater System	MeHg (μg/kg)	Estuary System	MeHg (μg/kg)
AC	1.43 ± 0.75	AC	1.03 ± 0.52
SAC	5.42 ± 1.92	SAC	1.16 ± 0.07
FeS	2.61 ± 0.65	FeS	2.07 ± 0.52
Control unit	3.97 ± 1.68	Control unit	1.99 ± 1.15

4. Conclusions

In this study, the aqueous batch experiments with the amendment of AC, SAC, and FeS sorbents were first carried out to comprehend the Hg removal efficiency in Hg-contaminated sediments influenced by salinity and DOM. The microcosms were then set up to examine the performance of these capping sorbents on Hg-contaminated sediment remediation. The experimental results showed that FeS on Hg removal was not significantly affected by salinity levels and maintained with high removal efficiency. The Hg removal efficiency of AC and SAC increased as salinity increased. In contrast, the Hg removal efficiency of sorbents decreased with the addition of DOM at different salinity levels because DOM competed with sorbents and may occupy the adsorption site, thus inhibited the Hg uptake by sorbents. The microcosm experiments showed that the THg immobilization abilities of three capping sorbents greatly varied as compared to that of control unit. The MeHg concentration of overlying water in the freshwater microcosm with no cap was higher than that in the estuary system. Therefore, Hg compounds in the freshwater system may be more bioavailable to microorganisms in methylated phase as compared to those in the estuary system. To summarize, the capping materials including AC, SAC, and FeS effectively decreased the concentration of overlying water MeHg in the freshwater system of microcosms. Because the production of MeHg in estuary system was low, the efficiency of materials on MeHg sorption was insignificant.

We suggest that future studies should be focused on scale-up design using large microcosms. Notably, because FeS showed the best Hg removal efficiency, resistance to salinity, and only slightly affected by DOM in aqueous adsorption experiments and AC showed as the best MeHg adsorption material, a "mixing cap" using both FeS and AC should be examined and the optimal mixing ratio should be obtained. A mixing cap of FeS and AC may also help preventing the leaching out of FeS from the cap layer, which was observed in our microcosm study.

It is also worth noting that for the vertical up-flow system, the accumulated Hg in capped layer may eventually breakthrough, which could cause sudden concentration shock that leads to instant risk of exposure. Long-term microcosm operation is critical and should be further conducted to obtain design parameters on subsequent pilot tests or full-scale application.

Supplementary Materials

Table S1. Freshwater and seawater ion species concentration, Table S2. Hg(II) speciation at various salinity levels, Table S3. The partitioning coefficients for Hg adsorptions at various salinity levels, Table S4. A one-way ANOVA or one-way ANOVA on ranks based on normality test, followed by a post hoc test ($p < 0.05$) used to determine the significance differences among various sorbents, Figure S1. Photos of the microcosms on (a) day 25 as the capping materials were initially applied and (b) on day 65, Figure S2. An-Shun site sediment texture, Figure S3. The temperature of microcosms (the symbol F refers to freshwater system and the symbol E refers to estuary system), Figure S4. The dissolved oxygen for the microcosms (the symbol F refers to freshwater system and the symbol E refers to estuary system), Figure S5. The pH value variation of microcosms (the symbol F refers to freshwater system and the symbol E refers to estuary system), Figure S6. The electricity conductivity of overlying water in (a) freshwater system and (b) estuary system, Figure S7. The oxidation reduction potential of overlying water in (a) freshwater system and (b) estuary system, Figure S8. DOM variation of microcosms for (a) freshwater system; (b) estuary system, Figure S9. The total Fe variation of microcosms in (a) freshwater system and (b) estuary system.

Author Contributions: Conceptualization, B.-L.C., Y.T., T.C.C., and H.-C.H.; methodology, B.-L.C., C.-J.H., Y.T., Y.-L.W., and H.-C.H.; formal analysis, B.-L.C. and C.C.; resources, H.-C.H.; writing—original draft preparation, B.-L.C. and C.-J.H.; writing—review and editing, Y.-L.W. and H.-C.H.; supervision, T.-C.C. and H.-C.H.; funding acquisition, H.-C.H. All authors have read and agreed to the published version of the manuscript.

Acknowledgments: We greatly appreciate the financial and technical supports from the Environmental Protection Administration, Taiwan and the Ministry of Science and Technology of Taiwan, Taiwan. The opinions expressed in this paper are not necessarily those of the sponsors.

References

1. Nriagu, J.O. *Biogeochemistry of Mercury in the Environment*; Elsevier/North-Holland Biomedical Press: Amsterdam, The Netherlands, 1979.
2. Compeau, G.C.; Bartha, R. Sulfate-reducing bacteria: Principal methylators of mercury in anoxic estuarine sediment. *Appl. Environ. Microbiol.* **1985**, *50*, 498–502. [CrossRef]
3. Gilmour, C.C.; Henry, E.A.; Mitchell, R. Sulfate stimulation of mercury methylation in freshwater sediments. *Environ. Sci. Technol.* **1992**, *26*, 2281–2287. [CrossRef]
4. Gilmour, C.C.; Podar, M.; Bullock, A.L.; Graham, A.M.; Brown, S.D.; Somenahally, A.C.; Johs, A.; Hurt, R.A.; Bailey, K.L.; Elias, D.A. Mercury methylation by novel microorganisms from new environments. *Environ. Sci. Technol.* **2013**, *47*, 11810–11820. [CrossRef] [PubMed]
5. Podar, M.; Gilmour, C.C.; Brandt, C.C.; Soren, A.; Brown, S.D.; Crable, B.R.; Palumbo, A.V.; Somenahally, A.C.; Elias, D.A. Global prevalence and distribution of genes and microorganisms involved in mercury methylation. *Sci. Adv.* **2015**, *1*, e1500675. [CrossRef] [PubMed]
6. Fitzgerald, W.F.; Lamborg, C.H.; Hammerschmidt, C.R. Marine biogeochemical cycling of mercury. *Chem. Rev.* **2007**, *107*, 641–662. [CrossRef] [PubMed]
7. Stein, E.D.; Cohen, Y.; Winer, A.M. Environmental distribution and transformation of mercury compounds. *Crit. Rev. Environ. Sci. Technol.* **1996**, *26*, 1–43. [CrossRef]
8. Zhang, C.; Zhu, M.Y.; Zeng, G.M.; Yu, Z.G.; Cui, F.; Yang, Z.Z.; Shen, L.Q. Active capping technology: A new environmental remediation of contaminated sediment. *Environ. Sci. Pollut. Res.* **2016**, *23*, 4370–4386. [CrossRef] [PubMed]
9. McDonough, K.M.; Murphy, P.; Olsta, J.; Zhu, Y.; Reible, D.; Lowry, G.V. Development and placement of a sorbent-amended thin layer sediment cap in the Anacostia River. *Soil Sediment Contam.* **2007**, *16*, 313–322. [CrossRef]

10. Gilmour, C.; Bell, T.; Soren, A.; Riedel, G.; Riedel, G.; Kopec, D.; Bodaly, D.; Ghosh, U. Activated carbon thin-layer placement as an in situ mercury remediation tool in a Penobscot River salt marsh. *Sci. Total Environ.* **2018**, *621*, 839–848. [CrossRef]
11. Ghosh, U.; Luthy, R.G.; Cornelissen, G.; Werner, D.; Menzie, C.A. In-situ sorbent amendments: A new direction in contaminated sediment management. *Environ. Sci. Technol.* **2011**, *45*, 1163–1168. [CrossRef]
12. Gomez-Eyles, J.L.; Yu, P.Q.; Beckingham, B.; Riedel, G.; Gilmour, C.C.; Ghosh, U. Evaluation of biochars and activated carbons for in situ remediation of sediments impacted with organics, mercury, and methylmercury. *Environ. Sci. Technol.* **2013**, *47*, 13721–13729. [CrossRef] [PubMed]
13. Ie, I.R.; Hung, C.H.; Jen, Y.S.; Yuan, C.S.; Chen, W.H. Adsorption of vapor-phase elemental mercury (Hg^0) and mercury chloride ($HgCl_2$) with innovative composite activated carbons impregnated with Na_2S and S^0 in different sequences. *Chem. Eng. J.* **2013**, *229*, 469–476. [CrossRef]
14. Puri, B.R.; Hazra, R.S. Carbon-sulphur surface complexes on charcoal. *Carbon* **1971**, *9*, 123–134. [CrossRef]
15. Li, Z.; Wu, L.; Liu, H.; Lan, H.; Qu, J. Improvement of aqueous mercury adsorption on activated coke by thiol-functionalization. *Chem. Eng. J.* **2013**, *228*, 925–934. [CrossRef]
16. Macıas-Garcıa, A.; Gomez-Serrano, V.; Alexandre-Franco, M.F.; Valenzuela-Calahorro, C. Adsorption of cadmium by sulphur dioxide treated activated carbon. *J. Hazard. Mater.* **2003**, *103*, 141–152. [CrossRef]
17. Ting, Y.; Chen, C.; Ch'ng, B.L.; Wang, Y.L.; Hsi, H.C. Using raw and sulfur-impregnated activated carbon as active cap for leaching inhibition of mercury and methylmercury from contaminated sediment. *J. Hazard. Mater.* **2018**, *354*, 116–124. [CrossRef]
18. Gong, Y.; Liu, Y.; Xiong, Z.; Kaback, D.; Zhao, D. Immobilization of mercury in field soil and sediment using carboxymethyl cellulose stabilized iron sulfide nanoparticles. *Nanotechnology* **2012**, *23*, 294007. [CrossRef]
19. Gong, Y.; Liu, Y.; Xiong, Z.; Zhao, D. Immobilization of mercury by carboxymethyl cellulose stabilized iron sulfide nanoparticles: Reaction mechanisms and effects of stabilizer and water chemistry. *Environ. Sci. Technol.* **2014**, *48*, 3986–3994. [CrossRef]
20. Han, D.S.; Orillano, M.; Khodary, A.; Duan, Y.; Batchelor, B.; Abdel-Wahab, A. Reactive iron sulfide (FeS)-supported ultrafiltration for removal of mercury (Hg (II)) from water. *Water Res.* **2014**, *53*, 310–321. [CrossRef]
21. Liu, J.; Valsaraj, K.T.; Devai, I.; DeLaune, R.E. Immobilization of aqueous Hg (II) by mackinawite (FeS). *J. Hazard. Mater.* **2008**, *157*, 432–440. [CrossRef]
22. Xiong, Z.; He, F.; Zhao, D.; Barnett, M.O. Immobilization of mercury in sediment using stabilized iron sulfide nanoparticles. *Water Res.* **2009**, *43*, 5171–5179. [CrossRef] [PubMed]
23. Paquette, K.; Helz, G. Solubility of cinnabar (red HgS) and implications for mercury speciation in sulfidic waters. *Water Air Soil Poll.* **1995**, *80*, 1053–1056. [CrossRef]
24. Benoit, J.M.; Gilmour, C.C.; Mason, R.P.; Heyes, A. Sulfide controls on mercury speciation and bioavailability to methylating bacteria in sediment pore waters. *Environ. Sci. Technol.* **1999**, *33*, 951–957. [CrossRef]
25. Ravichandran, M.; Aiken, G.R.; Ryan, J.N.; Reddy, M.M. Inhibition of precipitation and aggregation of metacinnabar (mercuric sulfide) by dissolved organic matter isolated from the Florida Everglades. *Environ. Sci. Technol.* **1999**, *33*, 1418–1423. [CrossRef]
26. Morse, J.W.; Arakaki, T. Adsorption and coprecipitation of divalent metals with mackinawite (FeS). *Geochim. Cosmochim. Acta* **1993**, *57*, 3635–3640. [CrossRef]
27. Wolthers, M.; Van der Gaast, S.J.; Rickard, D. The structure of disordered mackinawite. *Am. Mineral.* **2003**, *88*, 2007–2015. [CrossRef]
28. Hsi, H.C.; Rood, M.J.; Rostam-Abadi, M.; Chen, S.; Chang, R. Effects of sulfur impregnation temperature on the properties and mercury adsorption capacities of activated carbon fibers (ACFs). *Environ. Sci. Technol.* **2001**, *35*, 2785–2791. [CrossRef] [PubMed]
29. Hsu, C.J.; Chiou, H.J.; Chen, Y.H.; Lin, K.S.; Rood, M.J.; Hsi, H.C. Mercury adsorption and re-emission inhibition from actual WFGD wastewater using sulfur-containing activated carbon. *Environ. Res.* **2019**, *168*, 319–328. [CrossRef]
30. ASTM D6556–10. *Standard Test Method for Carbon Black—Total and External Surface Area by Nitrogen Adsorption*; ASTM International: West Conshohocken, PA, USA, 2012.
31. Lippens, B.C.; de Boer, J.H. Studies on pore systems in catalysts: V. the t method. *J. Catal.* **1965**, *4*, 319–323. [CrossRef]

32. Gee, G.W.; Bauder, J.W. Particle-size analysis. In *Methods of Soil Analysis, Part 1. Physical and Mineralogical Methods*, 2nd ed.; Klute, A., Ed.; Agronomy Monograph No. 9; American Society of Agronomy/Soil Science Society of America: Madison, WI, USA, 1986; pp. 383–411.
33. Nelson, D.W.; Sommers, L.E. Total carbon, organic carbon, and organic matter. In *Methods of Soil Analysis, Part 2. Chemical and Microbiological Properties*, 2nd ed.; Page, A.L., Ed.; Agronomy Series No. 9; American Society of Agronomy/Soil Science Society of America: Madison, WI, USA, 1982; pp. 539–579.
34. Lewis, P.A.; Klemm, D.J.; Lazorchak, J.M.; Norberg-King, T.J.; Peltier, W.H.; Heber, M.A. *Short-Term Methods for Estimating the Chronic Toxicity of Effluents and Receiving Waters to Freshwater Organisms*; US Environmental Protection Agency, Environmental Monitoring Systems Laboratory: Cincinnati, OH, USA, 1994.
35. Kester, D.R.; Duedall, I.W.; Connors, D.N.; Pytkowicz, R.M. Preparation of artificial seawater 1. *Limnol. Oceanogr.* **1967**, *12*, 176–179. [CrossRef]
36. Wang, Y.L.; Fang, M.D.; Chien, L.C.; Lin, C.C.; Hsi, H.C. Distribution of mercury and methylmercury in surface water and surface sediment of river, irrigation canal, reservoir and wetland in Taiwan. *Environ. Sci. Pollut. Res.* **2019**, *26*, 17762–17773. [CrossRef] [PubMed]
37. Kazemi, F.; Younesi, H.; Ghoreyshi, A.A.; Bahramifar, N.; Heidari, A. Thiol-incorporated activated carbon derived from fir wood sawdust as an efficient adsorbent for the removal of mercury ion: Batch and fixed-bed column studies. *Process Saf. Environ. Prot.* **2016**, *100*, 22–35. [CrossRef]
38. de Diego, A.; Tseng, C.M.; Dimov, N.; Amouroux, D.; Donard, O.F. Adsorption of aqueous inorganic mercury and methylmercury on suspended kaolin: Influence of sodium chloride, fulvic acid and particle content. *Appl. Organomet. Chem.* **2001**, *15*, 490–498. [CrossRef]
39. Ranganathan, K. Adsorption of Hg (II) ions from aqueous chloride solutions using powdered activated carbons. *Carbon* **2003**, *41*, 1087–1092. [CrossRef]
40. Thiem, L.; Badorek, D.; O'Connor, J.T. Removal of mercury from drinking water using activated carbon. *J. Am. Water Works Assoc.* **1976**, *68*, 447–451. [CrossRef]
41. Moreno, F.N.; Anderson, C.W.; Stewart, R.B.; Robinson, B.H.; Ghomshei, M.; Meech, J.A. Induced plant uptake and transport of mercury in the presence of sulphur-containing ligands and humic acid. *New Phytol.* **2005**, *166*, 445–454. [CrossRef]
42. Muller, K.A.; Brandt, C.C.; Mathews, T.J.; Brooks, S.C. Methylmercury sorption onto engineered materials. *J. Environ. Manag.* **2019**, *245*, 481–488. [CrossRef]
43. Schwartz, G.E.; Sanders, J.P.; McBurney, A.M.; Brown, S.S.; Ghosh, U.; Gilmour, C.C. Impact of dissolved organic matter on mercury and methylmercury sorption to activated carbon in soils: Implications for remediation. *Environ. Sci. Process. Impacts* **2019**, *21*, 485–496. [CrossRef]
44. Johs, A.; Eller, V.A.; Mehlhorn, T.L.; Brooks, S.C.; Harper, D.P.; Mayes, M.A.; Pierce, E.M.; Peterson, M.J. Dissolved organic matter reduces the effectiveness of sorbents for mercury removal. *Sci. Total Environ.* **2019**, *690*, 410–416. [CrossRef]
45. Skyllberg, U.; Bloom, P.R.; Qian, J.; Lin, C.M.; Bleam, W.F. Complexation of mercury (II) in soil organic matter: EXAFS evidence for linear two-coordination with reduced sulfur groups. *Environ. Sci. Technol.* **2006**, *40*, 4174–4180. [CrossRef]
46. Skyllberg, U.; Drott, A. Competition between disordered iron sulfide and natural organic matter associated thiols for mercury (II): An EXAFS study. *Environ. Sci. Technol.* **2010**, *44*, 1254–1259. [CrossRef] [PubMed]
47. Pandey, A.K.; Pandey, S.D.; Misra, V. Stability constants of metal–humic acid complexes and its role in environmental detoxification. *Ecotoxicol. Environ. Saf.* **2000**, *47*, 195–200. [CrossRef]
48. Ting, Y.; Ch'ng, B.L.; Chen, C.; Ou, M.Y.; Cheng, Y.H.; Hsu, C.J.; Hsi, H.C. A simulation study of mercury immobilization in estuary sediment microcosm by activated carbon/clay-based thin-layer capping under artificial flow and turbation. *Sci. Total Environ.* **2020**, *708*, 135068. [CrossRef] [PubMed]
49. Compeau, G.C.; Bartha, R. Effect of salinity on mercury-methylating activity of sulfate-reducing bacteria in estuarine sediments. *Appl. Environ. Microbiol.* **1987**, *53*, 261–265. [CrossRef] [PubMed]
50. Ullrich, S.M.; Tanton, T.W.; Abdrashitova, S.A. Mercury in the aquatic environment: A review of factors affecting methylation. *Crit. Rev. Environ. Sci. Technol.* **2001**, *31*, 241–293. [CrossRef]

Preliminary Analysis of the Diet of *Triturus carnifex* and Pollution in Mountain Karst Ponds in Central Apennines

Mattia Iannella [1], Giulia Console [1,*], Paola D'Alessandro [1], Francesco Cerasoli [1], Cristina Mantoni [1], Fabrizio Ruggieri [2], Francesca Di Donato [2] and Maurizio Biondi [1]

[1] Department of Life, Health and Environmental Sciences, Environmental Sciences Sect., University of L'Aquila, Via Vetoio, Coppito, 67100 L'Aquila, Italy; mattia.iannella@univaq.it (M.I.); paola.dalessandro@univaq.it (P.D.); francesco.cerasoli@univaq.it (F.C.); cristina.mantoni@graduate.univaq.it (C.M.); maurizio.biondi@univaq.it (M.B.)

[2] Department of Physical and Chemical Sciences (DSFC), University of L'Aquila, Via Vetoio, Coppito, 67100 L'Aquila, Italy; fabrizio.ruggieri@univaq.it (F.R.); francesca.didonato3@graduate.univaq.it (F.D.D.)

* Correspondence: giulia.console@graduate.univaq.it

Abstract: Mountain karst ponds are sensitive environments, hosting complex trophic networks where amphibians play a major role, often as top predators. The diet of the Italian crested newt (*Triturus carnifex*) is still poorly known for populations occupying mountain karst ponds. These are traditionally used as livestock's watering points, leading to water pollution due to excreta and wading behavior. The aim of this paper is to understand the relationship between *T. carnifex* diet composition, assessed through the stomach flushing technique, and physical and chemical characteristics in mountain ponds, focusing on parameters altered by livestock pressure, such as ammonium concentration and dissolved oxygen. The high diversity of prey items found within the newts' gut contents confirms the generalist diet even in mountain ponds. The number of prey taxa, their relative abundance and Shannon–Wiener diversity index show variations among the sampled sites, related to livestock organic pollution. Moreover, we report the very first European records of microplastic items in amphibians' stomach content, which also represent the first evidence for Caudata worldwide. Our findings suggest that livestock pressure directly influences *T. carnifex* diet and highlight that the emerging issue of plastics is a threat even in remote high-altitude environments.

Keywords: diet; *Triturus carnifex*; mountain karst ponds; pollution; microplastics

1. Introduction

High-altitude ecosystems are nowadays more and more studied because of their susceptibility to ongoing climate change and habitat loss [1–3]. Many species have been proven to shift their ranges to higher altitudes [4–6] or, when the habitat does not permit these migrations, to possibly go extinct [7–9].

Karst mountain ponds are sensitive water bodies because of their exposure to several highly variable environmental factors. They are subject to seasonal snow coverage, noticeable daily and annual thermal excursion, UV-radiation and water-level changes [10–13]. As a result, the biotic component is often unique as well, showing adaptations to temperature fluctuation, long diapause or hibernation and short breeding phenology [14–16]. The species belonging to these animal communities are often of conservation concern, because of their specialization to mountain environments and to low connectivity among ponds [10,17].

In the Apennines, ponds are usually located within grasslands of karstic plateaus, and are traditionally used as watering points for livestock [18–20]. Therefore, the occurrence of cattle around and inside the ponds determine the eutrophication of the water through the direct and indirect introduction of their excreta [21,22], as we also frequently observed during field campaigns. Even though most of these ponds fall within National and Regional Parks, they are still vulnerable and sensitive to many occurring environmental changes, because of their limited size [23,24]. Considering the alternation of turbid and clear water due to the alternative stable state of ponds [25], it is evident that the increase of non-managed livestock in ponds (especially horses and cows) determines nutrient enrichment, shifting towards turbid water conditions through sediment resuspension and alteration of chemicals, such as ammonium and oxygen concentrations [26,27]. Such physical and chemical modification affects both the macroinvertebrates and the vertebrates in ponds. In particular, the macroinvertebrate community structure changes and negatively influences the amphibians' community; this cascade effect depends on the environmental conditions of the ponds, because of a complex interaction between abiotic and biotic components [22,28,29]. Further, amphibians themselves show high sensitivity to climate [30,31], contaminants [32–34] and habitat changes [35,36].

Despite their highly threatened status [37,38] and their importance in food webs [39,40], amphibians' dietary ecology in high altitude lakes or ponds has been poorly investigated so far [41,42].

In this paper, we analyze how the ponds' alteration due to dense livestock presence could affect amphibians' diet, through the diet analysis of the Italian crested newt, *Triturus carnifex*, at mountain sampling sites. We study this aspect, which is still little known, focusing on both livestock-used ponds of different extents and artificial waters (rainwater collection cistern and troughs), characterized by different physical and chemical water conditions, in protected areas.

We explored, through a correspondence analysis, the association between physical and chemical variables and diversity in prey taxa in ponds with different degrees of livestock pollution. We also performed multiple statistical comparisons to test possible significant differences in the variability of newts' diet between natural and artificial sites and among the single sampling sites. We further assessed the presence of human-derived pollutants, such as plastics, which possibly affect freshwater ecosystems even in high altitudes.

2. Materials and Methods

2.1. Target Species

The Italian crested newt *Triturus carnifex* (Laurenti, 1768) was the target species of this study; it shows a disjunct distribution between the Italian and Balkan peninsulas [43]. In Italy, the species occurs from the sea level up to 1980 m [44]; in the Central Apennines, it reaches 1818 m a.s.l. in Lake Pantaniello [45,46].

The terrestrial environment preferred by the Italian crested newt ranges from deciduous beech forests to typically Mediterranean xeric meadows; the shelters must be fresh and moist, often found under downed woods, in forest litter and in the burrows of small mammals. The aquatic phase occurs in lakes, ponds, springs and puddles, preferably rich in aquatic vegetation for egg deposition.

Triturus carnifex is classified as "Near threatened" by the Italian IUCN red list, with a declining population trend [47]. The main threats are the destruction or alteration of its reproductive sites at different levels, the shift to intensive agricultural practices, and climate change, which often leads to the desiccation of their preferred habitats [48–50].

2.2. Study Area

The study area included the Central Apennines' mountain karst plains (900–1700 m a.s.l.) (Figure 1), characterized by a scarcity of surface water. In this area, we selected 9 natural and artificial sampling sites, classified as ponds, troughs and a cistern; all of them fell within two protected areas, the Gran Sasso and Monti della Laga National Park and the Sirente Velino Regional Park. In the former, the ponds named Fossetta (two neighboring ponds, coded as GSA and GSB), Fossa di Paganica (GSC) and Racollo (GSD) were studied. In the latter, the Tempra pond (SVA), the Piane di Iano pond (SVB), the

Pagliare di Tione no-roof rainwater collection cistern, hereafter named only "cistern" (SVC), the Piano Canale (SVD) and Fonte dell'Acqua (SVE) troughs were sampled. Relevant pieces of information about the sampling sites are reported in Table 1.

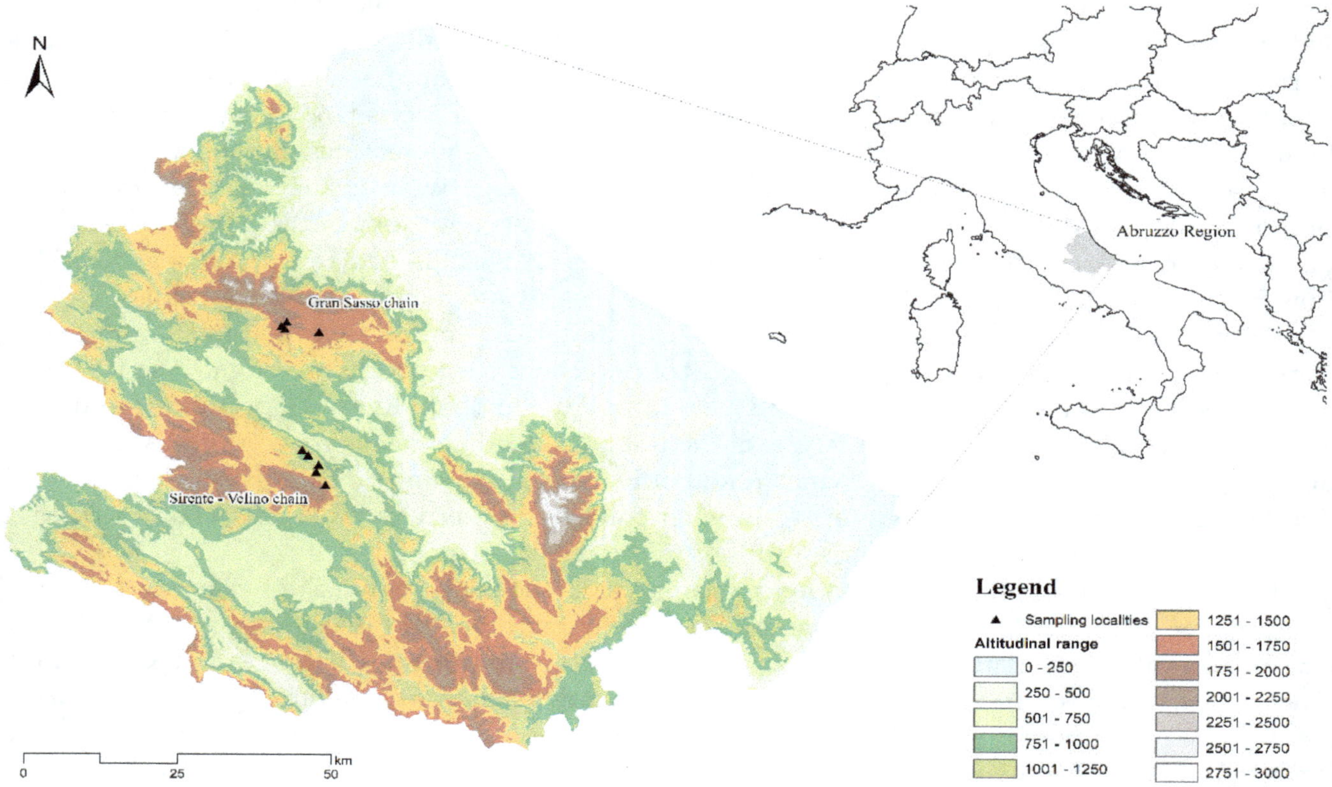

Figure 1. Study area with altitude classes and sampling localities (black triangles).

Table 1. Sampling sites considered in the analyses and the corresponding geographic information.

Name	Code	Type	Area (m^2)	Max. Depth (m)	Latitude	Longitude	Elevation (m a.s.l.)
Fossetta A	GSA	Pond	1870	1.2	42°24′17″ N	13°35′14″ E	1658
Fossetta B	GSB	Pond	358	1.2	42°24′19″ N	13°35′13″ E	1658
Fossa di Paganica	GSC	Pond	4500	1.5	42°24′55″ N	13°39′27″ E	1680
Racollo	GSD	Pond	7000	0.8	42°23′37″ N	13°39′27″ E	1570
Tempra	SVA	Pond	870	1.5	42°10′22″ N	13°38′30″ E	1183
Piane di Iano	SVB	Pond	780	1.2	42°11′21″ N	13°37′22″ E	960
Pagliare di Tione	SVC	Cistern	125	2.5	42°11′55″ N	13°36′41″ E	1080
Piano Canale	SVD	Trough	15	0.5	42°08′25″ N	13°39′17″ E	1342
Fonte dell'Acqua	SVE	Trough	26	0.5	42°08′21″ N	13°39′10″ E	1167

The GSA, GSB, GSD and SVB ponds show remarkable water level variations during summer; they are characterized by turbid waters (because of the silty bottom) and scarce vegetation (mainly composed by *Potamogeton*, *Ceratophyllum* and *Myriophyllum*) due to the constant presence of livestock (sheep, cows and horses), which enter the ponds for some meters from the banks (Figure 2). On the contrary, GSC and SVA show clearer waters, probably due to the wide extent (and the gravel bottom) of the former and due to the low livestock frequentation of the latter. In all these ponds, water is supplied only from snow melting and rainfall. SVC is not used by livestock as a watering point because of its limited accessibility due to perimeter walls; it has clear water and abundant aquatic vegetation. Finally, SVD and SVE show clear water and aquatic vegetation (including algae) as well; they are used as watering points, but livestock does not enter them.

Figure 2. Livestock near Fossa di Paganica pond, GSC (**a**); horses grazing aquatic plants (**b**) and cow cooling off during summer (**c**) in the Fossetta pond, GSA.

2.3. Sampling Methods

Newts (excluding larvae) were sampled starting from 2015 to 2019 (June–September, following the aquatic phase of the target species), employing both active and passive sampling methods. Dip netting sessions were performed, following a circular transect to cover the entire pond's perimeter. Funnel traps were also used as passive capture method, leaving them in the sampling sites for a 12-h period. The captured newts were stomach flushed using all the necessary equipment as described in [51]; the stomach contents were stored in Eppendorf safe-lock tubes with 70% ethanol.

A multi-parametric probe Eutech PD650 (EUTECH Instruments, Singapore, Singapore) was used to measure pH, dissolved oxygen (DO in mg/L), redox potential (mV) and the possible presence of ammonium (NH_4^+ concentration, mg/L), resulting from livestock excreta directly released within the ponds (or nearby the corresponding banks). All of the parameters were measured once for each sampling session.

2.4. Diet Analysis

The prey items were identified at the lowest taxon allowed by the level of digestion, generally as orders.

The sampled newts for each sampling site, the number of different prey groups and the total number of prey items found within the stomach contents of each newt are reported in Supplementary Material Table S1.

Diversification of predated taxa was calculated for each sampled newt within each sampling site through the Shannon–Wiener diversity index (H'), by means of the 'diversity' function of the 'vegan' [52] package in R [53]. The resulting H' values are reported in Supplementary Material Table S1.

Appropriate statistical tests were conducted in R to check for potential discrepancies in the diversification of prey groups among the sampling sites, as well as among the types of sampling sites (i.e., ponds, cistern and troughs) and their "class" (i.e., natural, comprising the ponds, and artificial, including the cistern and the troughs).

A Shapiro–Wilk test and QQ plots were used to check the normality of the residuals from a linear model built considering the H′ index as the response variable and the three above-mentioned grouping factors as covariates. Levene tests were performed to check homoscedasticity across the levels of the three grouping arrangements. Kruskall–Wallis tests were conducted to assess the significance of the considered grouping factors.

Subsequently, multiple pairwise Wilcoxon Rank Sum tests were conducted, using the 'pairwise.wilcox.test' function in R [53] with the Benjamini and Hochberg *p*-value adjustment method, to investigate differences among the single levels of the grouping factor(s) resulting as significant.

Moreover, we considered for each site the relative percentages of the different prey taxa per "average newt" (i.e., the ratio between the number of prey items, for each taxon, and the number of captured newts).

Correspondence analysis was performed through the 'ca' package [54] in RStudio [53], using both chemical parameters and relative prey abundance for each sampling sites as variables.

Together with prey items and other materials (e.g., parts of plants and sediments), we unexpectedly found microplastics in some stomach contents. All microplastics were observed and measured through a Leica M205C (Leica Microsystems, Wetzlar, Germany) binocular microscope. Photomicrographs were taken using a Leica DFC500 camera and the Zerene Stacker software (Zerene Systems LLC, Richland, WA, USA) version 1.04.

3. Results

The stomach contents of 239 Italian crested newts were collected from the nine sampling sites. The number of sampled newts and the total number of prey items for the different prey groups (for each site) are reported in Supplementary Material Tables S1 and S2.

Prey items were counted and classified in 19 different taxa, namely Crustacea (4750 prey items), Diptera (1800), Haplotaxida (328), Odonata (220), Hemiptera (191), Mollusca (135), Ephemeroptera (75), Coleoptera (53), Araneae (35), Acarina (17), Hymenoptera (13), Dermaptera (five), Caudata (four), Trichoptera (three), Tricladida (three), Lepidoptera (two), Julida (two), Neuroptera (one) and Mermithida (one). A total of 641 items could not be determined, due to the high level of deterioration; 161 fragments of plants, 93 particles of inorganic sediments probably deriving from the bottom, 59 microplastic items in the form of fibres, splinters and other undeterminable fragments were also found.

Crustacea (among which 47% were Ostracoda, 33% Branchiopoda and 19% Maxillopoda) and Diptera (among which 81% were Chironomidae and 9% Culicidae) showed the highest frequencies per average newt in almost all the sampling sites, with Crustacea exceeding 85% of the total prey items in GSD and SVB; differently, the highest Diptera frequency is found in SVA (Figure 3). GSC showed the highest number of taxa (14) per average newt, followed by SVD (13) and an equal number for SVA and SVE (11); the remaining five sampling sites showed a total number of prey taxa ≤ 10 (GSA = 10, SVC = 10, GSD = 8, GSB = 7, SVB = 6).

With respect to the analyses performed to assess the level of diversification within the diet of the single newts among the sampling sites, the performed Shapiro–Wilk test (W = 0.95, $p = 4.49 \times 10^{-7}$) and the inspected QQ plot led us to reject the hypothesis of the normality of the residuals from the linear model built upon the H′ index, while the Levene tests did not reject homoscedasticity for the sampling site (F = 0.97, $p = 0.46$), site type (F = 1.35, $p = 0.26$) and site class (F = 0.06, $p = 0.80$) grouping factors.

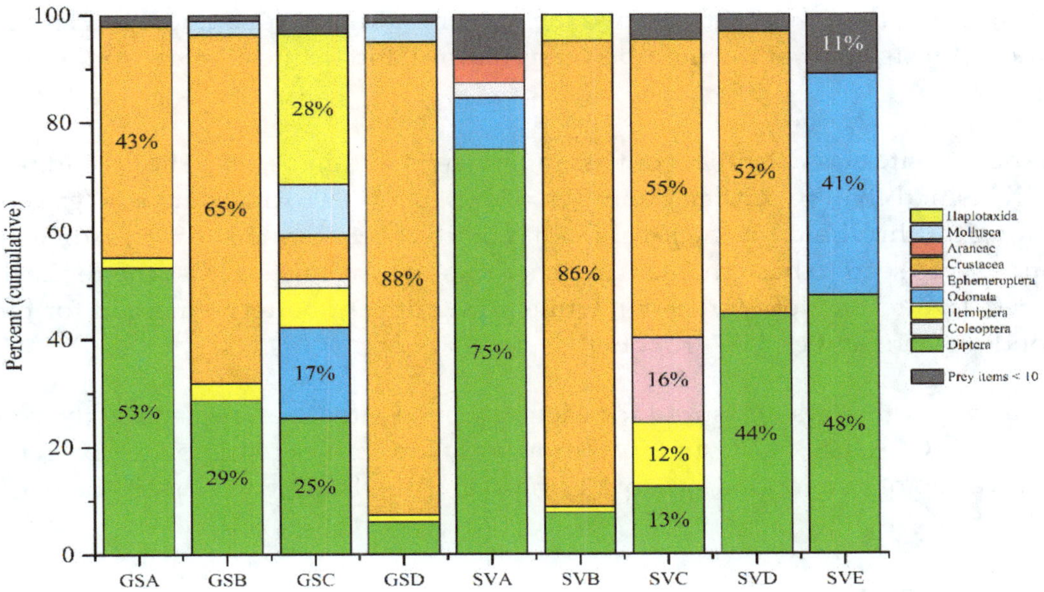

Figure 3. Frequency of prey items (taxa) for each sampled site. Taxa < 10 prey items were aggregated into a single class (grey); percentages < 10% are not reported within the figure for graphical purposes.

Given the normality assumption was not met, a Kruskall–Wallis test was used to check the significance of the three grouping factors—significant differences in the means of the single levels resulted in only considering the sampling sites as a grouping factor ($\chi^2 = 18.01$, df = 8, $p = 0.02$).

From the Wilcoxon Rank Sum tests successively performed on this grouping factor, GSB presented as significantly more diversified in terms of taxa predated by the single newts with respect to SVE ($p = 0.04$); it is also worth noticing the low median H′ value of GSA, GSD, SVB and SVE with respect to the remaining sampling sites (Figure 4a).

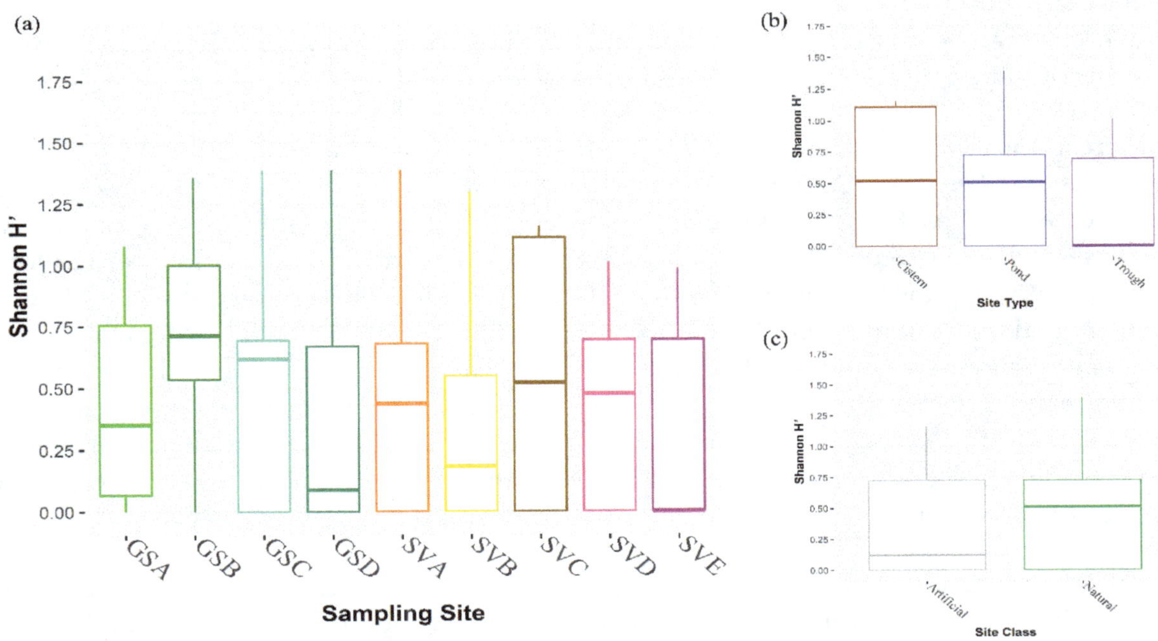

Figure 4. Boxplots of Shannon–Wiener H′ index values across sampling sites (**a**), site types (**b**), and site classes (**c**).

Notwithstanding, the site type ($\chi^2 = 4.29$, df = 2, $p = 0.12$) and site class ($\chi^2 = 2.32$, df = 8, $p = 0.13$) were shown to be non-significant grouping factors by the Kruskall–Wallis tests, considering the differences among the levels of these two factors emerging from boxplots in Figure 4b,c, single Mann–Whitney U tests were performed between natural and artificial sites, and among the three site types.

With respect to site class, the diets of the single newts within natural sites could be considered to be more diversified than in artificial ones (U = 5798.5, p = 0.06, alternative = 'greater') under a less constraining 0.1 threshold for the p-value; differently, considering site types, diets of the single newts in ponds were seen to be significantly richer than in (averages across the replicates) troughs (U = 4781, p = 0.02, alternative = 'greater'), while no significant pattern emerged for the other two pairs (i.e., ponds–cistern and cistern–troughs).

The chemical parameters measured for each sampling site (averages across the replicates) are reported in Table 2. GSA, GSB, GSD and SVB show very low values of dissolved oxygen, along with relatively high values of ammonium (up to 0.26 mg/L in SVB), leading to classify them as "low water quality sites".

Table 2. Chemical parameters measured for each sampling site.

Name	Code	pH	Redox Potential (mV)	DO (mg/L) (DO (%sat), Temp. (°C))	NH_4^+ (mg/L)
Fossetta A	GSA	7.3	−6.9	3.0 (38%, 20.8 °C)	0.12
Fossetta B	GSB	7.6	−17.7	3.6 (47%, 21.1 °C)	0.17
Fossa di Paganica	GSC	8.3	38.1	4.5 (62%, 24.0 °C)	0.00
Racollo	GSD	7.6	−7.9	2.3 (32%, 23.2 °C)	0.03
Tempra	SVA	9.5	−5.1	7.0 (90%, 24.2 °C)	0.06
Piane di Iano	SVB	7.5	−28.7	3.7 (45%, 17.9 °C)	0.26
Pagliare di Tione	SVC	8.5	16.2	7.2 (87%, 21.4 °C)	0.00
Piano Canale	SVD	8.1	27.2	7.8 (96%, 20.3 °C)	0.00
Fonte dell'Acqua	SVE	7.9	24.7	7.4 (93%, 23.4 °C)	0.00

SVA and GSC can be classified as "halfway sites", while SVC, SVD and SVE, which represent the artificial sites, could be grouped as "high water quality sites" (Table 2).

The correspondence analysis (Figure 5) resulted in a neat prevalence of Dim 1 (80.4%) along which natural sites (GSA, GSB, GSD, SVA and SVB) and artificial ones (SVC, SVD and SVE) are clearly separated, except for the GSC pond, which groups with the latter. Ammonium is distant from DO and redox potential and placed near to the ponds group; differently, pH does not seem to form any relevant group with the other variables. Mollusca, Crustacea, Hemiptera, Coleoptera and Diptera are placed in the negative portion of Dim 1, near the ponds group; on the contrary, Odonata, Trichoptera and Caudata are the main taxa placed close to the artificial sites. Tricladida, Mermithida, Haplotaxida, Julida, Dermaptera, Ephemeroptera and Hymenoptera are located between the two groups of sampling sites, in proximity to the artificial sites. Acarina, Araneae and Lepidoptera are distant from both the ponds and artificial sites groups, while Neuroptera is completely separated along the positive portion of the Dim 2 (not shown), probably due to their little contribution in diet diversity.

Microplastics were found in the stomach contents of newts inhabiting four sampling sites, namely GSC, GSD, SVD and SVE. A sample for each type of microplastic found is represented in Figure 6. Two net-like plastic fragments were found in the stomach flush of a newt sampled in GSD, measuring 3.7 mm and 3.1 mm, respectively (Figure 6a). Two textile-like fibres (out of many others found in all the sampling sites) are shown in Figure 6b, measuring 1.8 mm (blue) and 1.2 mm (red). Finally, a blue microplastic fragment (1.44 mm in size), was found in the stomach contents of a newt sampled in the SVE trough along with a little (0.1 mm) yellow fragment (Figure 6c).

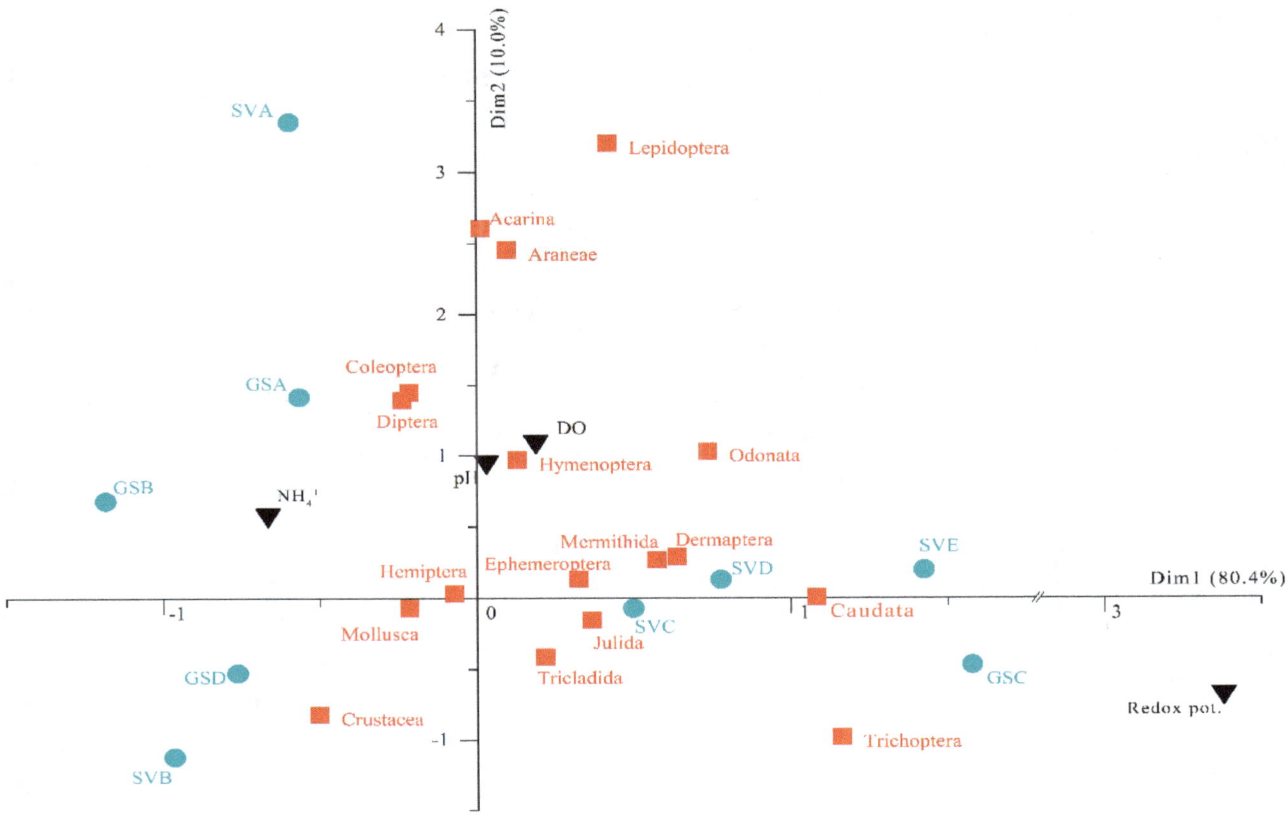

Figure 5. Correspondence analysis for both prey items and chemical parameters measured in all sampling sites. Blue circles: sampling sites, red squares: prey taxa and black triangles: physical and chemical parameters.

To characterize these items, ATR-FT-IR analyses were carried out using a Perkin Elmer (USA) Spectrum Two™ FTIR (PerkinElmer, Waltham, MA, USA) instrument equipped with the Attenuated Total Reflectance (ATR) accessory with a single bounce diamond crystal. The microplastics were analyzed at a 1 cm^{-1} resolution within the 4000–400 cm^{-1} range, with air used for the background spectrum. A consistent force was applied using the pressure monitoring system integrated with the instrument to maximize the spectrum intensity; the optimal number of scans was determined by collecting the spectra of the plastic fragments. An increase in the quality of the signal-to-noise ratio between 10 and 50 scans at a spectral resolution of 1 cm^{-1} was observed. After 50 scans, the improvement in spectral quality was less evident, thus, all samples were collected at 50 scans.

Analyses by ATR-FT-IR spectroscopy enabled the identification of microplastics by detecting absorbance bands. In Figure 7, the IR spectra of a fragment of polyethylene resulting from the analysis of the blue item pictured in Figure 6c are reported, corresponding to the bands of absorbance intrinsic of the stretching of C–H (3000–2770 cm^{-1}) and the bending of C–H (1500–1450 cm^{-1}) bonds. Additionally, ATR-FT-IR measurements enabled the detection of bands attributed to the scissoring of C–H bonds (750–700 cm^{-1}).

The IR analysis of the fragments pictured in Figure 6a shows the characteristic bands of a medical pressure-sensitive adhesive (spectrum in Supplementary Material Figure S1a). The fibres reported in Figure 6b are probably part of mountain technical clothes. The analysis of the corresponding IR spectrum (Supplementary Material Figure S1b) shows the presence of the typical bands of polyacrylic polymers, although the signal-to-noise ratio is moderately low.

Finally, the chop of Figure 6c is likely to be a part of a plastic bottle cap, considering its polyethylene composition revealed by the IR spectra.

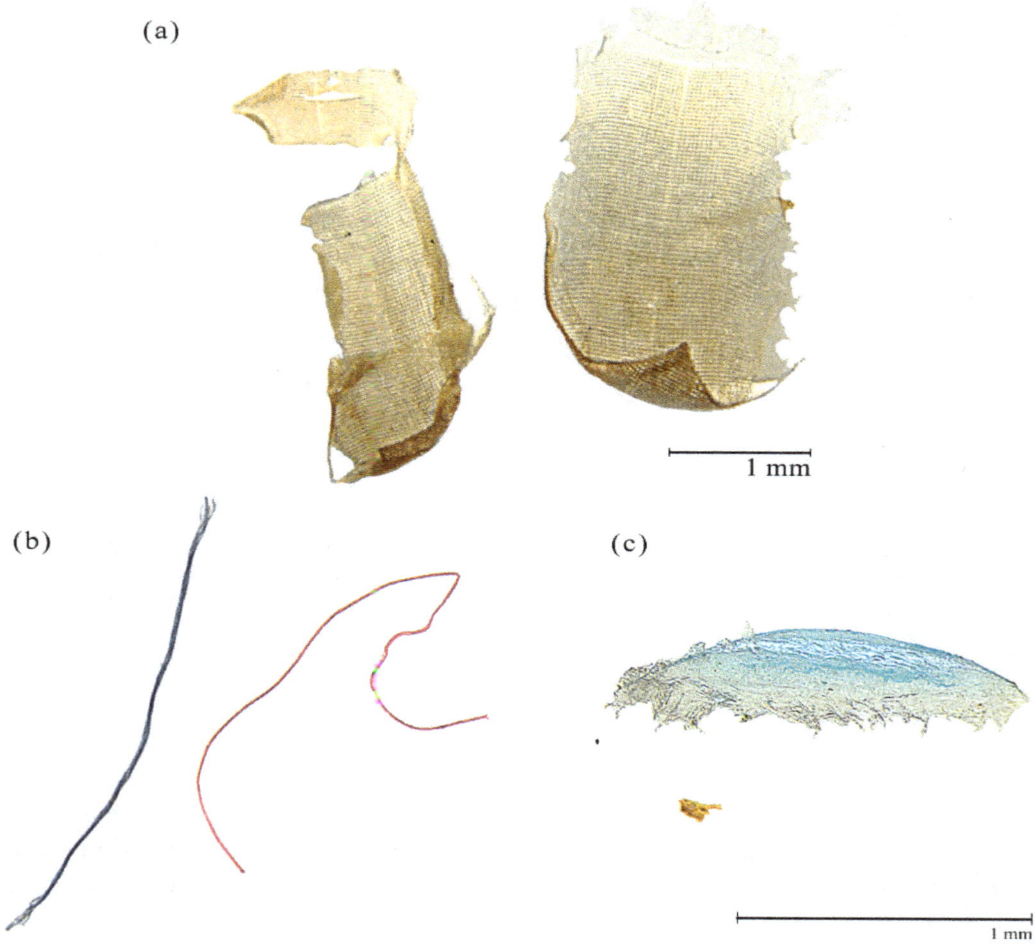

Figure 6. Microplastics found within *Triturus carnifex* stomach contents. (**a**) the biggest fragments found, probably deriving from a medical patch; (**b**) blue and red fibres, probably deriving from mountain clothes; (**c**) chop, probably deriving from a plastic bottle and very small yellow fragment (bottom) from an undeterminable origin.

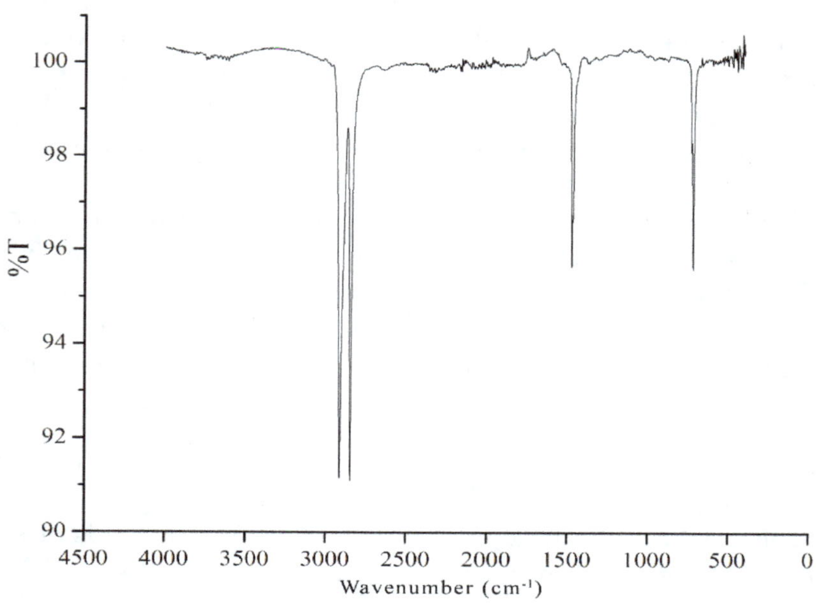

Figure 7. IR spectrum of the polyethylene chop (displayed in Figure 6c, above) found within an Italian crested newt's stomach contents.

4. Discussion

The Italian crested newt exhibits a generalist diet [44,55,56], consuming a considerable variety of preys [57]. In our samples, aquatic invertebrates presented the great majority of the collected preys; the most predated taxa were Crustacea, followed by Diptera, with a neat prevalence of small-sized prey items for both, while the others 17 taxa found in the stomach contents were less represented.

The study area has been used for centuries for grazing, and its management has been subject to national regulation for at least 100 years [58]; while in the past the studied territory was mainly pastured by sheep, nowadays cattle and horses represent the main grazers. This determines a change in grazing management, since usually only sheep are looked after by the shepherd during the whole day (personal field observations). The differences in the chemical and physical conditions observed in the sampled sites may be the direct consequence of such kind of pasture management: in fact, while sheep do not usually wade the ponds, cows and horses are frequently observed in them. Thus, ponds are subject to livestock wading and defecation, determining an alteration of water quality, while drinking troughs or tanks does not usually undergo these pressures [29,59]. This could explain the notable differences between the natural sampled sites and the artificial ones, in terms of abiotic conditions.

The correspondence analysis permitted us to segregate two main groups. Sites showing good physical and chemical parameters (i.e., the three artificial sites and the GSC pond) are associated along the positive half of Dim 1 with taxa with narrow ecological tolerances to disturbance and organic pollution (e.g., Odonata, Ephemeroptera and Trichoptera). Differently, GSD and SVB are placed on the negative portion of Dim 1 and Dim 2 and are mainly associated with Crustacea, which showed predominance in the diet of newts from these sites, leading to low prey diversity (Figures 3 and 4a). These two sites are part of a larger group comprising also GSA and GSB (as shown by the correspondence plot), in which all sites are associated with ammonium, probably reflecting the ease of livestock wading in these ponds. SVA, which shows low ammonium and high DO concentrations when compared to the aforementioned sites, is distant from this group along Dim 2; this can be explained by the low livestock density around this site (personal observation).

The controversial impact of grazing on amphibians' communities is still debated, as reviewed by Howell, Mothes, Clements, Catania, Rothermel and Searcy [29]; some studies highlighted the negative effects of grazing on amphibians, e.g., [60,61], while others documented a positive or intermediate response, due to the species-specific requirements of the target amphibians [62–64].

Based on our research, ponds, which represent typical habitats for the newts in grasslands of karst plains and show high biodiversity in uncontaminated conditions [13], are negatively influenced by livestock pressure, especially when they can be easily waded. Indeed, we found livestock contamination altering the chemical and physical water parameters in ponds (e.g., low redox potential, high ammonium and low DO); this can be related to the absence of highly sensitive taxa and the consequent low number of different predated taxa per site.

Another important aspect is that the only water supply for these ponds derives from snow melting and rain, resulting in a low rate of water replacement, especially during late summer, when the pasture pressure increases.

On the other hand, the artificial sampling sites, notwithstanding their small extent, present a high number of prey taxa, probably due to the constant water flow, which positively contributes to the high quality of water in these sites. This is further confirmed by the presence of some taxa which exhibit a reduced tolerance to the disturbances found in ponds, as also shown by the correspondence analysis.

Fossa di Paganica (GSC) is the only pond which does not follow this trend, showing conditions similar to the artificial sampling sites; in fact, it presents the highest number of prey taxa among all sites, and good chemical parameters. This could be explained by the large dimensions of the pond, which allow for the buffering of the negative effects of cattle and horses wading and defecating. Furthermore, the diffused presence of macrophytes found in GSC could favour macroinvertebrate diversity, as already observed in other karst ponds [65]. This is reflected by the high number of prey taxa found in this site, which could anyway be an underestimation of the total macroinvertebrate diversity.

Notwithstanding the evidence of livestock contamination of ponds and the corresponding influence on the number of prey taxa, the analysis of the Shannon H′ variations among the sites showed that prey diversification for each newt is occasionally high if compared to the number of prey taxa per sampling site (e.g., GSB and SVA, see Figure 4a).

Previous studies showed that the low diversity of available prey may hamper larval development, due to low intake of dietary antioxidants [66,67] leading to reduced immune response capability and mass growth [68]. Thus, we speculate that newts, also considering their generalist diet, may be led to diversify their preys as much as possible, to integrate a possible lack of nutrients and avoid the negative effects mentioned before.

Furthermore, amphibian feeding is also conditioned by interspecific competition [69], which can be changed by the noticeable disequilibria found in most of the sampled sites, considering the absence of invertebrate predators (sometimes also acting as preys), such as Coleoptera, Hemiptera and Odonata.

In this work, we present the first European record of microplastics in amphibians' stomach contents and the first one ever recorded for Caudata.

While several studies about microplastics in marine ecosystems have been published in recent decades [70], field studies examining their presence in freshwater ecosystems started only recently [71]. Despite both Caudata and Anuran being among the most threatened animal groups worldwide, the only papers assessing the presence of microplastics in gut content referred to Anuran species [71,72]. Moreover, the previous records of microplastics in Anuran stomach contents were collected in human-altered freshwater habitats, such as farmlands surrounding river deltas [72] and plantations [71]; on the contrary, we found evidences of microplastic ingestion in scarcely populated, high-elevation territories, even covered by protected areas.

The chemical composition of the collected microplastics, emerging from the performed ATR-FT-IR spectroscopy analyses, shows that they are associated to human activities carried out in mountain environments, such as pastoralism and tourism; indeed, the majority of microplastic items are textile fibres (Figure 6b). The presence of these fibres in sampling sites may be due to waste displacement by the wind from the neighbouring roads and mountain retreats, or to livestock-related items, such as horse bridles or cow tags.

While the effects of microplastic ingestion by amphibians have been already assessed in laboratory studies with contrasting results [73,74], there is no information about the potential negative impacts on amphibians' communities in freshwater habitats.

5. Conclusions

High-altitude environments are nowadays proven to be very sensitive to ongoing environmental changes; the alteration of their freshwater habitats leads to rapid and dramatic loss of biodiversity. While, on one hand, pasture has shaped some peculiar environments for centuries, on the other hand it may represent a serious threat if not correctly managed. Human presence has, in this case, a dual role; it may cause a loss of diversity (e.g., unmanaged pasture) but it may also lead to the persistence of amphibian populations, because of the construction of artificial watering points, which currently host highly diverse communities. This is the very first assessment of the livestock-related detrimental effects on newts' diets in mountain karstic ponds; ongoing targeted sampling campaigns will enrich the knowledge about these threats, which widely affect mountain ecosystems.

The unexpected discovery of microplastics in many stomach contents reinforces the alarm about the ubiquitous plastic pollution of natural habitats, even in high-altitude protected territories; it is thus worth conducting future targeted research about this emerging issue to properly adapt conservation strategies.

Author Contributions: Conceptualization, M.I. and M.B.; methodology, M.I., G.C. and F.C.; software, M.I., G.C. and P.D.; formal analysis, G.C., C.M. and P.D.; data curation, G.C. and C.M.; chemical analyses, F.R. and F.D.D.; writing—original draft preparation, all authors; writing—review and editing, all authors; funding acquisition, M.I. and M.B. All authors have read and agreed to the published version of the manuscript.

Acknowledgments: We are particularly thankful to the Scientific Department of Gran Sasso—Monti della Laga. National Park and to the Scientific and Naturalistic Office of the Sirente-Velino Regional Park for the technical support provided during the collection of some field data (permissions n. 0000219/15 UT-RAU-SCNZ448 and N. VII-02-02-P6213).

References

1. Rogora, M.; Frate, L.; Carranza, M.; Freppaz, M.; Stanisci, A.; Bertani, I.; Bottarin, R.; Brambilla, A.; Canullo, R.; Carbognani, M.; et al. Assessment of climate change effects on mountain ecosystems through a cross-site analysis in the Alps and Apennines. *Sci. Total Environ.* **2018**, *624*, 1429–1442. [CrossRef]
2. Dirnböck, T.; Essl, F.; Rabitsch, W. Disproportional risk for habitat loss of high-altitude endemic species under climate change. *Glob. Chang. Biol.* **2011**, *17*, 990–996. [CrossRef]
3. Brambilla, M.; Pedrini, P.; Rolando, A.; Chamberlain, D.E. Climate change will increase the potential conflict between skiing and high-elevation bird species in the Alps. *J. Biogeogr.* **2016**, *43*, 2299–2309. [CrossRef]
4. Di Musciano, M.; Carranza, M.; Frate, L.; Di Cecco, V.; Di Martino, L.; Frattaroli, A.; Stanisci, A. Distribution of plant species and dispersal traits along environmental gradients in central Mediterranean summits. *Diversity* **2018**, *10*, 58. [CrossRef]
5. Cerasoli, F.; Iannella, M.; Biondi, M. Between the hammer and the anvil: How the combined effect of global warming and the non-native common slider could threaten the European pond turtle. *Manag. Biol. Invasions* **2019**, *10*, 428. [CrossRef]
6. Chamberlain, D.; Brambilla, M.; Caprio, E.; Pedrini, P.; Rolando, A. Alpine bird distributions along elevation gradients: The consistency of climate and habitat effects across geographic regions. *Oecologia* **2016**, *181*, 1139–1150. [CrossRef] [PubMed]
7. Cerasoli, F.; Thuiller, W.; Guéguen, M.; Renaud, J.; D'Alessandro, P.; Biondi, M. The role of climate and biotic factors in shaping current distributions and potential future shifts of European Neocrepidodera (Coleoptera, Chrysomelidae). *Insect Divers. Conserv.* **2019**. [CrossRef]
8. Brunetti, M.; Magoga, G.; Iannella, M.; Biondi, M.; Montagna, M. Phylogeography and species distribution modelling of *Cryptocephalus barii* (Coleoptera: Chrysomelidae): Is this alpine endemic species close to extinction? *ZooKeys* **2019**, *856*, 3. [CrossRef] [PubMed]
9. Wiens, J.J. Climate-Related local extinctions are already widespread among plant and animal species. *PLoS Biol.* **2016**, *14*. [CrossRef] [PubMed]
10. Oertli, B.; Biggs, J.; Céréghino, R.; Grillas, P.; Joly, P.; Lachavanne, J.B. Conservation and monitoring of pond biodiversity: Introduction. *Aquat. Conserv. Mar. Freshw. Ecosyst.* **2005**, *15*, 535–540. [CrossRef]
11. Niedrist, G.; Psenner, R.; Sommaruga, R. Climate warming increases vertical and seasonal water temperature differences and inter-annual variability in a mountain lake. *Clim. Chang.* **2018**, *151*, 473–490. [CrossRef]
12. Sommaruga, R. The role of solar UV radiation in the ecology of alpine lakes. *J. Photochem. Photobiol. B Biol.* **2001**, *62*, 35–42. [CrossRef]
13. Hinden, H.; Oertli, B.; Menetrey, N.; Sager, L.; Lachavanne, J.B. Alpine pond biodiversity: What are the related environmental variables? *Aquat. Conserv. Mar. Freshw. Ecosyst.* **2005**, *15*, 613–624. [CrossRef]
14. Miaud, C.; Guyetant, R.; Faber, H. Age, size, and growth of the alpine newt, *Triturus alpestris* (Urodela: Salamandridae), at high altitude and a review of life-history trait variation throughout its range. *Herpetologica* **2000**, *56*, 135–144.
15. Alonso, M.; Garcia-De-Lomas, J. Systematics and ecology of *Linderiella baetica* n. sp. (Crustacea, Branchiopoda, Anostraca, Chirocephalidae), a new species from southern Spain. *Zoosystema* **2009**, *31*, 807–828. [CrossRef]
16. Mushet, D.M.; Euliss Jr, N.H.; Chen, Y.; Stockwell, C.A. Complex spatial dynamics maintain northern leopard frog (*Lithobates pipiens*) genetic diversity in a temporally varying landscape. *Herpetol. Conserv. Biol.* **2013**, *8*, 163–175.
17. Lund, J.O.; Wissinger, S.A.; Peckarsky, B.L. Caddisfly behavioral responses to drying cues in temporary ponds: Implications for effects of climate change. *Freshw. Sci.* **2016**, *35*, 619–630. [CrossRef]

18. Catorci, A.; Gatti, R.; Vitanzi, A. Relationship between phenology and above-ground phytomass in a grassland community in central Italy. In *Nature Conservation. Environmental Science and Engineering (Environmental Science)*; Gafta, D., Akeroyd, J., Eds.; Springer: Heidelberg, Germany, 2006; pp. 309–327.
19. Filibeck, G.; Cancellieri, L.; Sperandii, M.G.; Belonovskaya, E.; Sobolev, N.; Tsarevskaya, N.; Becker, T.; Berastegi, A.; Bückle, C.; Che, R. Biodiversity patterns of dry grasslands in the Central Apennines (Italy) along a precipitation gradient: Experiences from the 10th EDGG Field Workshop. *Bull. Eurasian Dry Grassl. Group* **2018**, *36*, 25–41.
20. Primi, R.; Filibeck, G.; Amici, A.; Bückle, C.; Cancellieri, L.; Di Filippo, A.; Gentile, C.; Guglielmino, A.; Latini, R.; Mancini, L.D. From Landsat to leafhoppers: A multidisciplinary approach for sustainable stocking assessment and ecological monitoring in mountain grasslands. *Agric. Ecosyst. Environ.* **2016**, *234*, 118–133. [CrossRef]
21. Ruggiero, A.; Solimini, A.G.; Carchini, G. Limnological aspects of an Apennine shallow lake. *Ann. Limnol. Int. J. Limnol.* **2004**, *40*, 89–99. [CrossRef]
22. Solimini, A.G.; Bazzanti, M.; Ruggiero, A.; Carchini, G. Developing a multimetric index of ecological integrity based on macroinvertebrates of mountain ponds in central Italy. In *Pond Conservation in Europe*; Oertli, B., Céréghino, R., Biggs, J., Declerck, S., Hull, A., Miracle, M.R., Eds.; Springer: Dordrecht, The Netherlands, 2007; pp. 109–123.
23. Calhoun, A.J.; Mushet, D.M.; Bell, K.P.; Boix, D.; Fitzsimons, J.A.; Isselin-Nondedeu, F. Temporary wetlands: Challenges and solutions to conserving a 'disappearing'ecosystem. *Biol. Conserv.* **2017**, *211*, 3–11. [CrossRef]
24. Hunter, M.L., Jr.; Acuña, V.; Bauer, D.M.; Bell, K.P.; Calhoun, A.J.; Felipe-Lucia, M.R.; Fitzsimons, J.A.; González, E.; Kinnison, M.; Lindenmayer, D. Conserving small natural features with large ecological roles: A synthetic overview. *Biol. Conserv.* **2017**, *211*, 88–95. [CrossRef]
25. Ruggiero, A.; Solimini, A.; Carchini, G. The alternative stable state concept and the management of Apennine mountain ponds. *Aquat. Conserv. Mar. Freshw. Ecosyst.* **2005**, *15*, 625–634. [CrossRef]
26. Solimini, A.G.; Ruggiero, A.; Anello, M.; Mutschlechner, A.; Carchini, G. The benthic community structure in mountain ponds affected by livestock watering in nature reserves of Central Italy. *Int. Ver. Theor. Angew. Limnol. Verh.* **2000**, *27*, 501–505. [CrossRef]
27. Ruggiero, A.; Solimini, A.; Carchini, G. Pratica dell'alpeggio e conservazione ambientale negli stagni montani dell'Appennino Centrale. In Proceedings of the Atti del Convegno Nazionale Conservazione Dell'ambiente e Rischio Idrogeologico, Perugia, Italy, 11–12 December 2002; pp. 687–694.
28. Wissinger, S.A.; Oertli, B.; Rosset, V. Invertebrate communities of alpine ponds. In *Invertebrates in Freshwater Wetlands*; Batzer, D., Boix, D., Eds.; Springer: Cham, Switzerland, 2016; pp. 55–103.
29. Howell, H.J.; Mothes, C.C.; Clements, S.L.; Catania, S.V.; Rothermel, B.B.; Searcy, C.A. Amphibian responses to livestock use of wetlands: New empirical data and a global review. *Ecol. Appl.* **2019**. [CrossRef]
30. Iannella, M.; Cerasoli, F.; Biondi, M. Unraveling climate influences on the distribution of the parapatric newts *Lissotriton vulgaris meridionalis* and *L. italicus*. *Front. Zool.* **2017**, *14*, 55. [CrossRef]
31. Iannella, M.; D'Alessandro, P.; Biondi, M. Evidences for a shared history for spectacled salamanders, haplotypes and climate. *Sci. Rep.* **2018**, *8*, 16507. [CrossRef]
32. Hayes, T.B.; Stuart, A.A.; Mendoza, M.; Collins, A.; Noriega, N.; Vonk, A.; Johnston, G.; Liu, R.; Kpodzo, D. Characterization of atrazine-induced gonadal malformations in African clawed frogs (*Xenopus laevis*) and comparisons with effects of an androgen antagonist (cyproterone acetate) and exogenous estrogen (17β-estradiol): Support for the demasculinization/feminization hypothesis. *Environ. Health Perspect* **2006**, *114*, 134–141.
33. Buck, J.C.; Scheessele, E.A.; Relyea, R.A.; Blaustein, A.R. The effects of multiple stressors on wetland communities: Pesticides, pathogens and competing amphibians. *Freshw. Biol.* **2012**, *57*, 61–73. [CrossRef]
34. Patar, A.; Giri, A.; Boro, F.; Bhuyan, K.; Singha, U.; Giri, S. Cadmium pollution and amphibians–Studies in tadpoles of *Rana limnocharis*. *Chemosphere* **2016**, *144*, 1043–1049. [CrossRef]
35. Ferreira, M.; Beja, P. Mediterranean amphibians and the loss of temporary ponds: Are there alternative breeding habitats? *Biol. Conserv.* **2013**, *165*, 179–186. [CrossRef]
36. Ficetola, G.F.; Rondinini, C.; Bonardi, A.; Baisero, D.; Padoa-Schioppa, E. Habitat availability for amphibians and extinction threat: A global analysis. *Divers. Distrib.* **2015**, *21*, 302–311. [CrossRef]
37. Stuart, S.N.; Chanson, J.S.; Cox, N.A.; Young, B.E.; Rodrigues, A.S.; Fischman, D.L.; Waller, R.W. Status and trends of amphibian declines and extinctions worldwide. *Science* **2004**, *306*, 1783–1786. [CrossRef] [PubMed]

38. May, R.M. Ecological science and tomorrow's world. *Philos. Trans. R. Soc. B Biol. Sci.* **2010**, *365*, 41–47. [CrossRef]
39. Best, M.L.; Welsh, J.; Hartwell, H. The trophic role of a forest salamander: Impacts on invertebrates, leaf litter retention, and the humification process. *Ecosphere* **2014**, *5*, 1–19. [CrossRef]
40. Davic, R.D.; Welsh, H.H. On the ecological roles of salamanders. *Annu. Rev. Ecol. Evol. Syst.* **2004**, *35*, 405–434. [CrossRef]
41. Salvidio, S.; Costa, A.; Crovetto, F. Individual trophic specialization in the Alpine newt increases with increasing resource diversity. *Ann. Zool. Fenn.* **2019**, *56*, 17–24. [CrossRef]
42. Vignoli, L.; Bombi, P.; D'Amen, M.; Bologna, M.A. Seasonal variation in the trophic niche of a heterochronic population of *Triturus alpestris apuanus* from the south-western Alps. *Herpetol. J.* **2007**, *17*, 183–191.
43. Schabetsberger, R.; Jehle, R.; Maletzky, A.; Pesta, J.; Sztatecsny, M. Delineation of terrestrial reserves for amphibians: Post-breeding migrations of Italian crested newts (*Triturus c. carnifex*) at high altitude. *Biol. Conserv.* **2004**, *117*, 95–104. [CrossRef]
44. Vanni, S.; Andreone, F.; Tripepi, S. *Istituto Superiore per La Protezione e La Ricerca Ambientale*; Anfibi d'Italia. Lanza, B., Nistri, A., Vanni, S., Eds.; Ministero dell'Ambiente e della Tutela del Territorio e del Mare: Rome, Italy, 2007; Volume 29.
45. Naviglio, L. Aspetti naturalistici del Lago Pantaniello. *Nat. Montagna.* **1984**, *31*, 49–57.
46. Ferri, V.; Di Tizio, L.; Pellegrini, M. *Atlante Degli Anfibi d'Abruzzo*; Ianieri–Talea: Pescara, Italy, 2007; p. 199.
47. Andreone, F.; Corti, C.; Ficetola, F.; Razzetti, E.; Romano, A.; Sindaco, R. *Triturus carnifex*. The IUCN Red List of Threatened Species. 2009: e.T59474A11947714. Available online: https://www.iucnredlist.org/species/59474/11947714 (accessed on 5 December 2018).
48. Beebee, T.J.C.; Griffiths, R.A. The amphibian decline crisis: A watershed for conservation biology? *Biol. Conserv.* **2005**, *125*, 271–285. [CrossRef]
49. D'Amen, M.; Bombi, P. Global warming and biodiversity: Evidence of climate-linked amphibian declines in Italy. *Biol. Conserv.* **2009**, *142*, 3060–3067. [CrossRef]
50. Collins, J.P.; Crump, M.L.; Lovejoy Iii, T.E. *Extinction in Our Times: Global Amphibian Decline*; Oxford University Press: New York, NY, USA, 2009.
51. Solé, M.; Rödder, D. Dietary assessments of adult amphibians. In *Amphibian Ecology and Conservation: A Handbook of Techniques*; Oxford University Press: New York, NY, USA, 2010; pp. 167–184.
52. Oksanen, J.; Blanchet, F.G.; Friendly, M.; Kindt, R.; Legendre, P.; McGlinn, D.; Minchin, P.R.; O'Hara, R.; Simpson, G.L.; Solymos, P. 2019. Vegan: Community Ecology Package. R package version 2.5-4. Available online: https://cran.r-project.org/web/packages/vegan (accessed on 1 September 2019).
53. R Core Team. *R: A Language and Environment for Statistical Computing*; R Foundation for Statistical Computing: Vienna, Austria, 2017; Available online: http://www.R-project.org (accessed on 15 June 2017).
54. Nenadic, O.; Greenacre, M. Correspondence Analysis in R, with two-and three-dimensional graphics: The ca package. *J. Stat. Softw.* **2007**, *20*, 1–13.
55. Costa, A.; Salvidio, S.; Posillico, M.; Matteucci, G.; De Cinti, B.; Romano, A. Generalisation within specialization: Inter-individual diet variation in the only specialized salamander in the world. *Sci. Rep.* **2015**, *5*, 13260. [CrossRef]
56. Vignoli, L.; Bologna, M.A. Dietary patterns and overlap in an amphibian assemblage at a pond in Mediterranean central Italy. *Vie Milieu-Life Environ.* **2009**, *59*, 47–57.
57. Fasola, M.; Canova, L. Feeding habits of *Triturus vulgaris*, *T. cristatus* and *T. alpestris* (Amphibia, Urodela) in the northern Apennines (Italy). *Ital. J. Zool.* **1992**, *59*, 273–280.
58. Regio Decreto Legislativo. *Riordinamento E Riforma Della Legislazione in Materia di Boschi E di Terreni Montani*; Italy, Ed.; NORMATTIVA: Rome, Italy, 1923.
59. Buono, V.; Bissattini, A.M.; Vignoli, L. Can a cow save a newt? The role of cattle drinking troughs in amphibian conservation. *Aquat. Conserv. Mar. Freshw. Ecosyst.* **2019**, *29*, 964–975. [CrossRef]
60. Schmutzer, A.C.; Gray, M.J.; Burton, E.C.; Miller, D.L. Impacts of cattle on amphibian larvae and the aquatic environment. *Freshw. Biol.* **2008**, *53*, 2613–2625. [CrossRef]
61. Hoverman, J.T.; Gray, M.J.; Miller, D.L.; Haislip, N.A. Widespread occurrence of ranavirus in pond-breeding amphibian populations. *Eco. Health* **2012**, *9*, 36–48. [CrossRef]
62. Badillo-Saldaña, L.M.; Ramírez-Bautista, A.; Wilson, L.D. Effects of establishment of grazing areas on

diversity of amphibian communities in tropical evergreen forests and mountain cloud forests of the Sierra Madre Oriental. *Rev. Mex. Biodivers.* **2016**, *87*, 133–139. [CrossRef]

63. Moreira, L.F.; Moura, R.G.; Maltchik, L. Stop and ask for directions: Factors affecting anuran detection and occupancy in Pampa farmland ponds. *Ecol. Res.* **2016**, *31*, 65–74. [CrossRef]
64. Howard, A.; Munger, J.C. *Effects of Livestock Grazing on the Invertebrate Prey Base and on the Survival and Growth of Larvae of the Columbia Spotted Frog. Rana Luteiventris*; Bureau of Land Management, Idaho State Office: Boise, ID, USA, 2003.
65. Zelnik, I.; Gregorič, N.; Tratnik, A. Diversity of macroinvertebrates positively correlates with diversity of macrophytes in karst ponds. *Ecol. Eng.* **2018**, *117*, 96–103. [CrossRef]
66. Byrne, P.G.; Silla, A.J. Testing the effect of dietary carotenoids on larval survival, growth and development in the critically endangered southern corroboree frog. *Zoo. Biol.* **2017**, *36*, 161–169. [CrossRef] [PubMed]
67. Clugston, R.D.; Blaner, W.S. Vitamin A (retinoid) metabolism and actions: What we know and what we need to know about amphibians. *Zoo. Biol.* **2014**, *33*, 527–535. [CrossRef] [PubMed]
68. Szuroczki, D.; Koprivnikar, J.; Baker, R.L. Dietary antioxidants enhance immunocompetence in larval amphibians. *Comp. Biochem. Physiol. Part A Mol. Integr. Physiol.* **2016**, *201*, 182–188. [CrossRef]
69. Arribas, R.; Díaz-Paniagua, C.; Caut, S.; Gomez-Mestre, I. Stable isotopes reveal trophic partitioning and trophic plasticity of a larval amphibian guild. *PLoS ONE* **2015**, *10*. [CrossRef]
70. Hidalgo-Ruz, V.; Gutow, L.; Thompson, R.C.; Thiel, M. Microplastics in the marine environment: A review of the methods used for identification and quantification. *Environ. Sci. Technol.* **2012**, *46*, 3060–3075. [CrossRef]
71. Döring, B.; Mecke, S.; Kieckbusch, M.; O'Shea, M.; Kaiser, H. Food spectrum analysis of the Asian toad, *Duttaphrynus melanostictus* (Schneider, 1799) (Anura: Bufonidae), from Timor Island, Wallacea. *J. Nat. Hist.* **2017**, *51*, 607–623. [CrossRef]
72. Hu, L.; Chernick, M.; Hinton, D.E.; Shi, H. Microplastics in small waterbodies and tadpoles from Yangtze River Delta, China. *Environ. Sci. Technol.* **2018**, *52*, 8885–8893. [CrossRef]
73. De Felice, B.; Bacchetta, R.; Santo, N.; Tremolada, P.; Parolini, M. Polystyrene microplastics did not affect body growth and swimming activity in *Xenopus laevis* tadpoles. *Environ. Sci. Pollut. Res.* **2018**, *25*, 34644–34651. [CrossRef] [PubMed]
74. Da Costa Araújo, A.P.; de Melo, N.F.S.; de Oliveira Junior, A.G.; Rodrigues, F.P.; Fernandes, T.; de Andrade Vieira, J.E.; Rocha, T.L.; Malafaia, G. How much are microplastics harmful to the health of amphibians? A study with pristine polyethylene microplastics and *Physalaemus cuvieri*. *J. Hazard. Mater.* **2020**, *382*, 121066. [CrossRef] [PubMed]

10

Amendment of Husk Biochar on Accumulation and Chemical Form of Cadmium in Lettuce and Pak-Choi Grown in Contaminated Soil

Kuei-San Chen [1], Chun-Yu Pai [2] and Hung-Yu Lai [1,3,*]

1 Department of Soil and Environmental Sciences, National Chung Hsing University, Taichung 40227, Taiwan; s104039023@smail.nchu.edu.tw
2 Department of Agronomy, National Taiwan University, Taipei 10617, Taiwan; r08628311@ntu.edu.tw
3 Innovation and Development Center of Sustainable Agriculture, National Chung Hsing University, Taichung 40227, Taiwan
* Correspondence: soil.lai@nchu.edu.tw

Abstract: (1) Background: Cadmium (Cd) accumulated in vegetables not only affects their growth but can also enter the human body via food chains and lead to various illnesses. Plants can decrease the toxicity by changing the chemical forms of Cd, which include inorganic (F_E), water-soluble (F_W), pectate- and protein-integrated (F_{NaCl}), undissolved phosphate (F_{HAc}), oxalate (F_{HCl}), and residual forms (F_R). Among them, F_E and F_W chemical forms show higher mobility to translocate upward from roots to shoots compared with the others. (2) Methods: Different varieties or cultivars of lettuce and pak-choi were grown in Cd-contaminated soils amended with husk biochar (BC) to replenish nitrogen to the recommended amount and also to raise the soil pH value. (3) Results: More than 73% of the accumulated Cd in the edible organs was compartmentalized in F_E chemical form in both leafy vegetables regardless of treatments. In comparison with control, the application of BC decreased the Cd concentrations and bioconcentration factors in the roots and shoots of two leafy vegetables at different growth periods in general. The chemical form and bioaccessible fraction of Cd in the edible blanching tissues were used to calculate the risk of oral intake. The vegetable-induced hazard quotients of lettuce and pak-choi were acceptable, except for pak-choi grown in control without applying BC.

Keywords: bioaccessibility; cadmium; chemical form; husk biochar; risk assessment

---

1. Introduction

Many illegal factories arbitrarily discharge wastewater into irrigation channels, where farmers unknowingly draw contaminated water for paddy rice cultivation, causing many farmlands in Taiwan to be contaminated with potential toxic elements (PTEs) [1]. If the total concentration of PTEs in the soil is above the control standard stipulated by the Soil and Groundwater Pollution Remediation Act (SGWPR Act) of Taiwan, crops should be removed to prevent human consumption [2]. However, when the total concentration of PTEs in the soil is slightly above the monitoring standard and below the control standard, crop safety must be emphasized, especially concerning the pathway of oral intake of vegetables.

Cadmium (Cd), a non-essential PTE, is one of the most toxic substances to plants and animals because it is easily transferrable from soil to plant [3,4]. The accumulation of Cd in the edible tissues of crops poses serious concerns to the trophic risk of Cd transfer along a food chain [5]. Because of the carcinogenic and adverse effects of Cd on biological processes, the International Agency of Research on Cancer and the United States Environmental Protection Agency classified Cd into Group 1 and

Class B, respectively [6]. Excessive concentration of Cd damaged mesophyll cells and epidermal cells in wheats [7] and resulted in the reduction of photosynthesis and chlorophyll content [8,9]. Once the roots uptake PTEs, plants can decrease the toxicity of PTEs by cell wall deposition and vacuolar compartmentation [10–13]. Cell walls are mainly composed of pectin, protein, cellulose, and semi-cellulose [14], and functional groups in cell walls could restrict PTEs into the cytoplasm, which prevents the protoplast from being poisoned [15].

The chemical forms can be of help in understanding the tolerance and mechanisms of PTE detoxification in plants. These chemical forms include inorganic (F_E), water-soluble (F_W), pectate- and protein-integrated (F_{NaCl}), undissolved phosphate (F_{HAc}), oxalate (F_{HCl}), and residual (F_R) forms [16,17]. Among them, F_E and F_W have high mobility and thus are easily translocated to other plant organs [18]. Many studies revealed that F_{NaCl} was related to adjusting Cd stress in ramie, cabbage and American pokeweed [18–20]. There was also a positive linear correlation between Cd concentration in the F_{NaCl} and F_{HAc} forms, and that in the shoot, which revealed that chemical forms can be used to predict the accumulation of Cd in the shoot [21].

Many soil amendments (e.g., compost, phosphate fertilizer, and organic matter) have been used for remediation of PTE-contaminated soil [22,23]. For instance, Liu et al. [24] found that the co-application of vermicompost and selenium could alleviate the Cd accumulation in rice. The presence of organic matter redistributed the PTE to less-available forms and thus ameliorated its toxicity to plants [25]. The solubility of PTE is high in an acidic pH, whereas addition of lime and biochar (BC) can increase the soil pH and thus can decrease the bioavailability [26–28]. Biochars are porous, low density, and carbon-rich solid products from the pyrolysis of waste biomass. Xu et al. [29] revealed that adding BCs to contaminated paddy soil improved the transformation of Cd from the acid-soluble fraction to the oxidizing and residual fractions. BCs also had a great capacity to reduce the percentage of inorganic and water-soluble fractions in lettuce roots grown in Cd-contaminated soil [30]. Additionally, BCs mainly increased undissolved cadmium phosphate and thus increased the Cd accumulation in pokeweed root [31].

In this study, different varieties or cultivars of lettuce and pak-choi were grown in soil dominantly contaminated with Cd. Husk BC was applied to replenish nitrogen to the recommended amount and to decrease the soil acidity. The objective of this study was to understand the effect of the application of husk BC on the accumulation, translocation, and chemical forms of Cd in two leafy vegetables. In addition, the vegetable-induced hazard quotient (HQ_v) was calculated via the chemical form and artificial digestant extractable concentration of Cd in the blanched edible parts to assess the risk from oral intake.

2. Materials and Methods

Experimental soil samples were collected from the surface layer (0–30 cm) of a Cd-contaminated site in central Taiwan because the soil Cd resulted from the use of irrigation water mainly accumulated in the top 30 cm [32]. Soil samples were air-dried, ground, passed through a 10-mesh stainless steel sieve, and then homogenized before analysis. Their basic properties were then analyzed, including pH [33], electrical conductivity (EC) [34], content of organic carbon (OC) [35], cation exchange capacity (CEC) [36], texture [37], and total concentration of Cd, chromium (Cr), nickel (Ni), and zinc (Zn) [38]. Other soil samples were homogenized and used for a pot experiment. The total concentrations of nitrogen, phosphorus, and potassium in the BC were analyzed in accordance with the methods described by Bremner [39], Kuo [40], and Helmke and Sparks [41], respectively. Two treatments used in this study were CK (control without applying biochar) and BC, which applied husk BC to replenish nitrogen to the recommended amount. For lettuce and pak-choi, 46,920 and 98,000 kg/ha of BC was applied, respectively. In total, 1.0 kg of soil samples or mixture was added to each pot, and seeds of two pak-choi—*Brassica chinensis* L. var. Chinensis (PCC) and *Brassica chinensis* L. cv. Wrinkled leaf (PCW)—and two lettuce—*Lactuca sative* L. cv. Chinese (LSC) and *Lactuca sative* L. var. Sative (LSS)—were sown. The above vegetables were selected because of their importance in human diets [42] and they are the most commonly consumed leafy vegetables in Taiwan [43]. The pots were translocated to a phytotron (25.1 ± 0.4 °C, relative humidity 61.2 ± 5.7%). Soil moistures were determined every two to three days, and remained at 60–80% of water-holding capacity by replenished deionized water.

All leafy vegetables were harvested 35, 42, and 49 days (D35, D42, and D49) after sowing and divided into roots and shoots. At this time, the shoot heights were measured and the chlorophyll content of the largest extended leaf was determined using a Konica Minolta SPAD-502 and recorded as SPAD (soil plant analyzer development) readings. Fresh plant tissues were rinsed with tap water and then deionized water. To remove the adsorbed Cd, the roots were soaked in 20 mM of Na_2-EDTA for 15 min. The Cd compartmentalized in different chemical forms in the shoots was analyzed using a sequential extraction in accordance with the methods described by Lai [21]. Six chemical forms were extracted in the following sequence, with the corresponding agents: inorganic Cd (F_E) extracted by 80% alcohol, water-soluble Cd (F_W) extracted by deionized water, pectate- and protein-integrated Cd (F_{NaCl}) extracted by 1 M NaCl, undissolved Cd phosphate (F_{HAc}) extracted by 2% CH_3COOH, Cd oxalate (F_{HCl}) extracted by 0.6 M HCl, and residual Cd (F_R) digested with aqua regia. All the other plant tissues were oven-dried at 65 °C for 72 h and then weighed (dry weight) before grinding (Rong Tsong Precision Tech. Co., Taichung, Taiwan), and then digested with $HNO_3/HClO_4$ (v/v = 3/1). The Cd concentration in the digestant was determined using a flame atomic absorption spectrophotometer (FAAS, Perkin Elmer AAnalyst 200, Waltham, MA, USA).

Soil samples of different treatments were air-dried, ground, passed through a 10-mesh stainless steel sieve, and then homogenized before analysis. The available concentration of Cd in the soils was extracted with 0.1 N HCl [44] and 0.05 N EDTA [45]. A BCR sequestration extraction based on the European Community Bureau of Reference [46] was also conducted to understand the Cd distribution in the different fractions. These four fractions include the exchangeable fraction (F-I), organic matter-bounding fraction (F-II), oxide-bounding fraction (F-III), and residual fraction (F-IV).

Statistical analysis was performed using the Statistical Package for the Social Sciences (SPSS, Armonk, NY, USA). Analysis of variance (ANOVA) was used to test the effect of different treatments on the different growth exhibitions and soil properties. The least significant difference (LSD) test was used to compare the significant differences of pH, EC, available Cd concentration, Cd distribution in the different fractions, four growth exhibitions, Cd concentration in the roots and shoots, bioconcentration factor (BCF = ratio of root concentration to soil concentration), and transfer factor (TF = shoot concentration to root concentration) between treatments ($p < 0.05$).

3. Results

3.1. Soil Properties

The total concentrations of Cd, Cr, Ni, and Zn were in the ranges of 10–13, 51–60, 126–146, and 61–68 mg/kg, respectively. For the four PTEs analyzed, only Cd and Ni were beyond the control standard (Cd 5 mg/kg, Ni 130 mg/kg) of farmland based on the SGWPR Act of Taiwan and could cause the samples to be regarded as Cd- and Ni-contaminated soils. However, only Cd accumulation in the lettuce and pak-choi was discussed in this paper because of following reasons. First, the total concentration of Cd was beyond not only the Canadian soil quality guideline (1.4 mg/kg) but also the Soil Contamination Warming Limit (4 mg/kg) and Counterplan Limit (12 mg/kg) of Korea for agricultural lands. Moreover, Cd has higher bioaccessibility in comparison with Ni [47] and is also the only one PTE regulated by the Office of the Journal of the European Union [48] among four soil PTEs analyzed in this study. The BC used in this study had very strong alkalinity and high EC. Its pH and EC were 10.54 ± 0.06 and 1.82 ± 0.08 dS/m, respectively. Total concentrations of nitrogen, phosphorus, and potassium in the BC were 0.24 ± 0.05%, 0.61 ± 0.10%, and 1.59 ± 0.09%, respectively. For the four PTEs analyzed in this study, the BC only had 1.40 ± 1.99 and 10.42 ± 2.25 mg/kg of Cd and Zn, respectively, and the total concentrations of Cr and Ni were not detectable.

The application of BC was able to significantly increase the soil pH and EC in comparison with CK. The BC treatments raised or significantly ($p < 0.05$) raised pH from 6.6–6.8 and 6.9–7.0 to 6.8–7.1 and 7.6–8.0 for the lettuce and pak-choi cultivated soils, respectively (Table 1). Because of the high EC of BC, the soil EC also increased or significantly ($p < 0.05$) increased from 0.03–0.08 (CK) to 0.06–0.18 dS/m under the BC treatment.

Table 1. Effect of different treatments on the pH, electrical conductivity, two chemical agents extractable Cd concentrations, and Cd's distribution in the different fractions.

Treatment [2]	Soil Property and Cd Concentration [1]							
	pH	Electrical Conductivity	Extracting Agent		BCR Sequestration Extraction [3]			
			0.1 N HCl	0.05 N EDTA	F-I	F-II	F-III	F-IV
		dS/m	mg/kg		%			
PCC-CK	6.90 ± 0.15b	0.03 ± 0.00b	6.63 ± 0.24a	6.81 ± 0.25a	55.99 ± 0.39b	26.37 ± 0.23c	16.18 ± 0.22b	1.46 ± 0.04b
PCC-BC	7.93 ± 0.18a	0.12 ± 0.02a	6.98 ± 0.61a	7.03 ± 0.38a	64.67 ± 0.31a	26.61 ± 0.07c	7.63 ± 0.10c	1.08 ± 0.03c
PCW-CK	6.90 ± 0.20b	0.05 ± 0.01b	7.23 ± 0.68a	7.01 ± 0.39a	52.57 ± 0.34c	31.00 ± 0.10a	16.44 ± 0.17ab	ND
PCW-BC	7.62 ± 0.46a	0.17 ± 0.06ab	6.91 ± 0.56a	6.27 ± 0.34b	51.93 ± 0.29d	29.91 ± 0.15b	16.56 ± 0.22a	1.60 ± 0.03a
LSC-CK	6.69 ± 0.11a	0.07 ± 0.02a	7.70 ± 0.62a	7.41 ± 0.85a	61.34 ± 0.27a	20.86 ± 0.19c	17.80 ± 0.23b	ND
LSC-BC	6.81 ± 0.16a	0.16 ± 0.08a	7.00 ± 0.64a	6.82 ± 0.56a	37.70 ± 0.39d	45.26 ± 0.11a	17.04 ± 0.22c	ND
LSS-CK	6.77 ± 0.11a	0.05 ± 0.01a	6.90 ± 0.88a	7.62 ± 0.30a	38.90 ± 0.28c	36.51 ± 0.32b	21.84 ± 0.16a	2.76 ± 0.05
LSS-BC	7.02 ± 0.23a	0.06 ± 0.00a	6.55 ± 1.18a	6.64 ± 0.94a	57.32 ± 0.44b	20.88 ± 0.12c	21.80 ± 0.25a	ND

[1] Mean ± standard deviation; ND: not detectable; The same lowercase letter indicates no significant difference between treatments for the same soil property and the same leafy vegetable. [2] PCC, PCW, LSC, and LSS are *Brassica chinensis* L. var. Chinensis, *Brassica chinensis* L. cv. Wrinkled leaf, *Lactuca sative* L. cv. Chinese, and *Lactuca sative* L. var. Sative, respectively; CK: control without applying husk biochar; BC: applying 46,920 and 98,000 kg/ha of husk biochar for lettuce and pak-choi, respectively. [3] F-I, F-II, F-III, and F-IV are exchangeable fraction, organic matter-bounding fraction, oxide-bounding fraction, and residual fraction, respectively.

3.2. Soil Cd Fraction and Growth Exhibition

Even through the BC treatment raised the soil pH in comparison with CK, the soil Cd fractions only slightly changed except for those of the lettuce. Regardless of treatments and crops, F-I was the primary fraction of Cd in the soils, and accounted for 37–65% of the total concentration in general (Table 1). In comparison with CK, the BC treatment significantly ($p < 0.05$) changed the Cd in F-I and F-II of lettuce. For LSS treated with BC, the Cd proportion of F-I significantly ($p < 0.05$) increased, but F-II significantly ($p < 0.05$) decreased compared with CK. For LSC treated with BC and in contrast with CK, however, the Cd in the F-I significantly ($p < 0.05$) decreased and F-II significantly ($p < 0.05$) increased. The BC treatment did not significantly change the Ni proportion in the different fractions (detailed data not shown). F-IV was the primary fraction of Ni in general and accounted for 57–63% of the total Ni in the soils. Besides sequence extraction, 0.1 N HCl and 0.05 N EDTA were used to extract the available concentration of Cd and Ni in the soils. Approximately 58–68% and 56–69% of the total Cd in the soils could be extracted by 0.1 N HCl and 0.05 N EDTA, respectively. Because most of the Ni was presented in the F-IV fraction, 0.1 N HCl and 0.05 N EDTA only extracted 10–15% and 9–13% of the total Ni in the soils, respectively. In comparison with CK, the BC treatment did not significantly decrease the two chemical agents' extractable Cd and Ni concentrations in the soils in general.

Compared to CK treatments, the root length, shoot height, shoot fresh weight, and SPAD reading increased under BC treatment in general (Table 2). However, most of the differences were not statistically significant. These results revealed that BC treatment promoted the growth of PCC in general.

Table 2. The growth exhibitions (mean ± standard deviation) of pak-choi and lettuce grown in the contaminated soils amended with or without applying husk biochar [1].

Growth Exhibition	Pak-choi [2]				Lettuce [2]			
	PCC-CK	PCC-BC	PCW-CK	PCW-BC	LSC-CK	LSC-BC	LSS-CK	LSS-BC
Root length (cm)	4.61 ± 1.49a	8.26 ± 3.52a	3.59 ± 0.20a	4.66 ± 0.95a	7.59 ± 1.30a	6.96 ± 0.21a	7.71 ± 0.45a	8.09 ± 1.38a
Shoot height (cm)	10.32 ± 2.75a	12.19 ± 0.67a	12.19 ± 0.94a	11.56 ± 1.79a	8.74 ± 0.38bc	7.58 ± 0.88c	15.47 ± 0.39a	17.99 ± 2.19ab
Shoot fresh weight (g/plant)	0.48 ± 0.07a	1.03 ± 0.12a	0.51 ± 0.28a	0.78 ± 0.56a	0.52 ± 0.09c	0.90 ± 0.23b	0.35 ± 0.08c	1.51 ± 0.22a
SPAD reading [3]	10.48 ± 1.70ab	7.84 ± 0.60c	9.71 ± 0.92bc	11.96 ± 1.14a	14.32 ± 1.97a	14.41 ± 1.74a	12.42 ± 0.50a	14.91 ± 0.30a

[1] The same lowercase letter indicates no significant difference between treatments for the same leafy vegetable and the same growth exhibition. [2] The meanings of abbreviations are the same as in Table 1. [3] SPAD reading: soil plant analyzer development reading.

3.3. Cd Accumulation

The concentrations of PTEs in the lettuce and pak-choi at three different growth periods (D35, D42, and D49) are shown in Figures 1 and 2. The Cd concentrations in the roots of LSC cultivars grown in CK for 35–49 days were in the range of 3.1–11.0 mg/kg, which were 1.4 to 6.4 times higher than that accumulated in the roots of LSS (Figure 1). In the shoots of different growing periods of lettuce, LSC also accumulated higher concentrations of Cd compared with LSS. The Cd concentrations in the shoots of LSC and LSS grown in CK of D35–D49 were in the ranges of 2.1–6.9 and 1.8–3.1 mg/kg, respectively. The LSC accumulated higher concentrations of Cd both in roots and shoots compared with LSS. Relative to LSS, the accumulated Cd concentrations in the edible parts of LSC were 1.7–3.5 times higher, which revealed that LSC can be regarded as a high-Cd-accumulation cultivar of lettuce. The PCW cultivar accumulated higher concentrations of Cd in the roots and shoots in comparison with PCC in general (Figure 2). The accumulated Cd concentrations in the roots of PCW and PCC growing for D35–D49 were in the ranges of 6.6–32.3 and 2.9–12.7 mg/kg, respectively. In the shoots of PCW and PCC, the Cd concentrations of D35–D49 were 3.4–8.9 and 3.4–7.8 mg/kg, respectively.

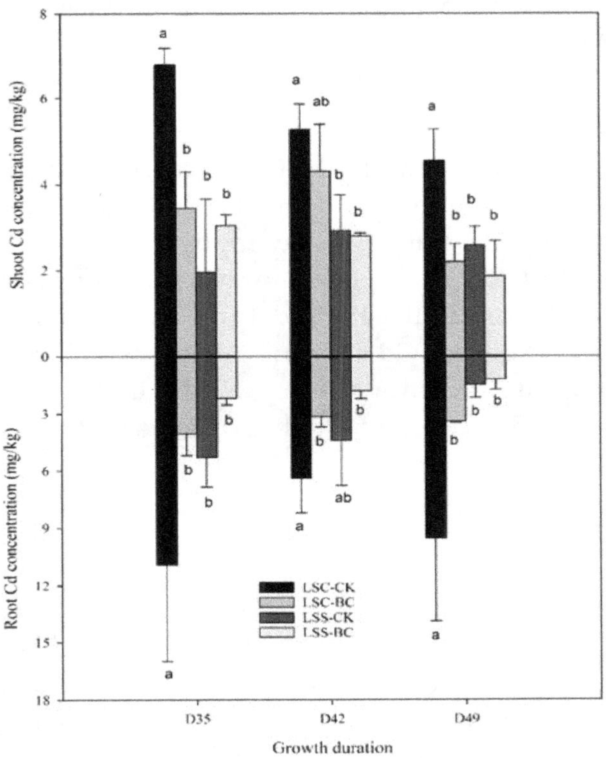

Figure 1. Effect of husk biochar treatment on the Cd accumulation in the roots and shoots of lettuce. The meanings of abbreviations are the same as in Table 1. The same lowercase letter indicates no significant difference between treatments for the same growth period.

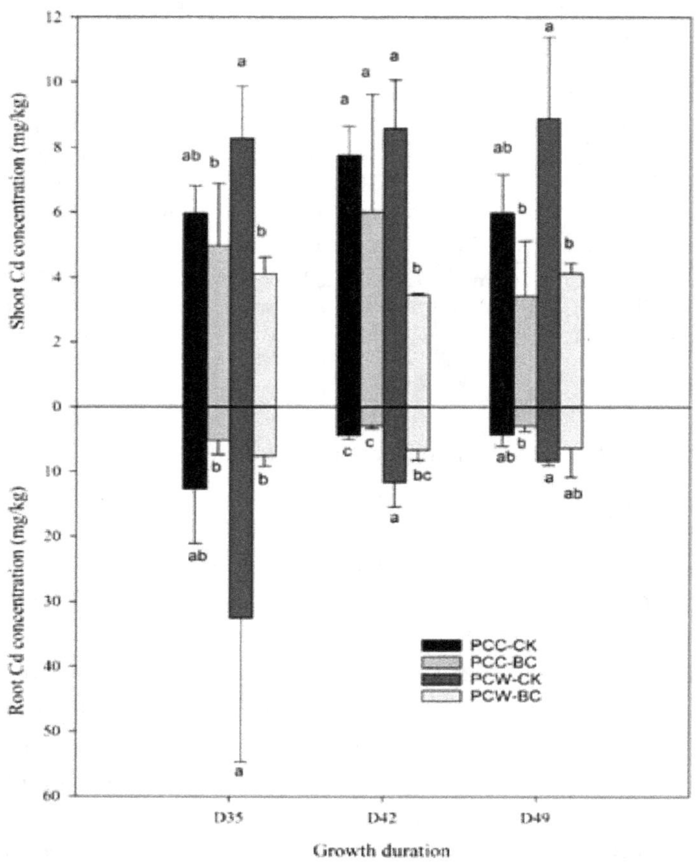

Figure 2. Effect of husk biochar treatment on the Cd accumulation in the roots and shoots of pak-choi. The meanings of abbreviations are the same as in Table 1. The same lowercase letter indicates no significant difference between treatments for the same growth period.

Except for LSS-D35, in comparison with CK, the BC treatments decreased or significantly ($p < 0.05$) decreased the accumulation of Cd in the roots and shoots of lettuce (Figure 1) and pak-choi (Figure 2). This phenomenon was especially true for the LSC variety, which had high accumulating capacity compared with LSS. Relative to CK, the Cd concentrations accumulated in the roots and shoots of LSC under BC treatments decreased 50–64% and 18–52%, respectively, in comparison with CK. Whether treating with BC or not, higher Cd concentrations were found in the roots and shoots of pak-choi compared with lettuce at different growth periods. Under the BC treatments, most of the Cd concentrations in the roots and shoots of pak-choi decreased 24–77% and 16–60%, respectively, in comparison with CK.

3.4. Chemical Form

According to the standard for the Tolerance of Heavy Metals in Plant Origin of Taiwan for leafy vegetables, 0.2 mg/kg·FW, most of the Cd concentrations in the edible parts of lettuce and pak-choi were beyond this standard if the water content of them was 90%. Whether treating with BC or not, Cd was mainly compartmentalized in the F_E chemical form in the shoots of lettuce and pak-choi. Approximately 94–98% and 82–98% of the Cd accumulated in the shoots of lettuce and pak-choi, respectively, were compartmentalized in the F_E at D42–D49 (Figure 3).

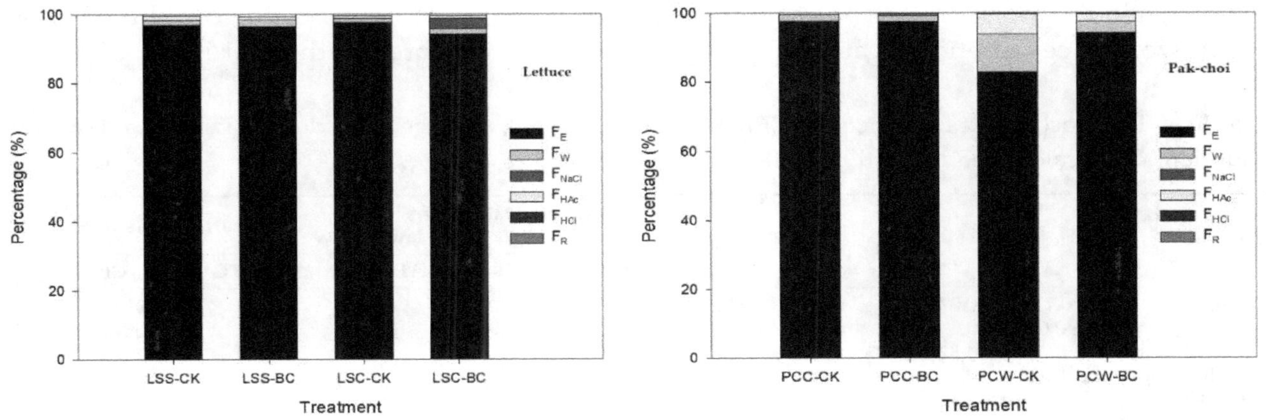

Figure 3. Effect of husk biochar treatment on the chemical form of Cd in the shoots of pak-choi and lettuce. The meanings of treatments are the same as in Table 1. F_E, F_W, F_{NaCl}, F_{HAc}, F_{HCl}, and F_R are inorganic, water-soluble, pectate- and protein-integrated, undissolved phosphate, oxalate, and residual chemical forms, respectively.

4. Discussion

For the three different growing periods, the accumulated Cd in the edible parts of LSC and PCW grown in CK were in the ranges of 4.5–6.9 and 8.2–8.9 mg/kg, respectively (Figures 1 and 2). These values were 1.7–3.5 and 1.1–1.5 times higher than LSS and PCC, respectively. Further results revealed that the accumulating capacity of Cd of LSC and PCW was stronger than LSS and PCC. Moreover, the lettuce and pak-choi accumulated almost constant concentration of Cd in the edible parts regardless of the growing period once planted in the Cd-contaminated soils. The experimental results of this study are in agreement with Lai and Chen [49].

BCF was used to evaluate the accumulation of Cd in the edible parts of two leafy vegetables at D49 (Table 3). BCF_R and BCF_S are the ratio of root and shoot concentration to soil concentration, respectively. The LSC accumulated more Cd than LSS, and the BCF_R of LSC at D49 was 6.1 and 2.6 times higher than LSS grown in CK and BC, respectively. All the BCF_R and BCF_S of lettuce grown in CK were less than 0.9, and there was a 25–68% and 32–58% decrease in BCF_R and BCF_S, respectively, under BC treatment in comparison with CK. Whether applying BC or not, the BCF_S of LSC was 1.1–1.8 times higher than that of LSS. This phenomenon reveals that LSC can accumulate higher Cd concentrations in the edible parts and also shows higher risk through oral consumption than LSS.

The PCW accumulated more Cd than PCC, and the BCF_R of PCW at D49 was 1.9 and 2.3 times higher than PCC grown in CK and BC, respectively. All the BCF_R and BCF_S of pak-choi grown in CK were less than 0.9, and there was a 27–55% and 45–50% decrease in BCF_R and BCF_S, respectively, under BC treatment compared with CK. Even though the BCF_R of PCW was higher than PCC, the BCF_S of PCW grown in different treatments was 57–75% that of PCC. This phenomenon reveals that the Cd accumulated in the roots of PCW was not efficiently upwardly transferred to shoots and thus had lower BCF_S compared with PCC.

Besides BCF, TF was used to evaluate the transfer of Cd from roots to edible parts of lettuce and pak-choi (Table 3). Approximately all the TF values of LSC and PCW, with higher BCF_R and BCF_S under different growth periods and treatments, were below unity in general. For LSS and PCC, however, TF values were higher than LSC and PCC, respectively, and were in the ranges of 0.3–1.8 and 0.4–2.1, respectively. Further results reveal that the Cd uptake by LSC and PCW grown in CK were mainly accumulated in the non-edible parts, regardless of growing periods. However, the TF values of LSS and PCC grown in BC at D42 and D49 reached 1.1–2.1. This result reveals that the accumulated Cd would transfer to shoots more easily than the other two varieties, even grown in BC. Except for LSS-D49, the BC treatment was efficient in increasing the TF of LSS and LSC in comparison with CK, regardless of growth periods. This result revealed that even the BC decreased the accumulated Cd concentration in the roots; on the contrary, BC increased the upward translocation of Cd from roots to shoots in two lettuce varieties because of Cd's high mobility compared with other PTEs.

Table 3. The bioconcentration factor (BCF), transfer factor (TF), average daily dose (ADD), and hazard quotient (HQ) calculated using three different methods [1].

Treatment [2]	BCF_R [3]	BCF_S [3]	TF [3]	Average Daily Dose [4] (ADD_v; µg/kg·BW/day)			Hazard Quotient (HQ_v) [4]		
				ADD_v-TC	ADD_v-CF	ADD_v-BF	HQ_v-TC	HQ_v-CF	HQ_v-BF
PCC-CK	0.40 ± 0.17ab	0.57 ± 0.11a	1.43 ± 0.67a	0.66	0.002	0.15	7.88	0.02	1.73
PCC-BC	0.25 ± 0.06b	0.29 ± 0.14a	1.15 ± 0.64a	0.37	0.001	0.08	4.42	0.02	0.97
PCW-CK	0.78 ± 0.06a	0.82 ± 0.23a	1.06 ± 0.23a	0.60	0.019	0.13	7.16	0.22	1.57
PCW-BC	0.57 ± 0.40ab	0.37 ± 0.03a	0.65 ± 0.49a	0.29	0.004	0.06	3.43	0.05	0.75
LSC-CK	0.89 ± 0.40a	0.42 ± 0.07a	0.47 ± 0.37b	0.34	0.002	0.08	4.07	0.03	0.90
LSC-BC	0.28 ± 0.01b	0.18 ± 0.04b	0.64 ± 0.13b	0.20	0.004	0.04	2.41	0.05	0.53
LSS-CK	0.14 ± 0.06b	0.25 ± 0.04b	1.70 ± 0.49a	0.26	0.002	0.06	3.04	0.03	0.67
LSS-BC	0.11 ± 0.05b	0.17 ± 0.07b	1.53 ± 0.79a	0.29	0.002	0.06	3.46	0.03	0.76

[1] The same lowercase letter indicates no significant difference between treatments for the same leafy vegetable. [2] The meanings of abbreviations are the same as in Table 1. [3] BCF_R: ratio of root conc. to soil conc.; BCF_S: ratio of shoot conc. to soil conc.; TF: ratio of shoot conc. to root conc. [4] ADDv and HQ_v are vegetable-induced ADD and HQ based on the total concentration, chemical form, and bioaccessible fraction of Cd, respectively.

In comparison with CK, the BC treatment did not have significant effects on changing the 0.1 N HCl and 0.05 N EDTA extractable Cd concentrations (Table 1), and also the chemical form of Cd compartmentalized in the shoots of lettuce and pak-choi (detail data not shown). However, the accumulated Cd concentration in the roots and shoots of lettuce and pak-choi grown under BC treatment at different growth periods were decreased or significantly decreased ($p < 0.05$) compared with CK (Figures 1 and 2).

Antoniadis et al. [50] reported that the vegetable-induced average daily dose (ADD_v) and vegetable-induced hazard quotient (HQ_v) can be calculated using Equations (1) and (2), respectively, where C_p is the Cd concentration (mg/kg) in the edible parts of vegetables. The mean individual daily vegetable consumption (MIDVC) in Taiwan during 2013–2016 was 0.133 kg/day based on the Report on the Nutrition and Health Survey and vegetable calorie counts, which can be used to calculate Cd intake daily per person from vegetables. The tolerable daily intake (TDI) of Cd set by the European Food Safety Authority (EFSA) was 0.36 µg/kg·BW/day. Nonetheless, food was the dominant source of Cd exposure of humans and accounts for approximately 90% of the intake [51]. Among all foods, approximately 26% was from vegetables [52], which means that the TDI from vegetables (TDI_v) is 0.084 µg/kg·BW/day.

$$ADD_v = \frac{C_p \times MIDVC}{kg \cdot BW} \quad (1)$$

$$HQ_v = \frac{ADD_v}{TDI_v} \quad (2)$$

Blanching is the most common method in Taiwan for cooking leafy vegetables, and it also decreases the concentration of PTE in thoroughly cooked vegetables. Based on the findings of Lam et al. [47], approximately 50% of the Cd accumulated in the water spinach was leached into boiling water. In this study, three methods based on total concentration (TC), chemical form (CF), and bioaccessible fraction (BF) of Cd in the edible parts of vegetables were used to calculate the HQ_v. The F_E and F_W were considered to have a higher mobility than other chemical forms and were easily leached into boiling water. Therefore, the sum of the proportion of the other four chemical forms, i.e., F_{NaCl}, F_{HAc}, F_{HCl}, and F_R, was used to calculate the HQ_v, coded as HQ_v-CF. Furthermore, approximately 32–55% (average is 44%) of the accumulated Cd in water spinach could be metabolized by in vitro digestive fluids [47], which reveals that approximately 44% of the Cd is bioaccessible and can be absorbed by the human body, coded as HQ_v-BF.

Regardless of treatments, the HQ_V-TC, HQ_V-CF, and HQ_V-BF values of usedleafy vegetables at D49 were in the ranges of 2.4–7.9, 0.01–0.3, and 0.5–1.8, respectively (Table 3). Because more than 73% of the accumulated Cd was compartmentalized in the F_E chemical form, which could be leached out of vegetable tissues during blanching, the HQ_V-CF values were less than 0.3 in general. The application of BC significantly decreased the HQ_V of pak-choi at D49, and the HQ_V-TC, HQ_V-CF, and HQ_V-BF was 20–89% in comparison with the CK. However, BC's effect on the HQ_V of lettuce was contrary to that of pak-choi because the HQ_V increased compared with CK. The HQ_V of lettuce and pak-choi used in this study was lower in comparison with water spinach grown in artificially Cd-spiked soils with a total concentration of 2.8–3.1 mg/kg [53]. According to the calculated results of HQ_V-CF and HQ_V-BF, except for pak-choi grown in CK, oral intake of these four leafy vegetables has a low risk even though the soil Cd concentration was 2 to 3 times beyond the control standard of farmland, i.e., 5 mg/kg, based on the SGWPR Act of Taiwan.

5. Conclusions

Experimental results evidenced that the application of BC was able to increase the soil pH and decrease the accumulation of Cd in the roots and shoots of leafy vegetables used in this study. The BCF also decreased under BC treatment compared with control. However, BC's effect on the upward transfer of Cd from root to shoot was dependent on crop species. Because some of the accumulated Cd in the vegetables will leach out of tissues during cooking, using total concentration of Cd in the vegetables cannot actually reflect the real dose of Cd absorbed by the human body. Based on the HQ_V calculation using the chemical form and bioaccessible fraction of Cd in the edible parts of blanched leafy vegetables, all of the vegetables grown in the study soil had low risk through oral intake, especially under BC treatment.

Author Contributions: Conceptualization and methodology, H.-Y.L.; sample analysis, K.-S.C. and C.-Y.P.; writing—original draft preparation, K.-S.C. and C.-Y.P.; writing—review and editing, H.-Y.L. All authors have read and agreed to the published version of the manuscript.

Acknowledgments: We would like to thank the students of the Soil Survey and Remediation Laboratory, Department of Soil and Environmental Sciences, National Chung Hsing University, for their assistance during the pot experiment and analysis.

References

1. Hseu, Z.Y.; Su, S.W.; Lai, H.Y.; Guo, H.Y.; Chen, T.C.; Chen, Z.S. Remediation techniques and heavy metal uptake by different rice varieties in metal-contaminated soils of Taiwan: New aspects for food safety regulation and sustainable agriculture. *Soil Sci. Plant Nutr.* **2010**, *56*, 31–52. [CrossRef]
2. SGWPR Act (Soil and Groundwater Pollution Remediation Act). Available online: https://sgw.epa.gov.tw/en/laws_policy/laws/458db6eb-5602-46b2-9471-745d58078aaf (accessed on 5 March 2020).
3. Di Toppi, L.S.; Gabbrielli, R. Response to cadmium in higher plants. *Environ. Exp. Bot.* **1999**, *41*, 105–130. [CrossRef]
4. Chen, D.; Guo, H.; Li, R.; Li, L.; Pan, G.; Chang, A.; Joseph, S. Low uptake affinity cultivars with biochar to tackle Cd-tainted rice—A field study over four rice seasons in Hunan, China. *Sci. Total Environ.* **2016**, *541*, 1489–1498. [CrossRef]
5. Ali, H.; Khan, E.; Ilahi, I. Environmental chemistry and ecotoxicology of hazardous heavy metals: Environmental persistence, toxicity, and bioaccumulation. *J. Chem.* **2019**, *2019*, 1–14. [CrossRef]
6. Barraza, F.; Schreck, E.; Lévêque, T.; Uzu, G.; López, F.; Ruales, J.; Prunier, J.; Marquet, A.; Maurice, L. Cadmium bioaccumulation and gastric bioaccessibility in cacao: A field study in areas impacted by oil activities in Ecuador. *Environ. Pollut.* **2017**, *229*, 950–963. [CrossRef] [PubMed]

7. Kovacevic, G.; Kastori, R.; Merkulov, L. Dry matter and leaf structure in young wheat plants as affected by cadmium, lead, and nickel. *Biol. Plant.* **1999**, *42*, 119–123. [CrossRef]
8. Liu, Y.T.; Chen, Z.S.; Hong, C.Y. Cadmium-induced physiological response and antioxidant enzyme changes in the novel cadmium accumulator, *Tagetes patula*. *J. Hazard. Mater.* **2011**, *189*, 724–731. [CrossRef]
9. Xu, D.Y.; Chen, Z.F.; Sun, K.; Yan, D.; Kang, M.J.; Zhao, Y. Effect of cadmium on the physiological parameters and the subcellular cadmium localization in the potato (*Solanum tuberosum* L.). *Ecotox. Environ. Safe.* **2013**, *97*, 147–153. [CrossRef]
10. Lasat, M.M.; Baker, A.J.M.; Kochian, L.V. Physiological characterization of root Zn^{2+} absorption and translocation to shoots in Zn hyperaccumulator and nonaccumulator species of Thlaspi. *Plant Physiol.* **1996**, *112*, 1715–1722. [CrossRef]
11. Lasat, M.M.; Fuhrmann, M.; Ebbs, S.D.; Cornish, J.E.; Kochian, L.V. Phytoremediation of a radiocesium-contaminated soil: Evaluation of cesium-137 bioaccumulation in the shoots of tree plant species. *J. Environ. Qual.* **1998**, *7*, 165–169. [CrossRef]
12. Küpper, H.; Zhao, F.J.; McGrath, S.P. Cellular compartmentation of zinc in leaves of the hyperaccumulator *Thlaspi caerulescens*. *Plant Physiol.* **1999**, *119*, 305–311. [CrossRef] [PubMed]
13. Ge, W.; Jiao, Y.Q.; Sun, B.L.; Qin, R.; Jiang, W.S.; Liu, D.H. Cadmium-mediated oxidative stress and ultrastructural changes in root cells of poplar cultivars. *S. Afr. J. Bot.* **2012**, *83*, 98–108. [CrossRef]
14. Haynes, R.J. Ion exchange properties of roots and ionic interactions within root apoplasm: Their role in ion accumulation by plants. *Bot. Rev.* **1980**, *46*, 75–99. [CrossRef]
15. Zhao, Y.F.; Wu, J.F.; Shang, D.R.; Ning, J.S.; Zhai, Y.X.; Shend, X.F.; Ding, H.Y. Subcellular distribution and chemical forms of cadmium in the edible seaweed, *Porphyra yezoensis*. *Food Chem.* **2015**, *168*, 48–54. [CrossRef] [PubMed]
16. Wu, F.B.; Dong, J.; Qian, Q.Q.; Zhang, G.P. Subcellular distribution and chemical form of Cd and Ca-Zn interaction in different barley genotypes. *Chemosphere* **2005**, *60*, 1437–1446. [CrossRef]
17. Su, Y.; Liu, J.L.; Lu, Z.W.; Wang, X.M.; Zhang, Z.; Shi, G.G. Effects of iron deficiency on subcellular distribution and chemical forms of cadmium in peanut roots in relation to its translocation. *Environ. Exp. Bot.* **2014**, *97*, 40–48. [CrossRef]
18. Fu, X.; Dou, C.; Chen, Y.; Chen, X.; Shi, J. Subcellular distribution and chemical forms of cadmium in *Phytplacca americana* L. *J. Hazard. Mater.* **2011**, *186*, 103–107. [CrossRef]
19. Wang, X.; Liu, Y.G.; Zeng, G.M.; Chai, L.Y.; Song, X.C.; Min, Z.Y.; Xiao, X. Subcellular distribution and chemical forms of cadmium in *Bechmeria nivea* (L.) Gaud. *Environ. Exp. Bot.* **2008**, *62*, 389–395. [CrossRef]
20. Qiu, Q.; Wang, Y.; Yang, Z.; Yuan, J. Effects of phosphorus supplied in soil on subcellular distribution and chemical forms in two Chinese flowering cabbage (*Brassica parachinensis* L.) cultivars differing in cadmium accumulation. *Food Chem. Toxicol.* **2011**, *49*, 2260–2267. [CrossRef]
21. Lai, H.Y. Subcellular distribution and chemical forms of cadmium in *Impatiens walleriana* in relation to its phytoextraction potential. *Chemosphere* **2015**, *138*, 370–376. [CrossRef]
22. Garau, G.; Castaldi, C.; Santona, L.; Deiana, P.; Melis, P. Influence of red mud, zeolite and lime on heavy metal immobilization, culturable heterotrophic microbial populations and enzyme activities in a contaminated soil. *Geoderma* **2007**, *142*, 47–57. [CrossRef]
23. Cao, X.; Dermatas, D.; Xu, X.; Shen, G. Immobilization of lead in shooting range soils by means of cement, quicklime, and phosphate amendments. *Environ. Sci. Pollut. Res.* **2008**, *15*, 120–127. [CrossRef] [PubMed]
24. Liu, N.; Jiang, Z.; Li, X.; Liu, H.; Li, N.; Wei, S. Mitigation of rice cadmium (Cd) accumulation by joint application of organic amendments and selenium (Se) in high-Cd-contaminated soils. *Chemosphere* **2020**, *241*, 125106. [CrossRef] [PubMed]
25. Shuman, L. Organic waste amendments effect on zinc fraction of two soils. *J. Environ. Qual.* **1999**, *28*, 1442–1447. [CrossRef]
26. Egene, C.E.; Van Poucke, R.; OK, Y.S.; Meers, E.; Tack, F.M.G. Impact of organic amendments (biochar, compost and peat) on Cd and Zn mobility and solubility in contaminated soil of the Campine region after three years. *Sci. Total Environ.* **2018**, *626*, 195–202. [CrossRef]
27. Ahmad, M.; Rajapaksha, A.U.; Lim, J.E.; Zhang, M.; Bolan, N.; Mohan, D.; Vithanage, M.; Lee, S.S.; Ok, Y.S. Biochar as a sorbent for contaminant management in soil and water: A review. *Chemosphere* **2014**, *99*, 19–33. [CrossRef]

28. Moreno-Jiménez, E.; Esteban, E.; Carpena-Ruiz, R.O.; Lobo, M.C.; Rénalos, J.M. Phytostabilisation with Mediterranean shrubs and liming improved soil quality in a pot experiment with a pyrite mine soil. *J. Hazard. Mater.* **2012**, *201–202*, 52–59. [CrossRef]
29. Xu, C.; Chen, H.X.; Xiang, Q.; Zhu, H.H.; Wang, S.; Zhu, Q.H.; Huang, D.Y.; Zhang, Y.Z. Effect of peanut shell and wheat straw biochar on the availability of Cd and Pb in a soil-rice (*Oryza sativa* L.) system. *Environ. Sci. Pollut. Res.* **2017**, *25*, 1147–1156. [CrossRef]
30. Wang, Y.M.; Tang, D.D.; Zhang, X.H.; Uchimiya, M.; Yuan, X.Y.; Li, M.; Chen, Y.Z. Effects of soil amendments on cadmium transfer along the lettuce-snail food chain: Influence of chemical speciation. *Sci. Total Environ.* **2019**, *649*, 801–807. [CrossRef]
31. Zhang, X.Y.; Zhang, Y.M.; Liu, X.Y.; Zhang, C.Y.; Dong, S.D.; Liu, Q.; Deng, M. Cd uptake by *Phytolacca americana* L. promoted by cornstalk biochar amendments in Cd-contaminated soil. *Int. J. Phytorem.* **2019**, *22*, 251–258. [CrossRef]
32. Chen, Z.S.; Lee, D.Y. Evaluation of remediation techniques on two cadmium-polluted soils in Taiwan. In *Remediation of Soils Contaminated with Metals*; Iskander, I.K., Adriano, D.C., Eds.; Science Reviews: Northwood, UK, 1997; pp. 209–223.
33. Thomas, G.W. Soil pH and soil acidity. In *Methods of Soil Analysis. Part. 3 Chemical Methods*; Sparks, D.L., Page, A.L., Helmke, P.A., Loeppert, R.H., Soltanpour, P.N., Tabatabai, M.A., Johnston, C.T., Sumner, M.E., Eds.; SSSA Inc./ASA Inc.: Madison, WI, USA, 1996; pp. 475–490.
34. Rhoades, J.D. Salinity: Electrical conductivity and total dissolved solids. In *Methods of Soil Analysis. Part 3. Chemical Methods*; Sparks, D.L., Page, A.L., Helmke, P.A., Loeppert, R.H., Soltanpour, P.N., Tabatabai, M.A., Johnston, C.T., Sumner, M.E., Eds.; SSSA Inc./ASA Inc.: Madison, WI, USA, 1996; pp. 417–435.
35. Nelson, D.W.; Sommers, L.E. Total carbon, organic carbon, and organic matter. In *Methods of Soil Analysis. Part 3. Chemical Methods*; Sparks, D.L., Page, A.L., Helmke, P.A., Loeppert, R.H., Soltanpour, P.N., Tabatabai, M.A., Johnston, C.T., Sumner, M.E., Eds.; SSSA Inc./ASA Inc.: Madison, WI, USA, 1996; pp. 961–1010.
36. Sumners, M.E.; Miller, W.P. Cation exchange capacity and exchange coefficients. In *Methods of Soil Analysis. Part 3. Chemical Methods*; Sparks, D.L., Page, A.L., Helmke, P.A., Loeppert, R.H., Soltanpour, P.N., Tabatabai, M.A., Johnston, C.T., Sumner, M.E., Eds.; SSSA Inc./ASA Inc.: Madison, WI, USA, 1996; pp. 1201–1229.
37. Gee, G.W.; Bauder, J.W. Particle-size analysis. In *Methods of Soil Analysis. Part 1. Physical and Mineralogical Method*, 2nd ed.; Klute, A., Ed.; SSSA Inc./ASA Inc.: Madison, WI, USA, 1986; pp. 383–412.
38. EPA/Taiwan. *Method Code No: NIEA S321.65B*; Environmental Protection Administration of Taiwan ROC: Taipei, Taiwan, 2018.
39. Bremner, J.M. Nitrogen–Total. In *Methods of Soil Analysis. Part 3. Chemical Methods*; Sparks, D.L., Page, A.L., Helmke, P.A., Loeppert, R.H., Soltanpour, P.N., Tabatabai, M.A., Johnston, C.T., Sumner, M.E., Eds.; SSSA Inc./ASA Inc.: Madison, WI, USA, 1996; pp. 1085–1121.
40. Kuo, S. Phosphorus. In *Methods of Soil Analysis. Part 3. Chemical Methods*; Sparks, D.L., Page, A.L., Helmke, P.A., Loeppert, R.H., Soltanpour, P.N., Tabatabai, M.A., Johnston, C.T., Sumner, M.E., Eds.; SSSA Inc./ASA Inc.: Madison, WI, USA, 1996; pp. 869–919.
41. Helmke, P.A.; Sparks, D.L. Lithium, Sodium, Potassium, Rubidium, and Cesium. In *Methods of Soil Analysis. Part 3. Chemical Methods*; Sparks, D.L., Page, A.L., Helmke, P.A., Loeppert, R.H., Soltanpour, P.N., Tabatabai, M.A., Johnston, C.T., Sumner, M.E., Eds.; SSSA Inc./ASA Inc.: Madison, WI, USA, 1996; pp. 551–574.
42. De Medici, D.; Komínková, D.; Race, M.; Fabbricino, M.; Součková, L. Evaluation of the potential for caesium transfer from contaminated soil to the food chain as a consequence of uptake by edible vegetables. *Ecotox. Environ. Safe* **2019**, *171*, 558–563. [CrossRef] [PubMed]
43. Yu, T.H.; Hsieh, S.P.; Su, C.M.; Huang, F.J.; Hung, C.C.; Yiin, L.M. Analysis of Leafy Vegetable Nitrate Using a Modified Spectrometric Method. *Int. J. Anal. Chem.* **2018**, *2018*, 1–6. [CrossRef] [PubMed]
44. Baker, D.E.; Amacher, M.C. Nickel, copper, zinc, and cadmium. In *Methods of Soil Analysis. Part 2. Chemical and Microbiological Properties*, 2nd ed.; Page, A.L., Millers, R.H., Keeney, D.R., Eds.; SSSA Inc./ASA Inc.: Madison, WI, USA, 1982; pp. 323–336.
45. Mench, M.J.; Didier, V.L.; Loffler, M.; Gomez, A.; Masson, P. A mimicked in-situ remediation study of metal-contaminated soils with emphasis on cadmium and lead. *J. Environ. Qual.* **1994**, *23*, 58–63. [CrossRef]
46. Ure, A.M. Methods of analysis of heavy metals in soils. In *Heavy Metals in Soils*, 2nd ed.; Alloway, B.J., Ed.; Blackie Academic and Professional: London, UK, 1995; pp. 58–102.

47. Lam, C.M.; Lai, H.Y. Effect of inoculation with arbuscular mycorrhizal fungi and blanching on the bioaccessibility of heavy metals in water spinach (*Ipomoea aquatica* Forsk.). *Ecotox. Environ. Safe* **2018**, *162*, 563–570. [CrossRef]
48. European Commission (EC). Commission Regulation (EU) No 420/2011 of 29 April 2011 amending Regulation (EC) No 1881/2006 setting maximum levels for certain contaminants in foodstuffs. *Off. J. Eur. Union* **2011**, *11*, 3–6.
49. Lai, H.Y.; Chen, B.C. The dynamic growth exhibition and accumulation of cadmium of pak choi grown in contaminated soils. *Int. J. Environ. Res. Public Health* **2013**, *10*, 5284–5298. [CrossRef]
50. Antoniadis, V.; Shaheen, S.M.; Boersch, J.; Frohne, T.; Laing, G.D.; Rinklebe, J. Bioavailability and risk assessment of potentially toxic elements in garden edible vegetables and soils around a highly contaminated former mining area in Germany. *J. Environ. Manag.* **2017**, *186*, 192–200. [CrossRef] [PubMed]
51. European Food Safety Authority. Scientific Opinion of the Panel on Contaminants in the Food Chain on a request from the European Commission of cadmium in food. *ESFA J.* **2009**, *980*, 1–139.
52. Kim, K.; Melough, M.M.; Vance, T.M.; Noh, H.; Koo, S.I.; Chun, O.K. Dietary cadmium intake and sources in the US. *Nutrients* **2019**, *11*, 2. [CrossRef]
53. Lam, C.M.; Chen, K.S.; Lai, H.Y. Chemical forms and health risk of cadmium in water spinach grown in contaminated soil with an increased level of phosphorus. *Int. J. Environ. Res. Public Health* **2019**, *16*, 3322. [CrossRef]

11

Marble Slurry's Impact on Groundwater

Leonardo Piccini [1,*], Tiziana Di Lorenzo [2], Pilario Costagliola [1] and Diana Maria Paola Galassi [3]

1 Department of Earth Science, University of Florence, 50121 Florence, Italy; pilario.costagliola@unifi.it
2 Research Institute on Terrestrial Ecosystems of the CNR, National Research Council of Italy, 50019 Florence, Italy; tiziana.dilorenzo@cnr.it
3 Department of Life, Health and Environmental Sciences, University of L'Aquila, 67100 L'Aquila, Italy; dianamariapaola.galassi@univaq.it
* Correspondence: leonardo.piccini@unifi.it

Abstract: Modern sawing techniques employed in ornamental stones' exploitation produce large amounts of slurry that can be potentially diffused into the environment by runoff water. Slurry produced by limestone and marble quarrying can impact local karst aquifers, negatively affecting the groundwater quality and generating a remarkable environmental and economic damage. A very representative case-study is that of the Apuan Alps (north-western Tuscany, Italy) because of the intensive marble quarrying activity. The Apuan Alps region extends over about 650 km^2; it hosts several quarries, known all over the world for the quality of the marble extracted, and a karst aquifer producing about 70,000 m^3/day of high-quality water used directly for domestic purposes almost without treatments. In addition, Apuan Alps are an extraordinary area of natural and cultural heritage hosting many caves (about 1200), karst springs and geosites of international and national interest. During intense rain events, carbonate slurry systematically reaches the karst springs, making them temporarily unsuitable for domestic uses. In addition, the deterioration of the water quality threatens all the hypogean fauna living in the caves. This paper provides preliminary insights of the hydrological and biological indicators that can offer information about the impact of the marble quarrying activities on groundwater resources, karst habitats and their biodiversity.

Keywords: groundwater; karst aquifer; pollution; quarry; Apuan Alps

---

1. Introduction

Karst areas constitute about 10% of the land surface worldwide and there is a widespread concern about the effects that human activities have upon karst environments [1,2]. Karst aquifers often host water resources that are very important, in terms of both quantity and quality, usually not requiring any treatment before being distributed through civil aqueducts. For this reason, karst groundwater sources are attaining a growing economic and social value. According to UNESCO's International Geoscience Programme (IGCP) 513, concerning "karst aquifers and water resources" and their preservation, about 25% of the world's population relies on drinking water supplies from karst aquifers. On the other hand, karst areas are extremely vulnerable and could be easily (and potentially irreversibly) damaged by human activities such as deforestation, agricultural practices, urbanization, tourism, military activities, water exploitation, mining and quarrying (e.g., [3], and reference therein). Climate change poses an additional risk to karst groundwater, exacerbating the major challenges in the next future.

An unexpectedly high diversity of living forms (micro, meio and macro-organisms) have been discovered in karst aquifers in the second half of the 20th century. Karst stygobionts are often geographically rare and phylogenetically isolated taxa, with ancient origins which date back to the Early Tertiary period [4]. The stygofauna show morphological, physiological and behavioral characteristics suitable for life in absence of light and scarcity of trophic resources (e.g., [5–8]). The stygofauna of karst habitats are represented by vertebrates as well as by small invertebrates, among which crustaceans are dominant [9,10]. Beyond the importance that such highly specialized fauna have in terms of global biodiversity, stygofauna provides a wide range of ecosystem services, including those related to the degradation of pollutants [11,12].

Known since the Roman age for the extraction of the precious white marble, the Apuan Alps, in north-western Tuscany, are a very peculiar mountain range and an extraordinary region of natural and cultural heritage in the Mediterranean basin, with many geological and biological features of national and international interest. This mountain range, whose maximum elevation is 1942 m a.s.l. at Pisanino Mount, consists of both metamorphosed and non-metamorphosed carbonate rocks belonging to three different tectonic units (Figure 1). Apuan metamorphic unit is mainly represented by meta-dolostone, marble and dolomitic marble (about 115 km^2 as outcrops) and by cherty limestones (about 20 km^2) ([13] and references therein).

In an area of about 650 km^2, the carbonate rock assemblages display very different structural and lithological traits. For this reason, they offer a wide variety of hydrogeological conditions displayed in many springs with very different hydrodynamics and hydro-chemical features [13]. The carbonate rocks contain important groundwater resources due to the high rainfall, and the high infiltration coefficient, that can reach 75% of total precipitation. Most of the major springs drain metamorphic carbonate aquifers, and supply drinking water mainly for inhabitants of the coastal plain (where Carrara and Massa towns are located).

The marble exploitation in the Apuan Alps is presently carried out in about 150 quarries that are frequently located at high altitudes (Figure 2a). Line drilling and sawing are the most modern techniques to produce large rectangular blocks suitable for being cut into smaller, regularly shaped products and slabs (Figure 2b). Sawing can be accomplished by a variety of tools, including wire, belt and chain saws. Pollutants resulting from quarrying, including the marble dust (locally known as "*marmettola*") that is a product of rock sawing, easily infiltrate by means of meteoric waters into the underlying karst aquifers through solution cavities. This condition is responsible for the high vulnerability of the groundwater bodies, which is particularly critical for metamorphic carbonate rocks (i.e., marbles and meta-dolostone), due to the absence of retention of pollutants in the unsaturated zone (epikarst and vadose zone) and to the high hydraulic conductivity in the saturated one [14]. As a matter of fact, this fine-grained material is easily mobilized by rainwater and dispersed into and through the subterranean karst systems [15].

In the last decade, the annual marble production in the Apuan Alps quarries has been about 1.2 Mt/year [16]. The extraction and squaring of marble blocks produce about 4% dust per unit [15], in respect to marble production, which so amounts to about 0.05 Mt/year, or, as volume, about 25,000 m^3/year, assuming a porosity of 35%. Most of this powder is disposed on quarry yards and stored in controlled repository areas but a substantial part of it is dispersed into the surrounding areas by wind and runoff before collecting. Furthermore, it is quite frequent that quarries intercept open fissures and solution pits through which marble powder enters the subterranean drainage usually driven by high-intensity rainfall events [17,18].

While the marble quarries in the Apuan Alps have been long and widely studied in many environmental respects, including marble formation volumes [19], rock mass stability [20] and restoration perspectives [21], the real effects of marble slurry pollution, and in particular its impact on groundwater and the processes governing its transport through the karst aquifers, have not yet been systematically examined, despite the fact that this aspect is crucial for the development of management measures aimed at preserving the water quality and the biodiversity of karst systems.

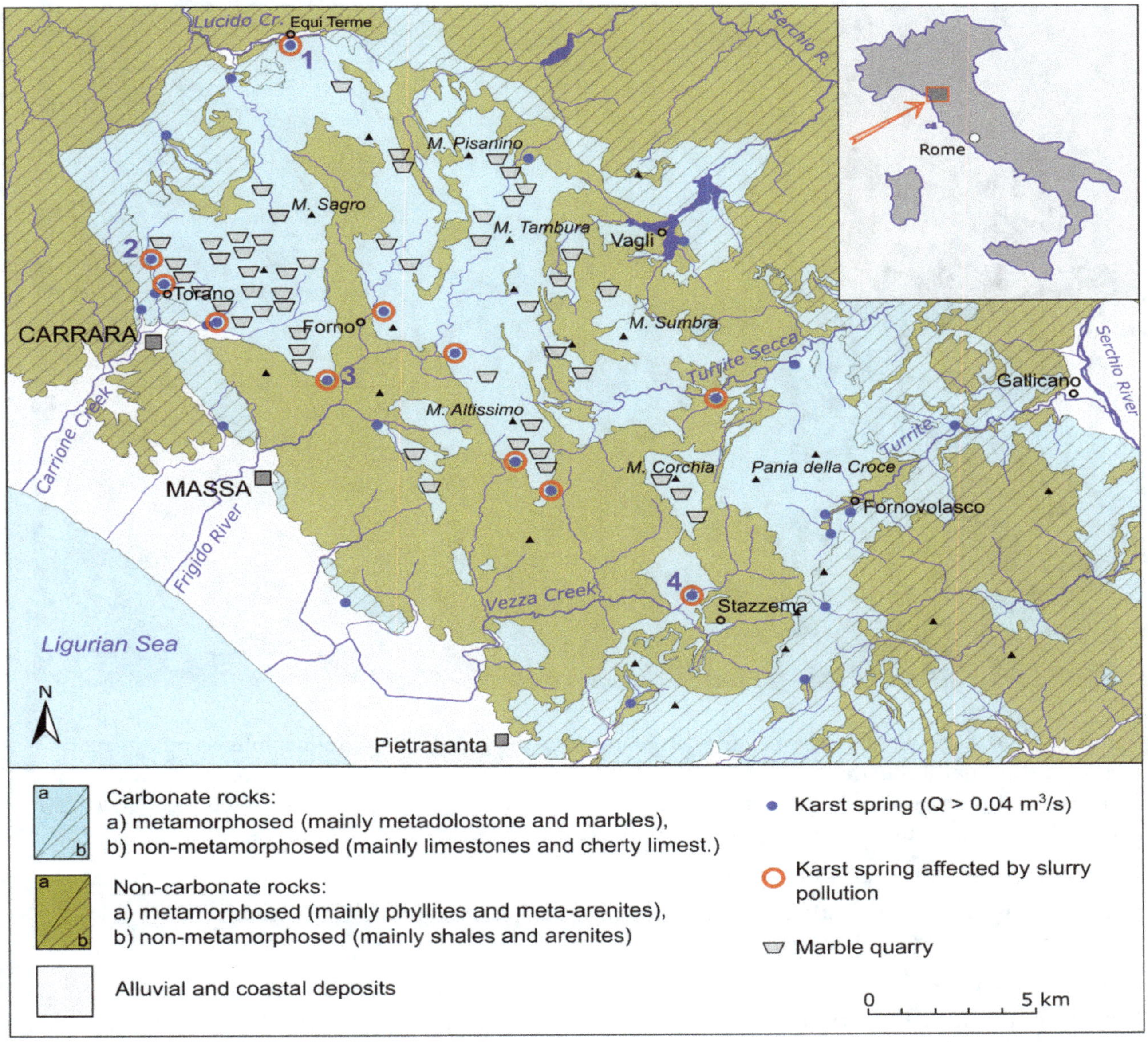

Figure 1. Hydrogeological sketch map of the Apuan Alps with the location of the major karst springs (average discharge >0.04 m^3/s) fed by carbonate aquifers and marble quarries. Numbers refer to monitored karst systems: (1) Equi springs; (2) Carbonera spring; (3) Cartaro spring; (4) Monte Corchia cave stream.

For these reasons, the Apuan Alps can be considered a relevant case-study area, where the impact of slurries produced by marble quarries on karst aquifers can be investigated in order to obtain an indication of general interest and worldwide application. The main purpose of this article is to provide the information available up to now regarding *marmettola* pollution, including its physical and chemical features, and the potential impacts that this particular waste product can cause on the Apuan Alps karst ecosystem, and more generally, in similar karst areas heavily affected by marble quarrying. The article also provides new data on both the hydro-physical properties of groundwater bodies and their biodiversity.

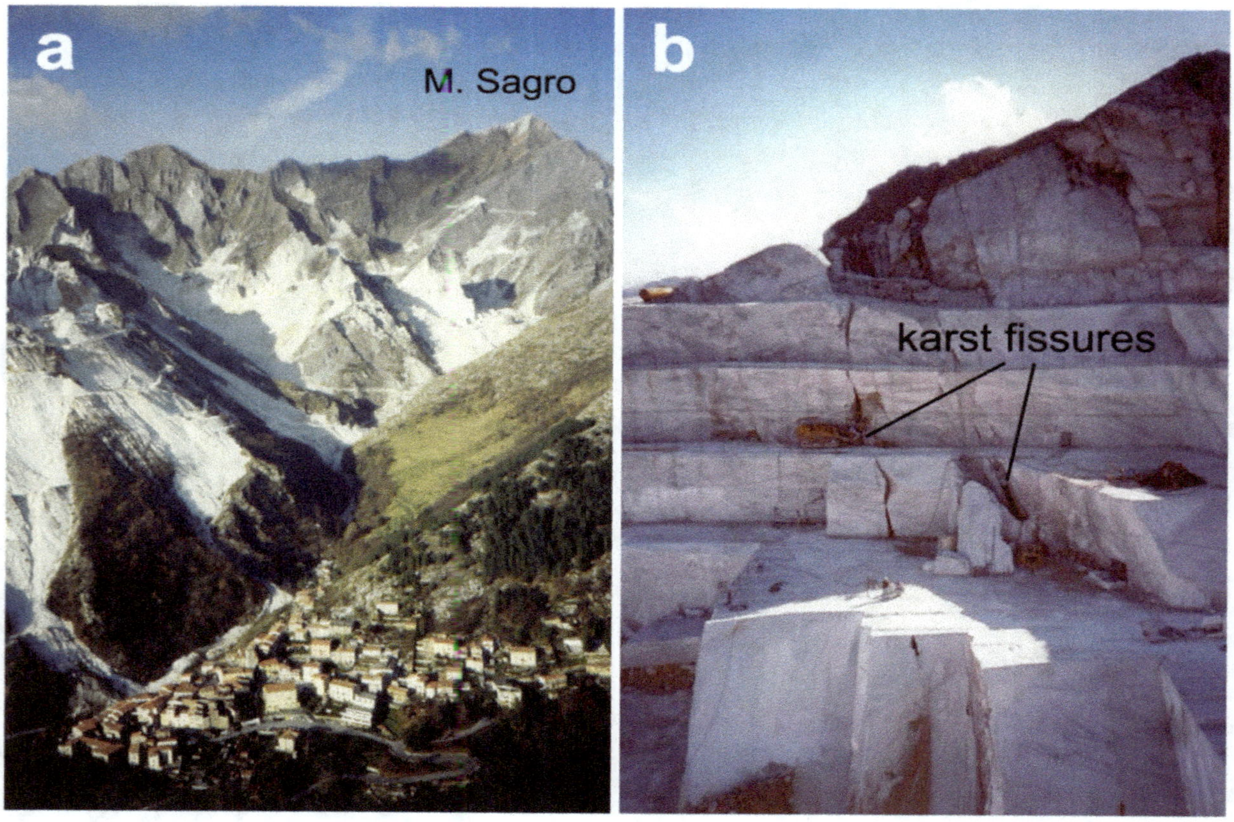

Figure 2. (**a**) A view of Colonnata valley, in Northern Apuan Alps, showing the intense quarrying that concerns high-altitude areas too. (**b**) Quarrying is performed through sawing large blocks with chain or diamond-wire saws; karst fissures or solution pit interception is very frequent (Focolaccia quarry, M. Tambura, central Apuan Alps).

2. Chemical and Physical Features of *Marmettola* and Karst Spring Sediments

Marmettola is a fine, white powder produced during the extraction and squaring of marble blocks by way of chain and diamond-wire saws. Minor sources are represented by the dust produced by the intense traffic of heavy trucks, bulldozers and tractors transporting and moving heavy marble blocks along quarry roads and service areas. The chemical composition of the powder reflects the mineralogical constituents of the different marble varieties. The grains are essentially composed by calcite (usually 95–99%), mainly in sand-to-silt sizes, with a small percentage (1–3%) of dolomite [15,22,23]. Silicates, primarily quartz, muscovite and/or chlorite and clay minerals, are often present in minor amounts [15,21].

In order to investigate textural features, we examined some slurry samples collected in 2017, coming from a quarry at Mount Sagro, by scanning electron microscope (SEM) analysis. Grains produced by chain saws were very angular in shape and with planar surfaces formed along calcite cleavage planes (Figure 3a). SEM image processing revealed that marble slurry is mainly composed by weight of a fine to very-fine sand (Figure 3b), whereas most of the particle sizes (expressed as equivalent circle diameter = ECD) range between 8 and 32 μm, if we consider the number of particles (Figure 3c). A consistent amount of the finer silt-sized fraction is possibly derived from consumption of larger grains during sawing [15]. The powder produced by diamond-wire cutting is generally finer (most grains <62.5 μm in size) but with grains showing an angular shape like those coming from chain saws [15].

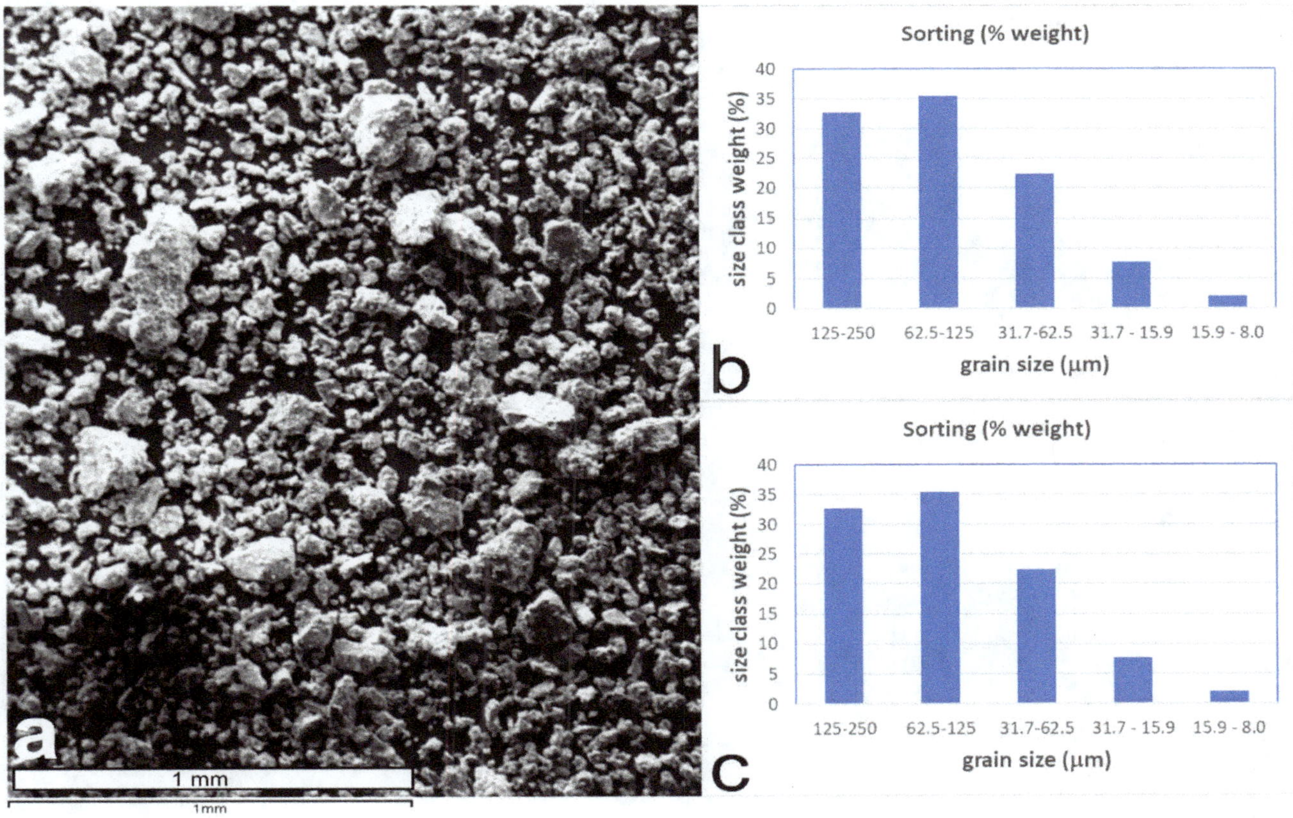

Figure 3. (**a**) SEM image of *marmettola* produced by chain sawing collected at a marble quarry in the Mount Sagro area (note the very angular shape of particles). (**b**) Weight sorting of the sample collected (weight percent for each grain size class). (**c**) Frequency sorting (number of grains) of the same sample.

By means of direct investigation of karst systems contaminated by marble slurry, such that of Monte Corchia cave, we realized that suspended sediments found in springs and cave-streams consisted of fine sand and mud having a light grey or ochre color. From a chemical and mineralogical point of view, these deposits had a higher amount of Mg and Si respect to *marmettola*, due to the occurrence of dolomite and silicates of natural origin. In a sample collected in 2012 at Tana dei Tufi, a captured spring located near Torano village (Carrara marble district) often affected by turbidity, SEM analysis showed that about 75% of the grains were carbonate, the remaining being silicate particles.

Grain morphometric analysis performed on SEM images revealed that grains' ECDs mainly ranged from 15.9 to 62.5 μm, with the silicate particles a little finer than carbonate ones (Figure 4a,b). Other morphometric parameters, such as the aspect ratio (length/breadth ratio of each) and the roundness (calculated as perimeter^2/4π * grain area), were similar in both carbonate and silicate grains (Figure 4c–f). SEM images confirmed also that most of this sediment consisted of calcite angular grains like those forming *marmettola*. Grains finer than 16 μm were poorly represented, probably because they were washed away as suspended load during floods. These preliminary data suggest that the transport through the karst network of the Apuan Alps produces an effective sorting and a mixing with natural sediments.

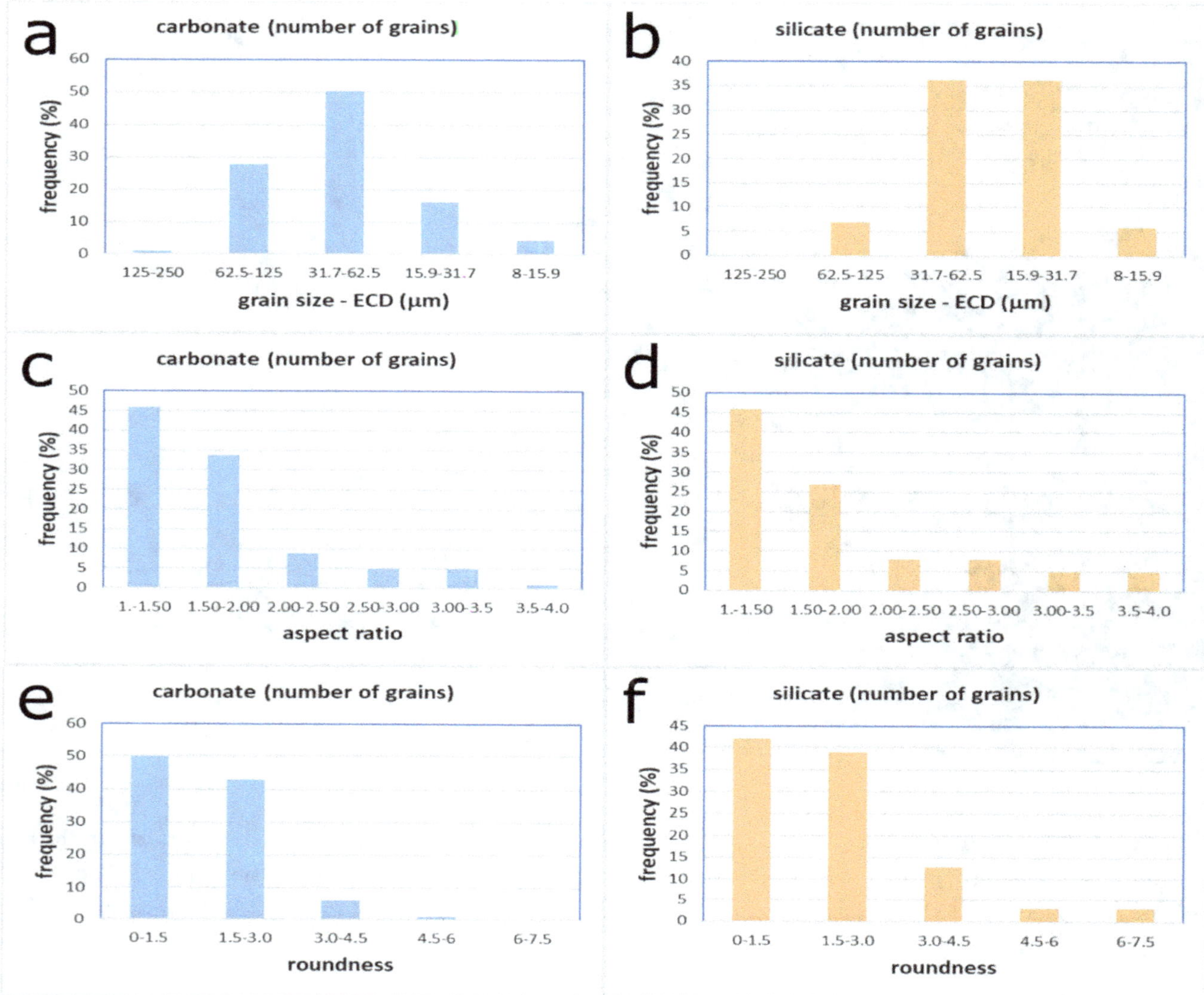

Figure 4. ECD (equivalent circular diameter; in µm), aspect ratio and roundness of carbonate and silicate fractions of a sediment collected at the Tana dei Tufi karst spring (Torano, Northern Apuan Alps) (see text for explanation).

3. Impact of Marble Slurry on Karst Groundwater

Once marble powder mixes up with water, it produces a fluid mud that rapidly disperses through the karst groundwater network [15,24]. This slurry negatively affects the groundwater quality, and its persistence in the karst system may produce remarkable environmental and economic damage [25,26]. In the Apuan Alps, the *marmettola* slurry has been observed in cave streams and karst springs (Figure 5).

Although several studies have described the environmental impacts of human activities on karst systems [1,27], there are relatively a few papers that specifically refer to the effects from quarrying on karst drainage systems. The relationship between quarrying activities and the environmental damage caused on carbonate rock has been well documented for over sixty years [28]. On the contrary, there are only a few reports that include major discussions of the environmental impacts on the karst aquifer of slurry produced by in-situ stone cutting and almost none referring to the impact on the karst stygofauna [29–31].

Figure 5. (**a**) *Marmettola* deposited in the occasionally flooded Buca di Equi cave. (**b**) Whitish, turbid water due to the marble slurry flowing out from the Equi karst spring observed several days after a strong flood in October 2013.

One of the most effective ways to investigate the dynamics of marble slurry transfer through karst subterranean networks is by monitoring of main spring hydro-physical parameters (flow rate, temperature, electrical conductivity and turbidity) coupled with sampling of suspended load. In the Apuan Alps, the turbidity observed at springs during floods can be often related to the suspended transport of marble slurry produced by quarries [15]. In fact, we have frequently observed that springs whose catchment areas do not embrace quarry districts have usually clear or slightly muddy water even during floods, apart from some springs that are fed by surface sinking water. On the contrary, turbidity due to carbonate slurry is systematically observed in water springs located in quarrying districts (Figure 5b).

In order to study the effects of the marble slurry on the karst aquifers of the Apuan Alps, a monitoring network has been recently (starting from April 2018) installed by the authority in charge for the environmental monitoring in Tuscany (Agenzia Regionale per la Protezione Ambientale Toscana-ARPAT). The monitoring settlement concerns three karst springs (Carbonera, Cartaro and Equi), and one cave stream (Monte Corchia Cave; see Figure 1 for location). Stream and spring monitoring is performed by automatic probes acquiring water level (resolution 1 mm), water temperature (resolution 0.1 °C), electrical conductivity (accuracy 1%) and turbidity (resolution 1—NTU (Nephelometric Turbidity Unit)); data are freely available at http://sira.arpat.toscana.it. Samples of *marmettola* were collected in quarries, whereas sediments were collected in a cave stream (Monte Corchia Cave) and karst springs, in order to compare the physical features of both marble sawing powder and springs suspended load.

One of the clearest examples of the marble slurry's impact on the Apuan Alps karst systems is provided by the springs located close to Equi Terme village (Massa) (see Figure 1 for location). Equi springs are fed by the Northern sector of the Apuan Alps karst aquifer, consisting mainly of marble and meta-dolostones. Marble quarries are the main human activity occurring in the catchment area. This multi-spring is characterized by high and very rapid fluctuations of water flow. During the dry season, the discharge can decrease down to 0.03–0.04 m^3/s, whereas major floods can increase it to a total flow rate of 15–20 m^3/s [32]. During floods and some days after, the water is highly turbid, but a slight whitish opalescence can persist for several days afterwards. Figure 6 shows the flow rate and turbidity diagram from April 2018 to April 2019. Floods are characterized by tight peaks followed by a rapid decrease of flow rate, indicating a quick transmission of hydraulic pressure through the karst network due to massive infiltration [33]. Turbidity peaks, which can reach 400 or more NTU, are not perfectly overlaid to the discharge peaks being shifted some days afterward.

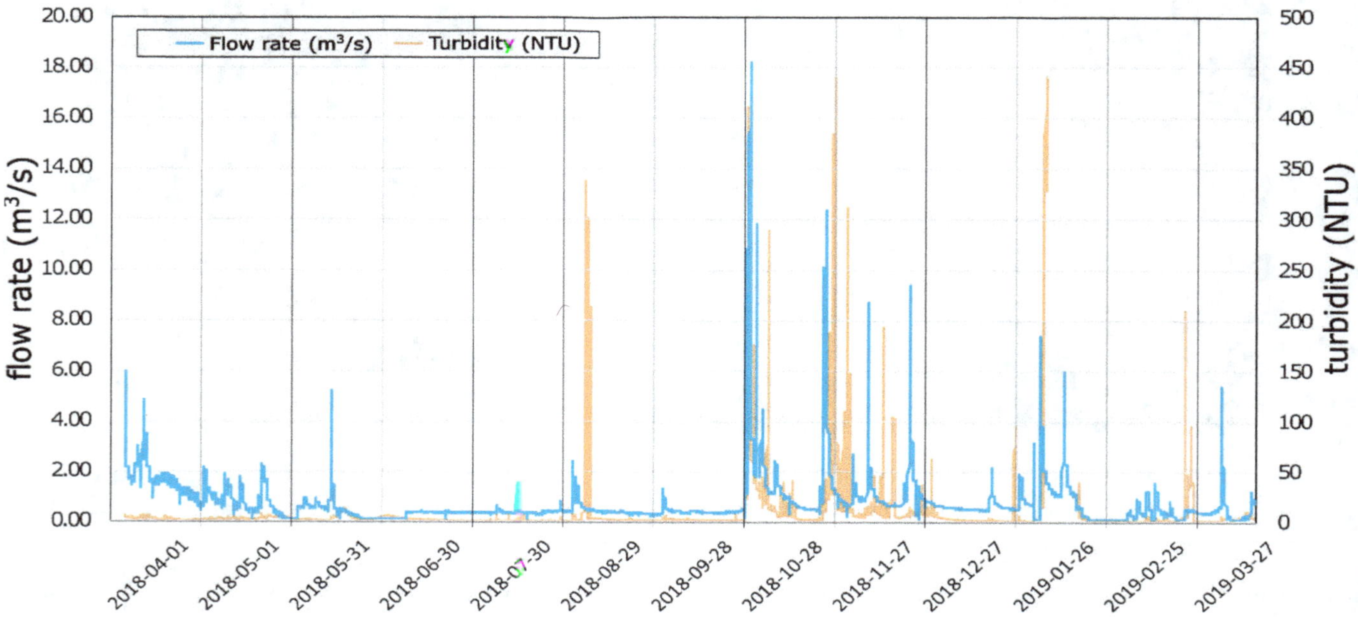

Figure 6. Flow rate and turbidity of Equi springs from April 2018 to April 2019 (dates are expressed as yyyy-mm-dd).

A detail graph (Figure 7) related to one minor flood (occurred on 1 September 2018 at Equi springs) following a long period of low flow rate, shows some features that clarify the origin of turbidity. The turbidity peak occurred around 48–50 h after the first flow peak, when the discharge decreased almost to the pre-flood level. This means that during minor floods the water remains clear until the *marmettola* slurry, washed away from the quarries and infiltrated in the catchment area, reaches the spring through a high-conductivity conduits network. In this case, the turbidity is not due to the mobilization of fine sediments contained in the phreatic sector of the aquifer but to the infiltration of surface runoff contaminated by marble slurry.

During major floods, the Equi springs displayed a different behavior of the turbidity parameters with respect to flow rate fluctuation (Figure 8). Turbidity showed a slight but significant increase just a few hours after the discharge rising, whereas the first substantial turbidity peak occurred with a delay of about 48 h after flood. Anyway, we observed that most of the solid load, measured as total suspended sediment (TSS), according to the methods proposed by Drysdale et al. [15], occurred during the maximum flow rate phase. Despite the high values of NTU, the turbidity peaks following with about two days of delay the first flood peak, contributed only for about the 40% to the amount of TSS, since discharge was low. Actually, the total TSS mobilized during this flood was exceptionally high, approaching 37 tons (37,000 kg).

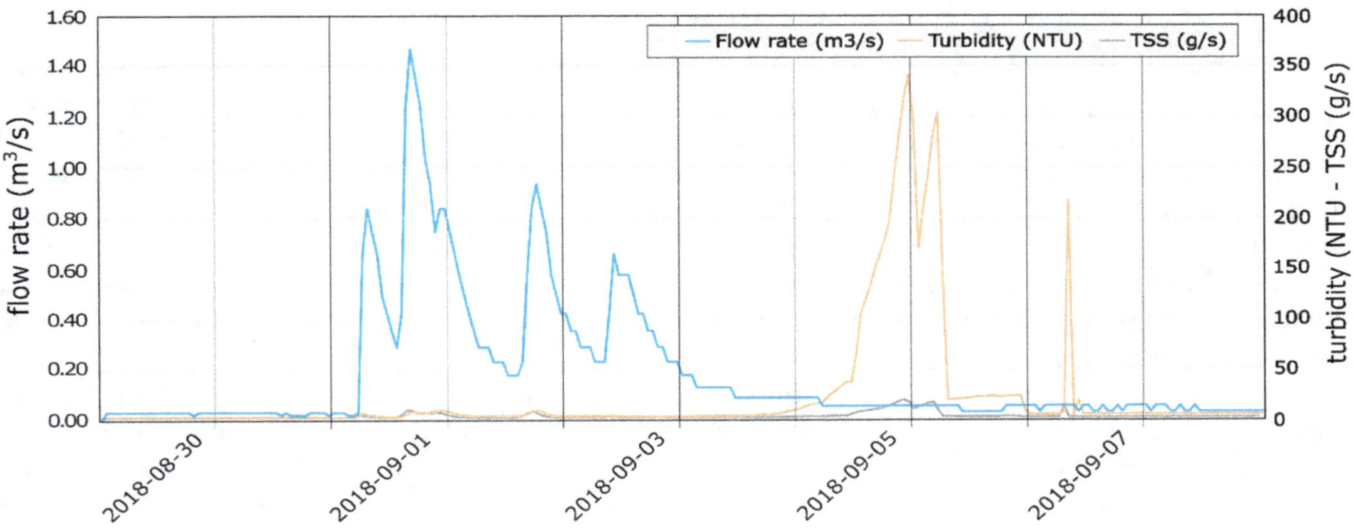

Figure 7. Flow rate and solid transport diagram of Equi springs during a minor flood on 1 September 2018 (dates are expressed as yyyy-mm-dd). The turbidity peak occurred more than four days after the flow peak.

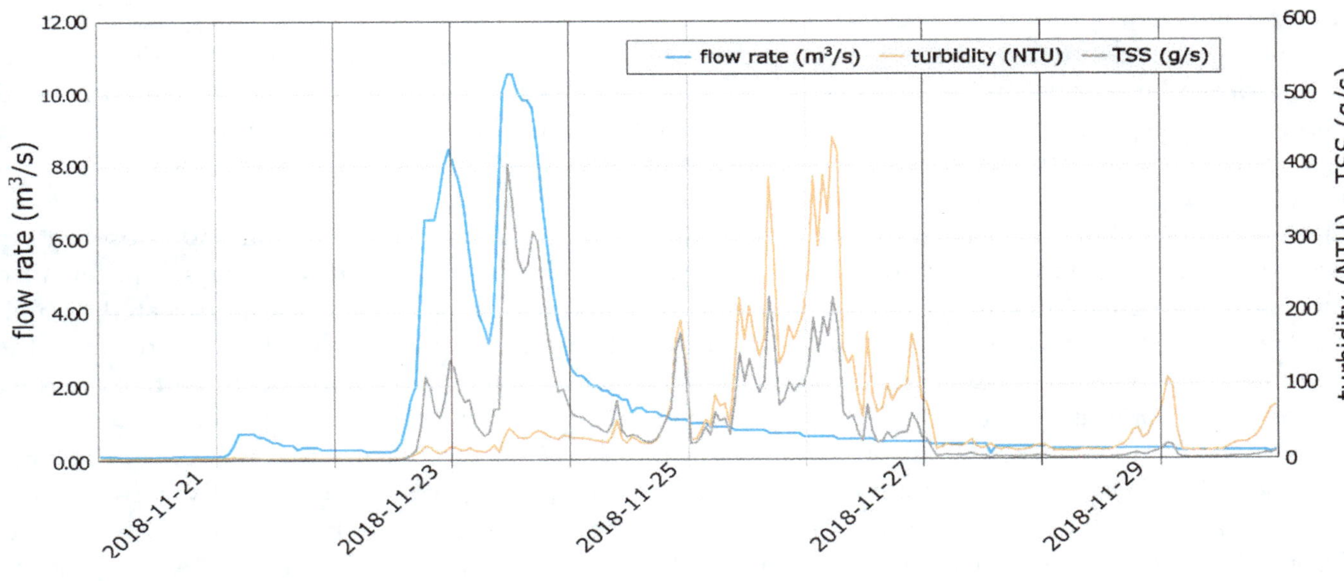

Figure 8. Flow rate and solid transport diagram of Equi springs during a major flood on 23 November 2018 (dates are expressed as yyyy-mm-dd). The major turbidity peak occurred about three days after the flow rate increase, but the TSS reached its greatest values during the second flood peak.

Comparing the different turbidity responses of these spring group during minor floods (Figure 7) with respect to major floods (Figure 8), we can infer that the TSS mobilized during the phase of major flow rate probably consists of sediments already stored in the phreatic system, which may be dragged away by water only in the case of a high flow velocity.

In conclusion, if we consider the effective TSS load instead of turbidity, we observe two major transport phases, the first during the flood peaks and the second one during the flood decremental phase (Figure 8). Considering the pattern of water and sediment fluxes, it may be argued that the second solid transport wave corresponded to the finest portion of *marmettola* washed away directly on quarry areas.

4. Impact of Marble Slurry on Karst Stygofauna

It is not an easy task assessing which are the main impacts of marble slurry on the organisms that live in karst aquifers. The main difficulty is the limited accessibility of these environments and the consequent impossibility of directly inspecting the habitats where stygobiotic species dwell [34]. Very appropriately, karst systems have been compared to black boxes in which one can assess exactly the input and the output but cannot directly know what is happening inside [35]. This similitude comes from hydrogeology but actually fits ecology very well. Karst stygofauna may be sampled in the epikarst (mostly caves) or at the karst outlets, through water filtration at the springs. In both cases, however, there is no direct inspection of the habitats except, perhaps, for gours, concretion pools, ponds and rivers, when present.

Despite the sampling difficulties, groundwater ecologists have collected enough information to outline the distributional patterns of many stygobiotic taxa. We know, for instance, that the distribution of stygofauna is not uniform in the aquifer, but that most of the species, which do not have adaptations to withstand fast flows, tend to colonize the habitats where the groundwater flow is low ([36–38] and references therein). Most of the species are collected from the small fractures and dissolution fissures (apertures from about one millimeter to a few centimeters) of the karst network, rather than in the fast drains [37,39–41]. Groundwater flow is therefore the main organizing factor for karst communities [42] and in extreme cases, such as in the event of an earthquake, the massive earthquake-induced flow discharge can trigger a huge washout of groundwater fauna with a destructive effect on the biological assemblages [36].

In normal conditions, however, groundwater flow acts together with other factors, such as behavioral adaptations. In fact, the occurrence of juveniles of some stygobiotic species at spring-fed karst outlets during a flood event [43] is suggestive of an adaptive strategy: some species release the eggs during the low groundwater level period and use the flood to disperse the juveniles so as to favor the colonization of new habitats [43]. Competition for trophic resources and predation are additional factors that can shape the distribution of communities within karst aquifers, as already observed in previous field and laboratory studies [44]. Finally, the chemical contamination of karst aquifers also influences the distribution of species within the karst network. An organic pollution (indicated by high nitrates and ammonium concentrations and particulate organic matter) may favor, for instance, the permanence of epigean species which, having high metabolic rates, exert a strong competitive pressure on the stygobiotic species, often replacing them completely within the community [45].

Unfortunately, there are no available data on the effect of marble quarry wastes on karst communities. Based on the observations made at Equi springs, we can realistically expect that a substantial quantity of *marmettola* settles in the smaller fractures of the karst systems, where the flow velocities are low and where, as mentioned above, most species are present. A complete or partial filling of these habitats likely induces chemical variations influencing the availability of oxygen, nutrients and habitat space reduction. This scenario recalls the phenomenon of clogging observed in hyporheic zones (transitional ecotones between surface water and groundwater bodies). In fact, the deposition of fine sediments in the hyporheic zone of the Alpine stream Noce Bianco in Northern Italy due to hydropeaking caused a reduction of the interstitial habitats available to stygobiotic taxa [46]. Similarly, Mori et al. [47] observed that the stygobiotic community of the hyporheic zones of the Sava River (Slovenia) was strongly affected after severe flooding with clogging of interstitial spaces by fine sediments. Microhabitat clogging may select species with certain physiological or trophic characteristics. For instance, in the reduced pore spaces of the hyporheic zones of the River Rhone catchment, only small-sized and slow-moving organisms, mainly copepods, ostracods and cladocerans, occurred [48]. This phenomenon has been confirmed recently by laboratory experiments. Korbel et al. [49] observed that stygobiotic crustaceans, such as the small-sized copepods and the large-sized amphipods and syncarids, are able to move easily in sand substrates (φ = 0.3–0.7 mm) and gravel (φ = 2–4 mm). However, only copepods were able to move in silts (φ < 60 μm), whereas the larger amphipods and syncarids remained on the sediment surface without being able to dig in.

The inability to cope with fine sediments for syncarids and amphipods increases when the groundwater level decreases. In microcosm experiments, none of these animals managed to dig in the mud layers when the overlying water was removed [49]. This phenomenon has serious consequences in terms of loss of ecosystem services. In fact, laboratory experiments have shown that syncarids and amphipods are crucial for maintaining the effective porosity of fine sediment-filled habitats. In the absence of these animals, porosity decreases significantly in two months, causing a further reduction of pore spaces [50]. The diameter of pore spaces in silts is predicted to be <25 μm [51] and the same is expected in quarry waste deposits because most of the particles we measured were from 8 to 32 μm in diameter. The intergranular spaces of *marmettola* deposits are, therefore, smaller than the smallest dimension of most of stygobiotic crustaceans. Copepods manage to dig in these deposits, and it has to be expected that fine sediments adhere tenaciously to the body of the animal (Figure 9). We do not know how this sort of "coating" will negatively affect the physiology of these organisms, but fissured aquifers filled with marble slurry are likely inhospitable habitats for most of karst stygofauna.

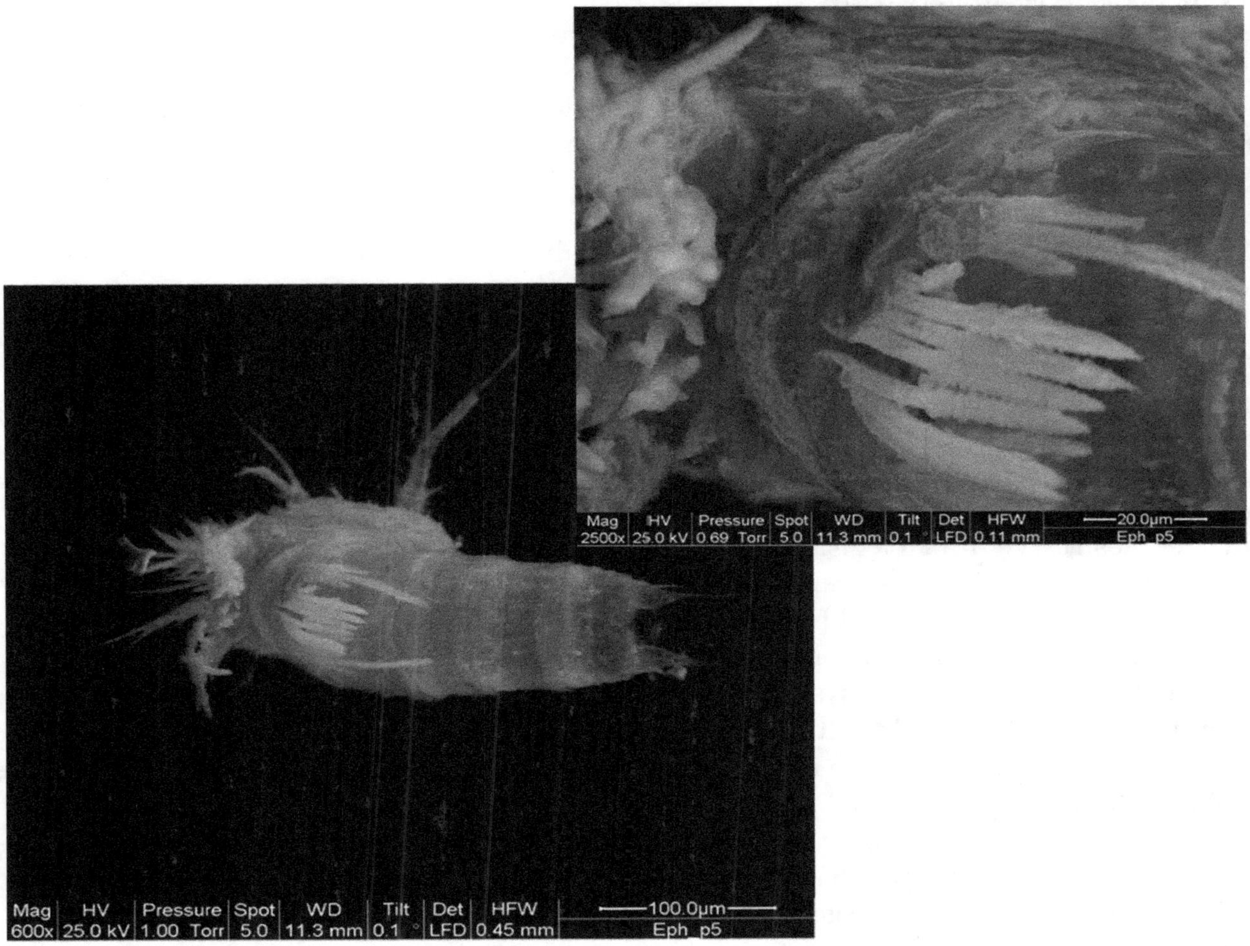

Figure 9. Silt-coating of an individual of *Elaphoidella phreatica* (Crustacea Copepoda) from a clay-sandy aquitard in Tuscany (Italy). The photos were obtained by scanning electron microscopy. The grains of silt adhere tenaciously to the body of the animal also covering the setae and appendages. In the detail of the photo, the coating of the fifth thoracic leg involved in mating behavior of these animals is visible.

In order to test this hypothesis, we carried out biological surveys in the Monte Corchia cave, located in one of the most intensely exploited areas with marble quarries in the Apuan Alps. The Monte Corchia karst system is located in the south-western part of the Apuan Alps and is developed inside the carbonate core of a syncline, almost completely enclosed by a low-permeable basement (Figure 1). Monte Corchia cave is about 65 km long and 1185 m deep and it is presently one the largest and deepest caves in Italy [52].

Specimens were collected from the vadose zone of the cave, located between 900 and 800 m a.s.l., which consists of several passages that have been part of the touristic path since 2001. We sampled six percolation water drips with permanent water flow and six concretional pools. The water in the concretional pools was collected by syringes at approximately quarterly intervals from June 2016 to October 2019. Percolation water drips were directed into a 500 mL plastic filtering bottles with partly removed walls covered by a net of 60 μm mesh a few cm above the bottom. The animals coming with the drips were retained in the small amount of water in the filtering bottles and collected on a quarterly basis. Altogether we collected about 700 specimens, 85% of which came from the drips. All the animals were copepods, as also observed in other studies (e.g., [35,53,54]). However, all the animals collected in our surveys belonged to a single taxon, a new undescribed species of the canthocamptid harpacticoid genus *Moraria*.

This result contrasts with the data of Meleg et al. [55], who reported 11 stygobiotic copepod species from drips and drip pools of five Romanian caves; of Galassi et al. [37], who reported up to eight copepod species from two drips of Frasassi cave; and especially, with the data (up to 33 species per cave) recorded in the Slovenian caves and reviewed by Pipan and Culver [53]. That of Monte Corchia Cave seems, therefore, to be an unusual situation, being similar only to two non-Dinaric caves in Slovenia; namely, Huda luknja, occurring in an isolated karst area, and Snežna jama, an ice cave in the Kamnik–Savinja Alps (two stygobiotic copepod species per cave) [53]. At present, data do not allow us to know whether the persistently low biodiversity observed in Monte Corchia Cave is due to the presence of *marmettola* and its clogging effects, or to other factors, such as the isolation of the cave that prevents colonization of groundwater similarly to the conditions of Huda luknja in Slovenia [54].

5. Discussion and Conclusions

This study did not aim at proposing strategies to contain the production of marble slurry or to prevent its dispersion in subterranean waterways but had the objective to denounce the fact that too many aspects of this environmental issue have not yet been adequately addressed. For example, the priority issue for the Italian authorities is the correct identification of *marmettola* as waste or as a by-product. This is not a matter of little importance, since the management regime of by-products is much less restrictive from a legislative point of view. Although the marble powder produced by the quarrying activities in the Apuan Alps is managed as waste, the cross-checks carried out by the competent authorities have highlighted that a relevant amount of it is dispersed in the environment. It would therefore be appropriate to apply an economic sanction that compensates for the environmental damage, including the loss of ecosystem services, consequently to *marmettola* dispersion. However, the quantification of this penalty, although expressly demanded by the Water Framework Directive in Europe, is difficult to determine now. The main impediment is due to the scarce knowledge of the impacts of marble slurry on the fauna dwelling in karst aquifers. With this review we aimed at disseminating information on the possible biological impacts, and secondly, on the poor biodiversity detected in some parts of the Apuan Alps karst system, such as in Monte Corchia cave. Stygofauna are "out of sight" and it is, therefore, easy to maintain these species as "out of mind" [56].

However, this short-sighted way to perceive karst ecosystems is an oversight before which we cannot remain silent. Fortunately, the turbidity due to *marmettola* detected at the springs is visible to the naked eye and has an immediate impact on the public, which is shocked by the idea of drinking "dirty" water. The quantification of the environmental impact in this case is easier, since it is possible to estimate the damage caused to the aqueduct filters by the high turbidity or the temporary closure of the distribution network. However, not only the springs presently captured for civil aqueducts but those that can be potentially captured in the future, must be monitored and possibly subjected to safeguarding measures.

Based on a first set of data, this study documented the worsening of the water quality during flood events due to the *marmettola* transport in the karst system of the Apuan Alps and a supposed negative effect of suspended solid deposition on the local groundwater community. The study highlighted that, although marble slurry does not constitute an immediate danger to public health, it is nevertheless responsible for a serious impact on the Apuan surface and subterranean waterways. We suspect that an unknown amount of *marmettola* has been already accumulated in the karst aquifers and it is discharged only in conjunction with high-intensity rain events. As already stated, the effects of such a physical pollution, apart the obvious deterioration of the quality of the water intended for human consumption, are still not known in detail. It can be expected that the marble slurry has potentially reduced the water exchanges among karst conduits and fractures and impacted the rock mass porosity, changing the hydrodynamics of the karst networks by also reducing their storage capacity. If so, it could be expected that the local environment will remain damaged (at some extent) for an unknown time period, even if the *marmettola* dispersion were to be completely stopped. Resilience will be hindered by the depletion of the groundwater biological communities that, otherwise, would be able to provide ecological services of great importance, most of them related to the aquifer self-purification.

In conclusion, the first results of this study have highlighted that the present-day footprint of quarrying activities in the Apuan Alps is greater than that indicated on maps or aerial photographs. It is necessary, therefore, to boost participatory processes involving economic stakeholders and environmental experts, and speleologists who are often the most informed about the conditions within karst systems too. Due to their long history of exploitation, the Apuan Alps represent a red flag that we hope will trigger biological and hydrogeological studies in other areas.

Author Contributions: Conceptualization, L.P., P.C. and T.D.L.; methodology, L.P., P.C. and T.D.L.; validation, L.P. and D.M.P.G.; formal analysis, L.P.; investigation, L.P., P.C. and T.D.L.; species identification: T.D.L. and D.M.P.G.: resources, D.M.P.G.; data curation, L.P., P.C., T.D.L. and D.M.P.G.; writing—original draft preparation, L.P.; writing—review and editing, L.P., P.C., T.D.L. and D.M.P.G.; visualization, L.P.; supervision, L.P. and D.M.P.G.; project administration, L.P. and D.M.P.G.; funding acquisition, D.M.P.G.

Acknowledgments: We wish to thank all the institutions that in different ways have encouraged and supported this research, with special mentions to: Agenzia Regionale per la Protezione dell'Ambiente, Toscana—ARPAT; Parco Regionale delle Alpi Apuane, Federazione Speleologica Toscana.

References

1. Williams, P.W. Environmental change and human impact on karst terrains: An introduction. *Catena* **1993**, *25*, 1–19.
2. Van Beymen, P.; Townsend, K. A disturbance Index for karst environments. *Environ. Manag.* **2005**, *36*, 101–116. [CrossRef] [PubMed]
3. Hamdan, I.; Margane, A.; Ptak, T.; Wiegand, B.; Sauter, M. Groundwater vulnerability assessment for the karstic aquifers of Tanour and Rasoun springs catchment area (NW-Jordan) using COP and EPIK intrinsic methods. *Environ. Earth Sci.* **2016**, *75*, 1474. [CrossRef]
4. Gibert, J.; Culver, D.C. Assessing and conserving groundwater biodiversity: An introduction. *Freshw. Biol.* **2009**, *54*, 639–648. [CrossRef]
5. Gibert, J.; Danielopol, D.L.; Stanford, J.A. *Groundwater Ecology*; Academic Press: New York, NY, USA, 1994; p. 571.
6. Coineau, N. Adaptations to interstitial groundwater life. In *Ecosystems of the World, 30: Subterranean Ecosystems'*; Wilkens, H., Culver, D.C., Humphreys, W.F., Eds.; Elsevier: Amsterdam, The Netherlands, 2000; pp. 189–210.
7. Di Lorenzo, T.; Di Marzio, W.D.; Spigoli, D.; Baratti, M.; Messana, G.; Cannicci, S.; Galassi, D.M.P. Metabolic rates of a hypogean and an epigean species of copepod in an alluvial aquifer. *Freshw. Biol.* **2015**, *60*, 426–435. [CrossRef]
8. Iepure, S.; Rasines-Ladero, R.; Meffe, R.; Carreno, F.; Mostaza, D.; Sundberg, A.; Di Lorenzo, T.; Barroso, J.L. The role of groundwater crustaceans in disentangling aquifer type features—A case study of the Upper Tagus Basin, central Spain. *Ecohydrology* **2017**, *10*, e1876. [CrossRef]

9. Galassi, D.M.P.; Stoch, F.; Fiasca, B.; Di Lorenzo, T.; Gattone, E. Groundwater biodiversity patterns in the Lessinian Massif of northern Italy. *Freshw. Biol.* **2009**, *54*, 830–847.

10. Di Lorenzo, T.; Cipriani, D.; Fiasca, B.; Rusi, S.; Galassi, D.M.P. Groundwater drift monitoring as a tool to assess the spatial distribution of groundwater species into karst aquifers. *Hydrobiologia* **2018**, *813*, 137–156. [CrossRef]

11. Boulton, A.J.; Fenwick, G.; Hancock, P.J.; Harvey, M.S. Biodiversity, functional roles and ecosystem services of groundwater invertebrates. *Invertebr. Syst.* **2008**, *22*, 103–116. [CrossRef]

12. Griebler, C.; Avramov, M. Groundwater ecosystem services: A review. *Freshwat. Sci.* **2014**, *34*, 355–367. [CrossRef]

13. Doveri, M.; Piccini, L.; Menichini, M. Hydrodynamic and geochemical features of metamorphic carbonate aquifers and implications for water management: The Apuan Alps (NW Tuscany, Italy) case study. In *Karst Water Environment*; The Handbook of Environmental Chemistry; Younos, T., Schreiber, M., Kosic Ficco, C., Eds.; Springer: Cham, Swizterland, 2019; Volume 68, pp. 209–249. [CrossRef]

14. Civita, M.; Forti, P.; Marini, P.; Meccheri, M.; Micheli, L.; Piccini, L.; Pranzini, G. *Pollution Vulnerability Map for the Aquifers of the Apuane Alps—A Brief Guide*; Gr. Naz. per la Difesa dalle Catastrofi Idrogeologiche; NRC: Firenze, Italy, 1991; p. 56.

15. Drysdale, R.N.; Pierotti, L.; Piccini, L.; Baldacci, F. Suspended sediments in karst spring waters near Massa (Tuscany), Italy. *Environ. Geol.* **2001**, *40*, 1037–1050.

16. Regione Toscana. Praer: Piano Regionale Delle Attività Estrattive. Available online: http://www.regione.toscana.it/-/praer-piano-regionale-delle-attivita-estrattive (accessed on 1 September 2019).

17. Gunn, J.; Hardman, D.; Lindesay, W. Problems of limestone quarrying in an adjacent to the Peak District National Park. *Annales de la Société géologique de Belgique* **1985**, *108*, 59–63.

18. Ekmekci, M. *Impact of Quarries on Karst Groundwater Systems. Hydrogeological Processes in Karst Terranes*; IAHS Publication: Wallingford, Oxfordshire, UK, 1993; Volume 207, pp. 3–6.

19. Mastrorocco, G.; Salvini, R.; Vanneschi, C. Fracture mapping in challenging environment: A 3D virtual reality approach combining terrestrial LiDAR and high definition images. *Bull. Eng. Geol. Environ.* **2018**, *77*, 691–707. [CrossRef]

20. Gulli, D.; Pellegri, M.; Marchetti, D. Mechanical behaviour of Carrara marble rock mass related to geo-structural conditions and in-situ stress. In Proceedings of the 15th Pan-American Conference on Soil Mechanics and Geotechnical Engineering (PCSMGE)/8th South American Congress on Rock Mechanics (SCRM), Buenos Aires, Argentina, 15–18 November 2015; pp. 253–260.

21. Gentili, R.; Sgorbati, S.; Baroni, C. Plant species patterns and restoration perspectives in the highly disturbed environment of the Carrara marble quarries (Apuan Alps, Italy). *Restor. Ecol.* **2011**, *19*, 32–42. [CrossRef]

22. Cantisani, E.; Fratini, F.; Malesani, P.; Molli, G. Mineralogical and petrophysical characterisation of white Apuan Marble. *Periodico di Mineralogia* **2005**, *74*, 117–140.

23. Cantisani, E.; Pecchioni, E.; Fratini, F.; Garzonio, C.A.; Malesani, P.; Molli, G. Thermal stress in the Apuan marbles: Relationship between microstructure and petrophysical characteristics. *Int. J. Rock Mech. Min. Sci.* **2009**, *46*, 128–137. [CrossRef]

24. Mantelli, F.; Montigiani, A.; Lotti, L.; Piccini, L.; Bianucci, P.; Malcapi, V. Le acque sotterranee del sistema carsico del M. Corchia: Valorizzazione, salvaguardia e rischi di inquinamento. In Proceedings of the 3 Convegno Nazionale "Protezione e gestione delle acque sotterranee per il III millennio", Parma, Italy, 13–15 October 1999; pp. 115–125.

25. Rizzo, G.; D'Agostino, F.; Ercoli, L. Problems of soil and groundwater pollution in the disposable of "marble" slurries in NW Sicily. *Environ. Geol.* **2008**, *55*, 929–935. [CrossRef]

26. Celik, M.Y.; Sabah, E. Geological and technical characterisation of Iscehisar (Afyon-Turkey) marble deposits and the impact of marble waste on environmental pollution. *J. Environ. Manag.* **2008**, *87*, 106–116. [CrossRef]

27. Drew, D.H.H.; Hotzl, H. *Karst Hydrogeology and Human Activities: Impact, Consequences, and Implications*; IAH International contributions to hydrogeology, 20; A.A. Balkema: Rotterdam, The Netherlands, 1999.

28. Foose, R.M. Ground-water behavior in the Hershey Valley, Pennsylvania. *Geol. Soc. Am. Bull.* **1953**, *64*, 623–645. [CrossRef]

29. Whitten, T. Applying ecology for cave management in China and neighbouring countries. *J. Appl. Ecol.* **2009**, *46*, 520–523. [CrossRef]
30. Souza-Silva, M.; Martins, R.P.; Ferreira, R.L. Cave conservation priority index to adopt a rapid protection strategy: A case study in Brazilian Atlantic rain forest. *Environ. Manag.* **2015**, *55*, 279–295. [CrossRef] [PubMed]
31. Sugai, L.S.M.; Ochoa-Quintero, J.M.; Costa-Pereira, R.; Roque, F.O. Beyond above ground. *Biodivers. Conserv.* **2015**, *24*, 2109–2112. [CrossRef]
32. Worthington, S.R.H. How preferential flow delivers pre-event groundwater rapidly to streams. *Hydrol. Proc.* **2019**, *33*, 2373–2380. [CrossRef]
33. Poggetti, E.; Lazzaroni, M.; Verole, M.; Piccini, L. Risultati e interpretazione idrodinamica del monitoraggio in continuo delle sorgenti carsiche di Equi Terme (Alpi Apuane). In Proceedings of the 22 Congresso Nazionale di Speleologia, Pertosa, Auletta (Salerno), Italy, 30 May–2 June 2015; pp. 369–374.
34. Korbel, K.; Chariton, A.; Stephenson, S.; Greenfield, P.; Hose, G.C. Wells provide a distorted view of life in the aquifer: Implications for sampling, monitoring and assessment of groundwater ecosystems. *Sci. Reports* **2017**, *7*, 40702. [CrossRef]
35. Mangin, A. L'approche systémique du karst, conséquences conceptuelles et méthodologiques. *Bulletin de la Société Méridionale de Spéléologie et de Préhistoire* **1984**, *24*, 131–145.
36. Galassi, D.M.P.; Lombardo, P.; Fiasca, B.; Di Cioccio, A.; Di Lorenzo, T.; Petitta, M.; Di Carlo, P. Earthquakes trigger the loss of groundwater biodiversity. *Sci. Rep.* **2014**, *4*, 6273. [CrossRef]
37. Galassi, D.M.P.; Fiasca, B.; Di Lorenzo, T.; Montanari, A.; Porfirio, S.; Fattorini, S. Groundwater biodiversity in a chemoautotrophic cave ecosystem: How geochemistry regulates microcrustacean community structure. *Aquat. Ecol.* **2017**, *51*, 75–90. [CrossRef]
38. Stoch, F.; Fiasca, B.; Di Lorenzo, T.; Porfirio, S.; Petitta, M.; Galassi, D.M.P. Exploring copepod distribution patterns at three nested spatial scales in a spring system: Habitat partitioning and potential for hydrological bioindication. *J. Limnol.* **2016**, *75*, 1–13. [CrossRef]
39. Brancelj, A. Microdistribution and high diversity of Copepoda (Crustacea) in a small cave in central Slovenia. *Hydrobiologia* **2002**, *477*, 59–72. [CrossRef]
40. Brancelj, A. Fauna of an unsaturated karstic zone in Central Slovenia: Two new species of Harpacticoida (Crustacea: Copepoda), *Elaphoidella millennii* n. sp. and E. tarmani n. sp., their ecology and morphological adaptations. *Hydrobiologia* **2009**, *621*, 85–104. [CrossRef]
41. Culver, D.C.; Christman, M.C.; Sket, B.; Trontelj, P. Sampling adequacy in an extreme environment: Species richness patterns in Slovenian caves. *Biodivers. Conserv.* **2004**, *13*, 1209–1229. [CrossRef]
42. Gibert, J.; Vervier, P.; Malard, F.; Laurent, R.; Reygrobellet, J.L. Dynamics of communities and ecology of karst ecosystems: Example of three karsts in Eastern and Southern France. In *Groundwater Ecology*; Gibert, J., Danielopol, D.L., Stanford, J.A., Eds.; Academic Press: San Diego, CA, USA, 1994; pp. 425–450.
43. Turquin, M.J. The tactics of dispersal of two species of Niphargus (perennial, Troglobitic amphipoda). In Proceedings of the 8th International Congress of Speleology, Bowling-Green, KY, USA, 18–24 July 1981; Volume 1, pp. 353–355.
44. Fattorini, S.; Lombardo, P.; Fiasca, B.; Di Cioccio, A.; Di Lorenzo, T.; Galassi, D.M.P. Earthquake-related changes in species spatial niche overlaps in spring communities. *Sci. Rep.* **2017**, *7*, 443. [CrossRef] [PubMed]
45. Malard, F.; Reygrobellet, J.L.; Mathieu, J.; Lafont, M. The use of invertebrate communities to describe groundwater flow and contaminat transport in a fractured rock aquifer. *Arch. Hydrobiol.* **1994**, *131*, 93–110.
46. Bruno, M.C.; Maiolini, B.; Carolli, M.; Silveri, L. Impact of hydropeaking on hyporheic invertebrates in an Alpine stream (Trentino, Italy). *Ann. Limnol.* **2009**, *45*, 157–170. [CrossRef]
47. Mori, N.; Simčič, T.; Žibrat, U.; Brancelj, A. The role of river flow dynamics and food availability for the hyporheic microcrustaceans: A reach scale study. *Fund. Appl. Limnol. Arch. Hydrobiol.* **2012**, *180*, 335–349. [CrossRef]
48. Descloux, S.; Datry, T.; Usseglio-Polatera, P. Trait-based structure of invertebrates along a gradient of sediment colmation: Benthos versus hyporheos responses. *Sci. Total Environ.* **2014**, *466*, 265–276. [CrossRef]
49. Korbel, K.L.; Stephenson, S.; Hose, G.C. Sediment size influences habitat selection and use by groundwater macrofauna and meiofauna. *Aquat. Sci.* **2019**, *81*, 39. [CrossRef]
50. Stumpp, C.; Hose, G.C. The impact of water table drawdown and drying on subterranean aquatic fauna in in vitro experiments. *PLoS ONE* **2013**, *8*, e78502. [CrossRef]

51. Stakman, W.P. Determination of pore size by the air bubbling pressure method. In Proceedings of the Wageningen Symposium "Water in the unsaturated zone"; IASH/AIHS: Unesco, Belgium, 1966; pp. 366–372. Available online: http://hydrologie.org/redbooks/a082/RB082.pdf (accessed on 1 November 2019).
52. Piccini, L.; Zanchetta, G.; Drysdale, R.N.; Hellstrom, J.; Isola, I.; Fallick, A.E.; Leone, G.; Doveri, M.; Mussi, M.; Mantelli, F.; et al. The environmental features of the Monte Corchia cave system (Apuan Alps, central Italy) and their effects on speleothem growth. *Int. J. Spel.* **2008**, *37*, 153–172. [CrossRef]
53. Pipan, T.; Culver, D.C. Organic carbon in shallow subterranean habitats. *Acta Carsologica* **2013**, *42*, 291–300. [CrossRef]
54. Pipan, T.; Navodnik, V.; Janzekovic, F.; Novak, T. Studies of the fauna of percolation water of Huda Luknja, a cave in isolated karst in Northeast Slovenie. *Acta Carsologica* **2008**, *37*, 141–151. [CrossRef]
55. Meleg, I.N.; Moldovan, O.T.; Iepure, S.; Fiers, F.; Brad, T. Diversity patterns of fauna in dripping water of caves from Transylvania. *Ann. Limnol. Int. J. Lim.* **2011**, *47*, 185–197. [CrossRef]
56. Wood, P.J.; Gunn, J.; Perkins, J. The impact of pollution on aquatic invertebrates within a subterranean ecosystem—Out of sight out of mind. *Archiv Fur Hydrobiol.* **2002**, *155*, 223–237. [CrossRef]

12

Evaluation of the Effect of Gold Mining on the Water Quality in Monterrey, Bolívar (Colombia)

Alison Martín [1,2], Juliana Arias [1], Jennifer López [1], Lorena Santos [1], Camilo Venegas [1], Marcela Duarte [3], Andrés Ortíz-Ardila [4], Nubia de Parra [3], Claudia Campos [1] and Crispín Celis Zambrano [2,*]

1 Department of Microbiology, Laboratorio de Indicadores de Calidad de Agua y Lodos (LIAL), Pontificia Universidad Javeriana, Carrera 7 No. 43–82, Bogotá 110231, Colombia; alison.martin@javeriana.edu.co (A.M.); julichaque@gmail.com (J.A.); jetalomo@gmail.com (J.L.); lorena28sant@gmail.com (L.S.); c.venegas@javeriana.edu.co (C.V.); campos@javeriana.edu.co (C.C.)

2 Department of Chemistry, Pontificia Universidad Javeriana, Carrera 7 No. 43–82, Bogotá 110231, Colombia

3 Independient Investigator, Bogotá 110111, Colombia; duartemarce@yahoo.com (M.D.); nubiapdeparra@gmail.com (N.d.P.)

4 Hydraulic and Environmental Engineering Department, Pontificia Universidad Católica de Chile, Av. Vicuña Mackenna 4860, Macul, Santiago 7820436, Chile; adortiz@uc.cl

* Correspondence: crispin.celis@javeriana.edu.co

Abstract: Gold mining uses chemicals that are discharged into rivers without any control when there are no good mining practices, generating environmental and public health problems, especially for downstream inhabitants who use the water for consumption, as is the case in Monterrey township, where the Boque River water is consumed. In this study, we evaluate Boque River water quality analyzing some physicochemical parameters such as pH, heavy metals, Hg, and cyanide; bioassays (*Lactuca sativa, Hydra attenuata*, and *Daphnia magna*), mutagenicity (Ames test), and microbiological assays. The results show that some physicochemical parameters exceed permitted concentrations (Hg, Cd, and cyanide). *D. magna* showed sensitivity and *L. sativa* showed inhibition and excessive growth in the analyzed water. Mutagenic values were obtained for all of the sample stations. The presence of bacteria and somatic coliphages in the water show a health risk to inhabitants. In conclusion, the presence of Cd, Hg, and cyanide in the waters for domestic consumption was evidenced in concentrations that can affect the environment and the health of the Monterrey inhabitants. The mutagenic index indicates the possibility of mutations in the population that consumes this type of water. Bioassays stand out as an alert system when concentrations of chemical contaminants cannot be analytically detected.

Keywords: bioassays; gold mining; health risk; mercury; microbiological indicators; mutagenicity; toxicity

---

1. Introduction

Gold mining in developing countries is the main source of income for 30 million miners globally. About 12% of global gold production is through illegal mining that provides a significant economic benefit to miners but also proves hazardous/harmful for the environment by causing impacts such as water source sedimentation, land cover degradation, deforestation, soil degradation, and chemical contamination with mercury, cyanide, nitric acid, and zinc [1–3]. In Colombia, despite the various alternatives to avoid the use of Hg in gold extraction, the use of the elemental Hg–Au amalgamation method in small-scale artisanal mining areas is extensive [4,5].

Within the gold mining protocols, mercury and cyanide play an important role. These materials are easy to use, available at a low-cost, and easily accessible. However, there is little awareness among the users or villagers about the use risk of cyanide and mercury in the gold extraction process [1,2]. This activity has led to serious pollution of terrestrial and aquatic ecosystems in emerging countries, impacting mining and fishing communities, and also these polluting elements can reach human beings [4–7].

According to the records of the Colombian Mining Association (ACM), gold production increased in 2020, going from 8.9 tons in 2019 to 9.5 tons in the first quarter of 2020, this represents a growth of 7% [8]. On the other hand, for gold extraction, 86% is considered illegal, taking place without a recognized mining title or without being registered. Medium-scale mining constitutes up to 26% and large-scale mining only takes up 2% of the total [9,10]. The population of the Bolívar department is 2,195,811 inhabitants, according to DANE's (National Statistics Administrative Department) projection for 2019 [11]. According to Carranza-Lopez et al. [4] the gold-mining districts (GMDs) at the department of Bolívar have extensive Hg contamination, and this situation requires special attention to reduce environmental and human health problems.

Municipalities of Montecristo, Santa Rosa del Sur, San Martín de Loba, Morales, San Pablo, Barranco de Loba, and Simití that are in Bolívar are where gold mining mainly takes place. Simití is known as the municipality that has the largest gold mining activity within the Bolívar department. It has an estimated population of 10,360 inhabitants in an area of 1345 km^2, the mining activity occurs in the Boque River, which flows in Simití. It starts on Serranía San Lucas, passes through Monterrey district, and flows into Magdalena River [12–14] (Figure 1).

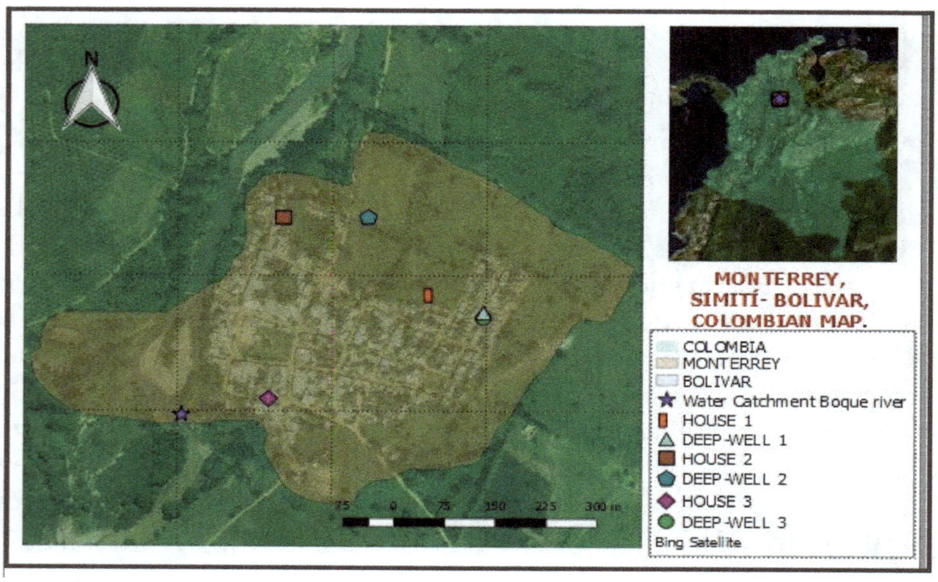

Figure 1. Location of sampling stations in Monterrey, Simití-Bolívar, Colombia. Source: authors.

Gold mining in Middle Magdalena has been carried out through artisanal practices, without considering the implications in the community and ecosystems due to the practice of non-regulated techniques affecting the environment, natural resources as well as health conditions and welfare of the population. Gold mining severely affects water resources, biodiversity, animals, flora, and fauna in its geographical area. In addition, the presence of certain types of mining settlements bring to pass certain types of domestic wastewater discharges without treatment to the Boque River, affecting the quality of the water and the inhabitants downstream [15–19]. The discharge of wastewater into a water body involves a large number and diversity of heavy toxic chemicals, many of which are unknown. These chemicals may react with each other, which can increase the toxicity level, which creates a negative impact on the structure and functioning of the natural ecosystem [4,20].

To determine the effect of gold extraction in the region, the evaluation of physicochemical parameters of the water is required. Nevertheless, the illegal settlements do not have sanitation systems, so microbiological contamination becomes an additional problem. However, even if in some of the above-mentioned situations the parameters could be between the legal requirements, it should be considered that trace heavy elements might have an impact on the population and the ecosystem after long periods of exposure. Thus, it is necessary to test different representatives of the trophic chain to identify the impact of the pollutants through bioassays tests [20–23].

Bioassays are described as alert mechanisms for long-term periods of exposure to chemical pollutants. These are used as indicators of substances that are harmful to living cells and tissues, useful even in the cases where physicochemical parameters fulfill the requirements of water quality [24]. Likewise, this possible bioaccumulation of chemical elements in the trophic chains can generate mutagenicity or toxicity, which is why it is important to be able to establish whether a complex system such as a water sample from a mining region has these undesirable characteristics, which can be detected by the Ames test or bioassays [25–27]. Some Latin American countries have made progress in the application of toxicity tests, while for Colombia, toxicity tests in natural environments are scarce compared to the evaluation of hazardous waste and industrial dumping [28,29]. On the other hand, the Ames test has proven to be effective for the identification of potentially carcinogenic or mutagenic chemicals, achieving its immediate adoption and its requirement by regulatory authorities around the world [30]. In the Ames test, *Salmonella typhimurium* (*S. typhimurium*) is used as an indicator of bacterial mutagenesis as a consequence of exposure to chemical contaminants [25,26].

Taking into consideration that the water of the Boque River is used in human consumption without treatment and it collects chemical pollutants from the mining activity, such as mercury and cyanide, the use of the Ames test in the evaluation of this water will permit the evaluation of its possible carcinogenic or mutagenic effect, making it a relevant issue for the inhabitants of Simití. For this reason, it is necessary to have data on bioassays and Ames test indicators in environmental samples, especially in mining, which has become one of the most important fonts of economic resources in Colombia and at the same time of damage not sufficiently evaluated to date. In order to have a complete evaluation of the water quality in relation to the possible presence of bacteria, viruses, and parasites and the risk to the inhabitants, it is necessary to use indicators of fecal contamination, with the most used indicators being total coliforms and *Escherichia coli* as bacterial indicators and somatic phages as viral indicators, which allow indication of the presence of pathogenic microorganisms in the water.

The aim of this research is to evaluate the impact generated by the exploitation of bad mining practices such as the use of dangerous chemical compounds in gold mining, which are drained into surface waters such as the Boque River in the South of Bolívar, Colombia, as well as the waste generated in the mining settlements. The assessment of the impact on the environment, living organisms, and human health will be done through the detection of heavy metals, microbiological indicators, and bioassays, which through a joint assessment will provide important aspects to protect the health of the inhabitants of Monterrey.

2. Materials and Methods

2.1. Study Area

The Boque River has an approximate area of 876 km^2 and merges into the Magdalena River. The Monterrey district belongs to the Municipality of Simití, Department of Bolívar (Colombia). The inhabitants who live within the Monterey township collect and use the water from the Boque River for different activities, this being the main source of water supply (Figure 1) [27].

2.2. Water Physicochemical Analysis, Heavy Metals, and Cyanide Detection

Some physicochemical parameters were analyzed such as pH (pH/T tester pHep®4, Hanna Instruments, RI, USA) [31], chemical oxygen demand (COD photometer Hanna Instruments, New England, RI, USA) [32], and total solids by the gravimetric method [33]. The detection of heavy metals was performed using a Varian SpectrAA 220 G Atomic Absorption Spectrometer(Varian-Agilent

Inst., Palo Alto, San Francisco, CA, USA), following previous publications: cadmium, chromium, zinc, and nickel [34]; mercury in a direct mercury analyzer (DMA-80, Milestone Inc., Sorisole, Italy) [35]; and cyanide in a portable photometer (Hanna Instruments, RI, USA) [36]. All reactive, analytical standards and reference materials were purchased from Merck (Merck KGaA, Darmstadt, Germany). The results obtained in the samples from Village Gato, Village Tigui, and the water catchment of the Boque River were compared with normative 0631/2015 [37], which establishes the parameters to be monitored and maximum permissible limits in the specific discharges of non-domestic wastewater (precious minerals and gold). While for houses and deep-well underground sites, they were evaluated in compliance with normative 2115/2007 [38], which regulates water for human consumption.

2.3. Bioassays

In the bioassays, *Lactuca sativa (L. sativa)* [39] and *Hydra attenuata (H. attenuata)* [40] were used as a biological indicators of water quality. After the follow-up of the results of the two first collections of water samples, a modification of the protocol was performed replacing *Daphnia magna (D. magna)* [41] instead of *H. attenuata* due to no evidence of sensitivity against the possible harmful substances that might be present in the water samples by *H. attenuata*. The effects on organisms can be inhibition, sublethality, and lethality volume/volume (v/v). The water samples were diluted in four different concentrations 25%, 50%, 75%, and 100% (v/v); reconstituted hard water (160–180 mg/L $CaCO_3$) was used as diluent for the *D. magna* and *H. attenuata*; while for *L. sativa*, distilled water was used. The response of *H. attenuata* was read using a binocular stereoscope (Leica). Before taking the readings, the containers were shaken in a circular way to reactivate the movement of *D. magna* and confirm their state. In the case of *L. sativa*, graph paper was used to measure the length of the radicle.

2.3.1. Endpoint and Toxic Response Model

L. sativa, half-maximal inhibitory concentration (IC_{50}): root growth reduction or inhibitory effects on lettuce seed germination and root growth after 5 days. *D. magna*, lethal concentration (LC_{50}): number dead/total number or lethal effects of water were observed after 48 h of exposure, and *H. attenuata*, median effective concentration (EC_{50}): density reduction or lethality test, produced by irreversible morphological changes after 96 h of exposure.

2.3.2. Toxicity

To calculate the lethal concentration (LC_{50}), the half-maximal inhibitory concentration (IC_{50}), and median effective concentration (EC_{50}), the method used was Environmental Protection Agency (EPA) Probit analysis model [42–44]. When results in EC_{50}/LC_{50}/IC_{50} cannot be reported by the statistical program, they are reported as the percentage of effect (%) in the lowest concentration at which the event is still present on the evaluated population [20].

2.4. Ames Test

The method was applied according to Ames [45] using *Salmonella typhimurium* (*S. typhimurium*) TA98 and TA100 strains to evaluate possible mutagenicity. The cultures were grown in Oxoid nutrient broth No. 2. The samples of water were diluted in four different concentrations 25%, 50%, 75%, and 100% (*v/v*). The mutagenic effect was evaluated from the number of revertant colonies per plate. The plates were prepared in triplicate for every test sample, and the result presented was the mean of triplicate observation. The mutagenic activity was detected after 120 h of exposure at 37 (±2) °C. The revertant colonies' readings were counted using an automatic colony counter (Industrial Scientific). For accuracy of the results, mutagenic index (MI) values greater than or equal to two (≥2) were considered mutagenic [46].

2.5. Statistic Analysis

To establish whether there is a relationship between evaluated physicochemical parameters and heavy metals with results of bioassays, a one-way analysis of variance (ANOVA) was performed with a significance level of $p < 0.05$.

2.6. Microbiological Analysis

The determination of total coliforms and *E. coli* as indicators of bacterial fecal contamination was performed according to the ISO 9308-1 standard method [47]. Cellulose acetate membranes of 0.45 μm × 47 mm (Sartorius) were used for the filtration. Dark blue/purple colonies on Chromocult agar (Merck) were presumed to be *E. coli*. The detection and quantification of somatic coliphages as indicators of viral fecal contamination was performed according to the ISO 10705-2 standard procedure and modified Scholten's agar (MSA) (OXOID) was used for the detection of coliphages [48].

The results obtained in the samples from Village Gato, Village Tigui, and the water catchment of the Boque River were compared with normative 1594/1984 [49], which regulates waters that can be treated by conventional systems for human consumption. While for houses and deep-well underground sites, they were evaluated in compliance with normative 2115/2007 [38], which regulates water for human consumption. Normative 2115/2007 only beholds the microbiological quality concerning total coliforms and *E. coli*. Although coliphage concentrations are not regulated within Colombian normatives, their detection is relevant since they confirm contamination of fecal origin and the possible presence of pathogenic viruses, both in drinking water and in water for human consumption.

3. Results

3.1. Physicochemical Parameters

Parameters such as COD, total solids, and pH did not exceed the limits of Colombian normative 0631/2015 [37], in the first three sample stations. However, the level of cadmium was excessive in the Village Gato station in the first sampling with a concentration of 0.05 mg/L. Chromium did surpass the limit in the first sampling in Village Tigui. Likewise, mercury in Village Gato exceeded the limit in the third sampling (0.0029 mg/L). Moreover, in the Village Tigui station, the measurements in the second sampling exceeded 0.0025 mg/L and those in the water catchment of the Boque River in the first sampling were also excessive with 0.0022 mg/L. Furthermore, the permitted concentration of cyanides in the first and second sampling with a concentration of 1.02 and 1.32 mg/L, respectively, was excessive at the Village Tigui sampling station. While in the water catchment of the Boque River station, the prescribed cyanide level was exceeded in the second sampling with a concentration of 1.57 mg/L (Appendix A—Table A1).

The values of heavy metals analyzed in the last two sampling stations were compared with normative 2115/2007 [38] based on waters for human consumption. On one hand, the level of cadmium exceeded the limits established by the regulations for the house station in the second and third sampling with concentrations of 0.03 and 0.01 mg/L, respectively. While the cyanide concentration was exceeded only in the second sampling (1.11 mg/L). On the other hand, the established values of cadmium exceeded only in the second sampling in deep-well underground, presenting a concentration of 0.01 mg/L. Mercury was detected in each of the samples for both the house and deep-well underground stations, but the concentrations did not exceed the limits established in normative 2115 of 2007 [38]. The other metals evaluated (Zn, Ni, and Cr) were not detected in any of the samples analyzed (Appendix A—Table A2).

3.2. Bioassays

Table 1 shows the different percentages of growth inhibition of *L. sativa* in three samples for five evaluated stations. As the results show, there was root inhibition in some sampling stations and overgrowth in others. In Village Gato, in the third sampling, there was excess growth at a concentration of 25%. Likewise, in Village Tigui, the greatest inhibitions registered were observed at a concentration of 25%. In general, among all the sampled stations where the greatest inhibition was observed the highest was in Village Tigui, followed by house, deep-well underground, Village Gato, and the smallest recorded in the water catchment of the Boque River.

Table 1. Bioassays results with *Lactuca sativa.*

Sampling Station (n = 15)	1S % (*v*/*v*) Effect	2S % (*v*/*v*) Effect	3S % (*v*/*v*) Effect
Village Gato	33% Inhibition to 75%	27% Inhibition to 100%	133% Growth to 25%
Village Tigui	44% Inhibition to 25%	24% Inhibition to 25%	4% Inhibition to 50%
Water catchment of the Boque River	0% Inhibition to 100%	27% Inhibition to 100%	110% Growth to 100%
House	34% Inhibition to 25%	19% Inhibition to 75%	6% Inhibition to 50%
Deep-well underground	35% Inhibition to 50%	36% Inhibition to 75%	6% Inhibition to 75%

S: sampling, the numbers 1S, 2S, and 3S correspond to the months of July, September, and December in which the sample was taken; n: is the number of samples.

In Table 2, it is observed that the *D. magna* indicator has different mortality percentages since the same four concentrations of the sample are evaluated as in the *L. sativa* bioassay. The highest was 23% mortality at a concentration of 50% at the water catchment of the Boque River station and 17% mortality at 75% at the Village Tigui sampling station, followed by the house and deep-well underground stations. Finally, the lowest concentration of mortality was obtained in the Village Gato station with 33% mortality at 100%. For *H. attenuata*, it was not possible to determine the EC_{50} and LC_{50} values because there were no morphological changes, indicating lethality or sublethality, in the three samplings carried out, reporting 0% sublethality at 100% (*v*/*v*) and 0% lethality at 100% (*v*/*v*).

Table 2. Results of bioassay with *Daphnia magna.*

Sampling Station (n = 15)	3S % (*v*/*v*) Effect
Village Tigui	17% mortality to 75%
Village Gato	33% mortality to 100%
Water catchment of the Boque River	23% mortality to 50%
House	47% mortality to 100%
Deep-well underground	43% mortality to 100%

S: sampling, the number 3S corresponds to the month of December in which the sample was taken; n: is the number of samples.

3.3. *Ames Test*

The results obtained with the Ames test using *S. typhimurium* TA98 and TA100 strains are presented in Table 3, where the mutagenic index (MI) is shown for each condition used in the assay. According to Table 3, for the second sampling of the house station, mutagenic values were observed for both strain TA98 and TA100 in each of the concentrations evaluated. While for Village Gato, with strain TA100, a value of 2.4 was observed in the 100% concentration (Table 3). For the other sampling sites, there was no mutagenic index.

Table 3. Mutagenic index, for each concentration analyzed in the five sampling stations with strain *Salmonella typhimurium* TA98 and *S. typhimurium* TA100.

Sampling Stations (n = 15)	Concentration *v/v* (%)	TA98 (MI)			TA100 (MI)		
		S1	S2	S3	S1	S2	S3
Village Gato	25	0.77	1.07	0.00	0.97	0.46	1.00
	50	0.86	0.87	0.20	1.09	0.68	1.20
	75	1.06	1.49	0.60	1.27	0.77	1.70
	100	1.12	1.93	1.00	1.51	1.14	2.40
Village Tigui	25	0.97	0.70	0.00	1.37	0.20	0.60
	50	0.95	0.74	0.00	1.39	0.34	0.90
	75	1.12	1.14	0.10	1.48	0.33	1.00
	100	1.21	1.82	0.40	1.81	0.56	1.20
Water catchment of the Boque River	25	1.17	0.39	0.30	0.94	0.27	0.60
	50	1.39	0.63	0.50	1.00	0.24	0.70
	75	1.41	0.47	0.60	1.21	0.35	1.10
	100	1.50	0.59	0.90	1.71	0.44	1.30
House	25	0.68	41.78	0.40	1.01	11.05	0.60
	50	1.00	48.49	0.60	1.17	12.39	1.00
	75	1.06	56.62	0.80	1.36	13.75	2.00
	100	1.21	58.31	1.10	1.70	15.09	2.50
Deep-well underground	25	88.0	0.39	0.20	1.02	0.37	0.80
	50	0.88	0.50	0.50	1.16	1.21	1.20
	75	1.00	0.50	0.80	1.23	0.42	1.70
	100	0.55	0.64	0.90	1.28	0.64	2.00

S: sampling, the numbers 1S, 2S, and 3S correspond to the month of July, September, and December in which the sample was taken; n: is the number of samples; MI: mutagenic index.

3.4. Statistical Analysis

The statistical analysis performed to determine a possible relationship between the results of the physicochemical parameters against the toxicity indicators showed that there is a relationship between the inhibition of *L. sativa* concerning to mercury with a significance of $p < 0.05$.

However, due to the low number of samples analyzed for *H. attenuata* (three samples) versus the number of samples for *D. magna* (15 samples), it was not possible to establish whether there was a correlation with the concentration of metals or with the results of *L. sativa*.

3.5. Microbiological Analysis

Table 4 shows the results of the concentration of fecal contamination indicators (total coliforms, *E. coli*, and somatic coliphages) for the different types of water of the five sampling stations. Table 4 shows that concentrations between 10^3 and 10^5 colony forming unit (CFU)/100 mL for total coliforms were obtained at the different stations. While in the case of *E. coli*, concentrations between 10^3 and 10^4 CFU/100 mL were obtained. Somatic coliphages were detected in samples taken at Village Gato and house stations. Colombia does not have regulations for the presence of this indicator, although this is necessary since the presence of somatic coliphages represents a risk to the health of the community.

While, for the water catchment of the Boque River, Village Tigui, and deep-well underground stations, the presence of phages in some samples was not detected ($<1.0 \times 10^3$). The results were compared with decree 1594/1984 [49], which establishes the concentration of total coliforms (2.0×10^4/100 mL) allowed in waters that will be treated by conventional systems: while normative 2115/2007 [38], establishes that the concentrations for total coliforms and *E. coli* for drinking water is 0 CFU/100 mL.

Table 4. Results of total coliforms, *Escherichia coli*, and somatic coliphages in waters of the Boque River and drinking waters.

Sampling Stations (n = 15)	Microbiological Indicators								
	Total Coliforms CFU/100 mL			*E. coli* CFU/100 mL			Somatic Coliphages PFU/100 mL		
	1S	2S	3S	1S	2S	3S	1S	2S	3S
Village Gato	7.0×10^3	1.1×10^5	4.1×10^5	1.0×10^3	4.0×10^3	8.0×10^4	1.4×10^3	1.0×10^2	4.5
Village Tigui	1.3×10^5	3.2×10^4	2.2×10^5	2.0×10^3	2.0×10^3	3.0×10^4	4.9×10^3	3.0×10^2	$<1.0 \times 10^3$
Water catchment of the Boque River	2.4×10^4	1.7×10^4	4.0×10^5	1.0×10^3	1.0×10^3	1.0×10^4	$<1.0 \times 10^2$	$<1.0 \times 10^3$	1.9
House	4.0×10^4	1.4×10^5	2.8×10^5	1.0×10^3	3.0×10^4	3.0×10^4	1.0×10^2	2.0×10^4	1.0×10^2
Deep-well underground	3.2×10^4	6.0×10^4	3.8×10^5	1.0×10^3	4.0×10^4	4.0×10^4	2.0×10^2	$<1.0 \times 10^3$	1.0

CFU/100 mL: colony forming units in 100 mL of analyzed water; PFU/100 mL: plaque forming units in 100 mL of analyzed water; n: number of samples analyzed. S: sampling, the numbers 1S, 2S, and 3S correspond to the month of July, September, and December in which the sample was taken; <: less than the limit of quantification; n: is the number of samples.

4. Discussion

4.1. Bioassays

4.1.1. *Hydra attenuata* and *Daphnia magna*

By applying the *H. attenuata* toxicity test, it was not possible to determine lethality or sub-lethality since there were no morphological changes in the three samples taken. An important factor that could influence why *H. attenuata* was not sensitive to contaminants present in this water, is that the toxicity of metals is modified by abiotic factors such as hardness, pH, and water temperature [50]. For example, if water hardness is high, the formation of metal complexes tends to increase, which in turn lowers the effect of toxic divalent metals [50,51].

H. attenuata has a higher sensitivity to toxic substances at acidic pH, compared to that at alkaline or neutral pH [52]. The pH value of the water sample from the Boque River is about 7 (Appendix A—Tables A1 and A2), which could influence *H. attenuata* not presenting sensitivity when heavy metals, cyanides, or other toxic substances are in the water.

Due to the results obtained with *H. attenuata* in the first two samples, *D. magna* was used in the last sampling, to find an animal indicator that presented a greater sensitivity to these types of contaminants. Table 2 shows different mortality percentages that were found, demonstrating the sensitivity of this organism to the contaminants present in the water of the Boque River that is consumed by the Monterrey population. Studies conducted by Forget et al. [53] with *D. magna*, show percentages of toxicity up to 70% against heavy metals. Castro-Català et al. [54] evaluated the toxicity of sediments and water in rivers with the presence of pesticides and heavy metals using *D. magna* as an animal indicator, showing that it can be sensitive to these types of samples, due to its high metabolic rate [55]. Likewise, Lattuada et al. [56], in southern Brazil, used *D. magna* as an indicator of toxicity in waters affected by coal mining, in which heavy metals such as Fe, Mn, Zn, Ni, Cd, and Pb were found. The results showed sensitivity by this indicator in this type of water and suggest the evaluation of toxicity in waters from gold mining. Moreover, studies conducted in China by Wu et al. [57] demonstrated that the most frequently encountered heavy metals in a region affected by gold mining were mercury and cadmium, as observed in the results found in the drinking water of the population of Monterrey (Appendix A—Tables A1 and A2).

4.1.2. *Lactuca Sativa*

The differences observed between chemical parameters and toxicity may be related to the fact that the samples were not collected simultaneously and that it is not the same water because along the river route and on the different sampling days, diverse factors can alter its quality. Likewise, dilution effects due to rain, sedimentation, the introduction of new pollutants, among others, can have an influence.

Additionally, the water entering the treatment plant can be more contaminated, taking into account that it travels through tanks that are not in operation or comes into contact with sludge that might have a higher concentration of contaminants, which may return to the column of water.

The increase in germination, compared to the positive control (overgrowth), is related to the presence of organic matter because they are essential nutrients for *L. sativa* seed germination, and if they are available in high concentrations, they will stimulate growth. On the other hand, mercury and cyanide at the Village Tigui station (Appendix A—Tables A1 and A2) had higher concentrations. The inhibition rates of 24% and 44% for samples 1 and 2 at the concentration of 25%, affect the growth of the seed as the higher concentration of pollutants results in greater inhibition. Castillo et al. [24] found inhibition in the germination of seeds in waters contaminated with mercury and argue that it can occur due to the harmful effects caused by mercury at the cellular level in the seed. These results coincide with other studies where *L. sativa* has been proposed as a useful tool to evaluate and compare the toxicity of industrial effluents that present heavy metal contamination [58].

Likewise, the level of cadmium (Appendix A—Tables A1 and A2), also exceeded the minimum values established by the regulations for water for human consumption; studies have been reported where the exposure of *L. sativa* to this metal causes toxic and harmful effects that decrease its growth as the concentration of Cd, and thereby its adsorption, increases [59–61]. Just as the presence of heavy metals and cyanide has toxic implications in the plant and animal model, in the same way, it will affect the health of the human being [62,63].

Cd is one of the most toxic elements to which man is exposed since the accumulation of this metal in the body is gradual and increases with age due to its long half-life, greater than 20 years [64]. This is why eating food or drinking water with very high levels of cadmium causes severe stomach irritation, which causes vomiting, diarrhea, and sometimes death [64]. Moreover, cyanide exceeded the allowed limits (0.05 mg/L) in one of the samples analyzed in one of the houses (1.11 mg/L). The guide values of the World Health Organization [65] establish that the concentration of cyanide toxic to humans is 0.07 mg/L. Exposure to this concentration or higher may cause inhibition of cell growth, thereby affecting the breathing process and the metabolism of nitrogen and phosphate. It also inhibits the activity of some metalloproteins, joining cofactors such as the heme group of hemoglobin [66].

Finally, cyanide has acute effects on human health such as irritation of the eyes, nose, and throat. High exposure causes intoxication with headache, weakness, nausea, strong heartbeat, coma, and even death. As for chronic effects, it causes nosebleeds and nose lesions and can cause enlargement of the thyroid gland, which can interfere with its regular function [67].

4.2. Ames Test

In some cases, there was a decrease in reverts with increasing doses, which may be due to the presence of toxic substances that prevent the growth of bacteria [68]. However, in most of the sampling stations, a direct relationship was observed between the number of revertants and the increase in the concentration of heavy metals. This demonstrates the high probability of the presence of substances such as heavy metals and organic compounds in the Boque River that cause base-pair mutations and changes in the DNA reading frame of bacteria.

When observing the reversion of the strains, it was evidenced that they exceed 2–40 times the value of the negative control for the TA98 strain in the second sample in one of the houses, and from 2.0 to 2.5 for the TA100 strain in the third sample for the house and underground well. According to Orozco and Zuleta [69], some samples can exceed 100 times the negative control and these results are related to the quality of the water.

Likewise, Meléndez et al. [70] investigated the mutagenic activity of drinking water before and after chlorination at the Villa Hermosa plant, Medellín, Colombia, finding that contamination and chlorination influence mutagenicity. They used the Ames test with strains TA100 and TA98. Sierra et al. [71] evaluated the mutagenic activity of the Cauca River water with the same strains with and without the enzyme activator S9, finding that the highest rate of mutagenicity was observed with strain TA98 without enzyme activator. However, the TA100 strain is characterized by presenting the *hisG46* mutation and has specific markers that give it greater sensitivity to the test; within these are the *uvr* mutation, the *uvrB* mutation, and the plasmid pKM101 [26,68].

Mesquidaz et al. [72] reported alarming figures in the mercury concentrations used in the gold extraction process in a mine in northern Colombia, which ranges from 50 to 100 tons in 2007. Furthermore, it is reported that, thanks to this pollutant present in water, the health of the population has been affected, since Hg was found in human hair at a concentration of 12.8 μg/g, a figure that is well above international standards. It has been shown that inhabitants of different municipalities in southern Bolívar where gold mining takes place have high levels of Hg contamination, and this situation requires special attention to reduce environmental and human health problems [4]. Mercury contamination has been linked to health problems, as direct absorption of mercury vapor released by incinerators in gold mining, or ingestion of mercury-containing wastes, causes hydrargyrism and poisoning. Mercury (Hg) is one of the heavy metals of greatest concern to populations that consume fish. This pollutant can be released from many sources and has various toxic effects in humans [73].

Some of the health problems caused are excessive salivation, shortness of breath and fatigue, bronchitis, tremors and irritability, personality changes (due to brain damage), a sensation of floating teeth and pain in them, kidney and respiratory damage that can lead to death from problems in the lungs and other organs of the body [74–76]. While breathing polluted air, elemental mercury can reach the brain, affecting nerve cells and the olfactory system. The main organs in which mercury accumulates are the brain and the kidney [77,78].

4.3. Statistical Analysis

The statistical analysis showed a relationship between the inhibition of *L. sativa* concerning mercury with a significance of $p < 0.05$; this inhibition in the germination of *L. sativa* with this metal was also reported in Chile, where the exposure of the seeds to Hg inhibited their growth [19]. The toxicity caused multiple harmful effects in the seed at the cellular level such as a change in permeability in the cell membrane and the affinity to react with phosphate groups and the sulfhydryl group (SH). When mercury interacts with the SH groups to form the S–Hg–S bonds, it disrupts the stability of the group can affect seed germination and seedling growth whose tissues are rich in SH groups [79].

On the other hand, it was not possible to establish a correlation between the vegetal and the animal model due to the number of samples collected. When comparing the results of the bioassays associating for *D. magna* and *L. sativa*, it was observed that the variability due to the sampling was not simultaneous for every sample and it could be possibly affected by a new spill in the river. Additionally, these are different organisms with different sensitivity to the contaminants present in the Boque River water and there exist other factors that can influence this response. For example, some bacteria can naturally modify mercury (Hg^{2+}) by ion methylation forming CH_3–Hg^+, which is more toxic and is incorporated into trophic chains, affecting the animal model more than the vegetable model [80].

4.4. Total Coliforms, Escherichia Coli, and Somatic Coliphages

The microbiological results confirm the high fecal contamination in all the sampling stations (Table 4). Total coliform concentrations exceeded the limits for Colombian regulations [49]. In the case of drinking water for human consumption (deep-well underground and house), total coliforms and *E. coli* were well above levels required by regulation for drinking water [38]. Likewise, in the case of drinking water for human consumption by the treatment system (Village Gato, Tigui, and water catchment of the Boque River), the concentrations allowed for total coliforms were exceeded [49].

Campos-Pinilla et al. [81] and Sánchez-Alfonso et al. [82] in studies carried out in the Bogotá River found a total coliform concentration between 10^3 and 10^6 CFU/100 mL and for *E. coli* between 10^3 and 10^5/100 mL. This coincides with the values found in this study, which range between 10^3 and 10^5 CFU/100 mL of total coliforms and for *E. coli* between 10^3 and 10^4 CFU/100 mL (Table 4). Likewise, studies conducted by Lucena et al. [83] and Sánchez-Alfonso et al. [82] in rivers show average concentrations of somatic coliphages between 10^2 and 10^4 plaque forming unit (PFU)/100 mL, similar to those found in this study with ranges from 1 and 10^3 PFU/100 mL. The concentration of microorganisms in river water varies depending on climatic factors, geographical area, and the amount of organic matter present in water bodies [84,85]. The mine exploitation site is a settlement space for the population that works in this activity legally or illegally, which generates a high level of household waste in the river causing contamination by the discharge of fecal matter and organic matter, which explains the concentration of indicators of fecal contamination. It is related to the absence of treatment systems and improper installation of septic tanks.

The detected concentrations of total coliforms and *E. coli* in all the drinking water samples and the detection of somatic coliphages in some samples of water used for human consumption confirm the fecal contamination and the possible presence of pathogenic viruses in the drinking water (Table 4). These concentrations of indicators are similar to those detected in river samples as reported by Lucena et al. [83], Campos-Pinilla et al. [81], and Sánchez-Alfonso et al. [82], which could increase the risk for residents.

5. Conclusions

The results obtained with the three toxicity indicators reveal that *H. attenuata* does not present sensitivity to toxic substances present in this type of water, so its use for this purpose is not recommended. On the other hand, *D. magna* showed sensitivity even in diluted samples as well as *L. sativa*, which showed growth inhibition and excessive growth in different concentrations of the analyzed water, inclusive of waters with pollutant concentrations below the detection level. The Ames test shows an increase in the revertants indicating the possibility of mutations in the population that consumes this type of water, which is correlated with the results of the mutagenicity test that showed a mutagenic effect in the five stations evaluated with both strains used in the study. The highest mutagenic index was found in the water sample taken from the house sampling station. The concentration of bacteria in the water exceeded the limits allowed by Colombian regulations, creating a health risk, also with an alert call to the presence of possible pathogenic viruses, and the risk that they imply for the inhabitants of Monterrey due to somatic coliphage levels determined.

This research recognizes the potential use of bioassays to evaluate the toxic effects generated by chemical wastes produced by gold mining and discharged into surface waters. The use of animal and plant models is recommended to evaluate said effects on the environment and public health and infer the damages that until now have not been sufficiently evaluated having as correlation factors physicochemical and microbiological parameters.

Finally, this research generated data that contribute to the knowledge of the effects caused to the environmental and public health by illegal and legal mining carried out with bad practices in emerging countries with inefficient controls of this type of activity. These assays used can help sanitation organizations in different countries to take preventive actions on this issue

Author Contributions: All authors contributed to all features of the paper. A.M., J.A., J.L., L.S., C.V., A.O.-A., and M.D. were involved in the sampling and analysis of bacteria, coliphages, bioassays, and physicochemical determinations. C.C. and C.C.Z. conceived the idea for the research and contributed to the development of the project by obtaining economic resources. C.C., N.d.P., and C.C.Z. provided consultation and interpretation of the results and contributed to writing the manuscript. C.V., A.M., and J.L. edited the manuscript. All authors have read and agreed to the published version of the manuscript.

Acknowledgments: The authors would like to acknowledgment the residents of Monterrey Sur de Bolívar, Colombia, and Programa de Desarrollo y Paz del Magdalena Medio by the logistical support in sampling.

Appendix A

Table A1. Results of physicochemical parameter analysis compared with the normative 0631/2015.

Physicochemical Parameters (n = 9)	Village Gato			Village Tigui			Water Catchment of the Boque River			Limit of the Regulations
Number of Sampling	**S1**	**S2**	**S3**	**S1**	**S2**	**S3**	**S1**	**S2**	**S3**	**Normative 0631/2015 [37]**
pH	7.36	7.49	7.33	7.55	6.05	6.57	7.61	7.56	7.63	6.0–9.0
COD (mg/L)	32.67	<0.001	5.52	22.46	<0.001	24.17	29.63	40.72	4.82	150
Total solids (g/10 mL)	0.0024	0.00165	0.0	0.0007	0.0058	0.0029	0.0001	0.00316	0.0005	50
Cyanide (mg/L)	<0.025	<0.025	0.025	1.02	1.32	<0.025	<0.025	1.57	<0.025	1.0
Cadmium (Cd) (mg/L)	0.05	0.03	0.02	$<1.0 \times 10^{-2}$	0.02	0.02	$<1.0 \times 10^{-2}$	0.02	0.01	0.05
Chrome (Cr) (mg/L)	$<1.0 \times 10^{-6}$	$<1.0 \times 10^{-6}$	$<1.0 \times 10^{-6}$	0.06	$<1.0 \times 10^{-6}$	0.04	$<1.0 \times 10^{-6}$	$<1.0 \times 10^{-6}$	$<1.0 \times 10^{-6}$	0.5
Mercury (Hg) (mg/L)	0.0008	0.0029	0.0003	0.001	0.0025	0.0008	0.0022	0.002	0.0008	0.002
Nickel (Ni) (mg/L)	$<1.0 \times 10^{-3}$	$<1.0 \times 10^{-3}$	$<1.0 \times 10^{-3}$	$<1.0 \times 10^{-3}$	$<1.0 \times 10^{-3}$	$<1.0 \times 10^{-3}$	$<1.0 \times 10^{-3}$	$<1.0 \times 10^{-3}$	$<1.0 \times 10^{-3}$	0.5
Zinc (Zn) (mg/L)	$<1.0 \times 10^{-3}$	$<1.0 \times 10^{-3}$	$<1.0 \times 10^{-3}$	$<1.0 \times 10^{-3}$	$<1.0 \times 10^{-3}$	$<1.0 \times 10^{-3}$	$<1.0 \times 10^{-3}$	$<1.0 \times 10^{-3}$	$<1.0 \times 10^{-3}$	3.0

mg/L: milligram per liter; S: sampling, the numbers 1S, 2S, and 3S correspond to the month of July, September, and December in which the sample was taken; <: less than the limit of quantification; n: is the number of samples; COD: chemical oxygen demand.

Table A2. Results of physicochemical parameters analysis compared with the normative 2115 of 2007.

Physicochemical Parameters (n = 6)	House			Deep-Well Underground			Limit of the Regulations
Number of sampling	**S1**	**S2**	**S3**	**S1**	**S2**	**S3**	**Normative 2115/2007 [38]**
pH	7.6	7.53	7.44	6.95	6.79	6.61	6.5–9.0
Cyanide (mg/L)	<0.025	1.11	<0.025	<0.025	<0.025	<0.025	0.05
Cadmium (Cd) (mg/L)	$<1.0 \times 10^{-2}$	0.03	0.03	$<1.0 \times 10^{-2}$	0.01	$<1.0 \times 10^{-2}$	0.003
Chrome (Cr) (mg/L)	0.02	$<1.0 \times 10^{-6}$	$<1 \times 10^{-6}$	$<1.0 \times 10^{-6}$	$<1.0 \times 10^{-6}$	$<1.0 \times 10^{-6}$	0.05
Mercury (Hg) (mg/L)	0.0004	0.0005	0.0003	0.0007	0.0003	0.0007	0.001
Nickel (Ni) (mg/L)	$<1.0 \times 10^{-3}$	$<1.0 \times 10^{-3}$	$<1.0 \times 10^{-3}$	$<1.0 \times 10^{-3}$	$<1.0 \times 10^{-3}$	$<1.0 \times 10^{-3}$	0.2
Zinc (Zn) (mg/L)	$<1.0 \times 10^{-3}$	$<1.0 \times 10^{-3}$	$<1.0 \times 10^{-3}$	$<1.0 \times 10^{-3}$	$<1.0 \times 10^{-3}$	$<1.0 \times 10^{-3}$	3.0

mg/L: milligram per liter; S: sampling, the numbers 1S, 2S, and 3S correspond to the month of July, September, and December in which the sample was taken; <: less than the limit of quantification; n: is the number of samples.

References

1. UNIDO Green Industry Initiative for Sustainable Industrial Development 2011. Available online: https://www.greengrowthknowledge.org/sites/default/files/downloads/resource/Green_Industry_Initiative_for_Sustainable_Development_UNIDO.pdf (accessed on 12 February 2020).
2. Ogola, J.S.; Mitullah, W.V.; Omulo, M.A. Impact of gold mining on the environment and human health: A case study in the Migori Gold Belt, Kenya. *Environ. Geochem. Health* **2002**, *24*, 141–157. [CrossRef]
3. Gafur, N.A.; Sakakibara, M.; Sano, S.; Sera, K. A case study of heavy metal pollution in water of Bone River by artisanal small-scale gold mine activities in eastern part of Gorontalo, Indonesia. *Water* **2018**, *10*, 1507. [CrossRef]
4. Carranza-Lopez, L.; Caballero-Gallardo, K.; Cervantes-Ceballos, L.; Turizo-Tapia, A.; Olivero-Verbel, J. Multicompartment mercury contamination in major gold mining districts at the department of Bolivar, Colombia. *Arch. Environ. Contam. Toxicol.* **2019**, *76*, 640–649. [CrossRef] [PubMed]
5. Pinedo-Hernández, J.; Marrugo-Negrete, J.; Díez, S. Speciation and bioavailability of mercury in sediments impacted by gold mining in Colombia. *Chemosphere* **2015**, *119*, 1289–1295. [CrossRef] [PubMed]
6. Von Behren, J.; Liu, R.; Sellen, J.; Duffy, C.N.; Gajek, R.; Choe, K.-Y.; DeGuzman, J.; Janes, M.K.; Hild, J.; Reynolds, P. Heavy metals in California women living in a gold mining-impacted community. *Int. J. Environ. Res. Public Health* **2019**, *16*, 2252. [CrossRef]
7. Ouboter, P.E.; Landburg, G.; Satnarain, G.U.; Starke, S.Y.; Nanden, I.; Simon-Friedt, B.; Hawkins, W.B.; Taylor, R.; Lichtveld, M.Y.; Harville, E.; et al. Mercury levels in women and children from interior villages in Suriname, South America. *Int. J. Environ. Res. Public Health* **2018**, *15*, 1007. [CrossRef] [PubMed]
8. Producción de oro en Colombia Creció 7% en el Primer Trimestre de 2020. Available online: https://www.elespectador.com/economia/produccion-de-oro-en-colombia-crecio-7-en-el-primer-trimestre-de-2020-articulo-920587/ (accessed on 22 May 2020).
9. Robles Mengoa, M.E.; Urán, A. Colombia: Legal loopholes behind illegal gold trade. In *Global Gold Production Touching Ground*, 1st ed.; Springer International Publishing: New York, NY, USA, 2020; pp. 151–161.
10. Güiza, L. Small scale mining in Colombia: Not such a small activity. *DYNA* **2013**, *80*, 109–117.
11. DANE Proyecciones de Población Departamentales Por Área 2015–2020. Available online: https://www.dane.gov.co/files/investigaciones/poblacion/proyepobla06_20/7Proyecciones_poblacion.pdf (accessed on 22 February 2020).
12. Ministerio de Minas y Energía Censo Minero Departamental 2010–2011. Available online: https://www.minenergia.gov.co/documents/10180/698204/CensoMinero.pdf/093cec57-05e8-416b-8e0c-5e4f7c1d6820 (accessed on 8 February 2020).
13. De la Hoz, V.J. Economía y conflicto en el Cono Sur del Departamento de Bolívar. *Banco de la República* **2009**, *110*, 1–109.
14. Alcaldía de Simití-Bolívar Plan de Desarrollo ¡Simití Pata Todos! 2016–2019. Available online: https://simitibolivar.micolombiadigital.gov.co/sites/simitibolivar/content/files/000132/6580_plan-de-desarrollo-municipal-20162019.pdf (accessed on 22 July 2020).
15. Grajales, A.K. *El Río Grande la Magdalena Como Eje Estructurante del Desarrollo territorial. Estudio de Caso: Magdalena Medio 1950–2010*; Empresas Publicas de Medellin (EPM): Medellin, Colombia, 2012; pp. 1–65.
16. Donato, D.B.; Nichols, O.; Possingham, H.; Moore, M.; Ricci, P.F.; Noller, B.N. A critical review of the effects of gold cyanide-bearing tailings solutions on wildlife. *Environ. Int.* **2007**, *33*, 974–984. [CrossRef]
17. Hilson, G. Abatement of mercury pollution in the small-scale gold mining industry: Restructuring the policy and research agendas. *Sci. Total Environ.* **2006**, *362*, 1–14. [CrossRef]
18. Hilson, G.; Murck, B. Progress toward pollution prevention and waste minimization in the North American gold mining industry. *J. Clean Prod.* **2001**, *9*, 405–415. [CrossRef]
19. Cordy, P.; Veiga, M.M.; Salih, I.; Al-Saadi, S.; Console, S.; Garcia, O.; Mesa, L.A.; Velásquez-López, P.C.; Roeser, M. Mercury contamination from artisanal gold mining in Antioquia, Colombia: The world's highest per capita mercury pollution. *Sci. Total Environ.* **2011**, *410*, 154–160. [CrossRef] [PubMed]
20. Bohórquez-Echeverry, P.; Duarte-Castañeda, M.; León-López, N.; Caicedo-Carrascal, F.; Vásquez-Vásquez, M.; Campos-Pinilla, C. Selection of a bioassay battery to assess toxicity in the affluents and effluents of three water-treatment plants. *Univ. Sci.* **2012**, *17*, 152–166. [CrossRef]

21. Keddy, C.J.; Greene, J.C.; Bonnell, M.A. Review of whole-organism bioassays: Soil, freshwater sediment, and freshwater assessment in Canada. *Ecotoxicol. Environ. Saf.* **1995**, *30*, 221–251. [CrossRef]
22. Miller, W.E.; Peterson, S.A.; Greene, J.C.; Callahan, C.A. Comparative toxicology of laboratory organisms for assessing Hazardous Waste Sites. *J. Environ. Qual.* **1985**, *14*, 569–574. [CrossRef]
23. Férard, J.F.; Burga Pérez, K.F.; Blaise, C.; Péry, A.; Sutthivaiyakit, P.; Gagné, F. Microscale ecotoxicity testing of Moselle river watershed (Lorraine Province, France) sediments. *J. Xenobiot.* **2015**, *5*, 1–7. [CrossRef]
24. Castillo, G.C.; Vila, I.C.; Ella, N. Ecotoxicity assessment of metals and wastewater using multitrophic assays. *Environ. Toxicol.* **2000**, *15*, 370–375. [CrossRef]
25. MacGregor, J.T.; Casciano, D.; Müller, L. Strategies and testing methods for identifying mutagenic risks. *Mutat. Res.* **2000**, *455*, 3–20. [CrossRef]
26. Maron, D.M.; Ames, B.N. Revised methods for the *Salmonella* mutagenicity test. *Mutat. Res.* **1983**, *113*, 173–215. [CrossRef]
27. González Scancella, T. *Evaluación del Sistema de Abastecimiento de Agua Potable y Disposición de Excretas de la Población del Corregimiento de Monterrey, Municipio de Simití, Departamento de Bolívar, Proponiendo Soluciones Integrales al Mejoramiento de Los Sistemas y la Salud de la Comunidad*; Ecology College Degree, Pontificia Universidad Javeriana: Bogotá, Colombia, 2013.
28. Barrera, J.A.; Espinosa, A.J.; Álvarez, J.P. Contaminación en el Lago de Tota, Colombia: Toxicidad aguda en Daphnia magna (Cladocera: Daphniidae) e Hydra attenuata (Hydroida: Hydridae). *Rev. biol. Trop.* **2019**, *67*, 11–23. [CrossRef]
29. Espinosa-Ramírez, A.J. El Agua, un Reto Para la Salud Pública. La Calidad del Agua y las Oportunidades Para la Vigilancia en Salud Ambiental. Ph.D. Thesis, Universidad Nacional de Colombia, Bogotá, Colombia, 2018.
30. Zeiger, E. The test that changed the world: The Ames test and the regulation of chemicals. *Mutat. Res.* **2019**, *841*, 43–48. [CrossRef] [PubMed]
31. American Public Health Association 2310 B. Titration method. In *Standard Methods for the Examination of Water and Wastewater*; American Public Health Association: Washington, DC, USA, 2012; p. 1946.
32. American Public Health Association 5220 D. Closed reflux, colorimetric method. In *Standard Methods for the Examination of Water and Wastewater*; American Public Health Association: Washington, DC, USA, 2012; p. 1946.
33. American Public Health Association 2540 B. Total solids dried at 103–105 °C. In *Standard Methods for the Examination of Water and Wastewater*; American Public Health Association: Washington, DC, USA, 2012; p. 1946.
34. American Public Health Association 3111. Metals by flame atomic absorption spectrometry. In *Standard Methods for the Examination of Water and Wastewater*; American Public Health Association: Washington, DC, USA, 2012; p. 1946.
35. American Public Health Association 3112B. Metals by cold-vapor atomic absorption spectrometry. In *Standard Methods for the Examination of Water and Wastewater*; American Public Health Association: Washington, DC, USA, 2012; p. 1946.
36. American Public Health Association 4500-CN-E. Colorimetric method. In *Standard Methods for the Examination of Water and Wastewater*; American Public Health Association: Washington, DC, USA, 2012; p. 1946.
37. Ministerio de Ambiente y Desarrollo Sostenible Resolución 0631 de 2015, Parámetros y los Valores Límites Máximos Permisibles en los Vertimientos Puntuales a Cuerpos de Aguas Superficiales y a Los Sistemas de al Cantarillado Público y se Dictan Otras Disposiciones. Available online: https://www.minambiente.gov.co/images/normativa/app/resoluciones/d1-res_631_marz_2015.pdf (accessed on 21 March 2020).
38. Ministerio de Protección Social, Ministerio de Ambiente, Vivienda y Desarrollo Territorial Resolución 2115 de 2007, Características, Instrumentos Básicos y Frecuencias del Sistema de Control y Vigilancia Para la Calidad del Agua Para Consumo Humano. Available online: https://www.minambiente.gov.co/images/GestionIntegraldelRecursoHidrico/pdf/Legislaci%C3%B3n_del_agua/Resoluci%C3%B3n_2115.pdf (accessed on 21 March 2020).
39. Dutka, B. Short-term root elongation toxicity bioassay. In *Methods for Toxicological Analysis of Waters, Wastewaters and Sediments, Rivers Research Branch*; Canada Centre for Inland Waters, National Water Research Institute: Burlington, ON, Canada, 1989; pp. 120–122.
40. Trottier, S.; Biaise, C.; Kusui, T.; Johnson, E.M. Acute toxicity assessment of aqueous samples using a microplate-based *hydra attenuata* assay. *Environ. Toxicol. Water Qual.* **1997**, *12*, 265–271. [CrossRef]

83. Lucena, F.; Méndez, X.; Morón, A.; Calderón, E.; Campos, C.; Guerrero, A.; Cárdenas, M.; Gantzer, C.; Shwartzbrood, L.; Skraber, S.; et al. Occurrence and densities of bacteriophages proposed as indicators and bacterial indicators in river waters from Europe and South America. *J. Appl. Microbiol.* **2003**, *94*, 808–815. [CrossRef]
84. Augustyn, Ł.; Babula, A.M.; Joniec, J.; Stanek-Tarkowska, J.; Hajduk, E.; Kaniuczak, J. Microbiological indicators of the quality of river water, used for drinking water supply. *Pol. J. Environ. Stud.* **2016**, *25*, 511–519. [CrossRef]
85. Jofre, J.; Lucena, F.; Blanch, A.R.; Muniesa, M. Coliphages as model organisms in the characterization and management of water resources. *Water* **2016**, *8*, 199. [CrossRef]

13

The Toxicity and Uptake of As, Cr and Zn in a Stygobitic Syncarid (Syncarida: Bathynellidae)

Grant C. Hose [1,*], Katelyn Symington [1], Maria J. Lategan [2] and Rainer Siegele [3]

1 Department of Biological Sciences, Macquarie University, Sydney, NSW 2109, Australia; katelyn.symington@planning.nsw.gov.au
2 Department of Molecular Sciences, Macquarie University, Sydney, NSW 2109, Australia; josielategan@bigpond.com
3 Australian Nuclear Science and Technology Organisation, Locked Bag 2001, Kirrawee DC, NSW 2232, Australia; rns@ansto.gov.au
* Correspondence: grant.hose@mq.edu.au

Abstract: Ecotoxicological data for obligate groundwater species are increasingly required to inform environmental protection for groundwater ecosystems. Bathynellid syncarids are one of several crustacean taxa found only in subsurface habitats. The aim of this paper is to assess the sensitivity of an undescribed syncarid (Malacostraca: Syncarida: Bathynellidae) to common groundwater contaminants, arsenic(III), chromium(VI) and zinc, and examine the bioaccumulation of As and Zn in these animals after 14-day exposure. Arsenic was the most toxic to the syncarid (14-day LC50 0.25 mg As/L), followed closely by chromium (14-day LC50 0.51 mg Cr/L) and zinc (14-day LC50 1.77 mg Zn/L). The accumulation of Zn was regulated at exposure concentrations below 1 mg Zn/L above which body concentrations increased, leading to increased mortality. Arsenic was not regulated and was accumulated by the syncarids at all concentrations above the control. These are the first published toxicity data for syncarids and show them to be among the most sensitive of stygobitic crustaceans so far tested, partly due to the low hardness of the groundwater from the aquifer they inhabit and in which they were tested. The ecological significance of the toxicant accumulation and mortality may be significant given the consequent population effects and low capacity for stygobitic populations to recover.

Keywords: groundwater ecology; stygofauna; stygobite; aquifer; syncarida

---

1. Introduction

Groundwater is vital to society globally. It provides drinking water to over 2 billion people worldwide [1], and, in many areas, is the only reliable water source. Increasing demand for water and risk of contamination from surface activities such as agriculture, urban development and mining threaten the integrity of aquifers, their suitability as a water source, and the biota that inhabit these subterranean ecosystems [2].

Groundwater ecosystems support a unique biota, consisting of microbial assemblages, frequently crustacean-dominated invertebrate assemblages and, somewhat rarely, vertebrates such as fish [3]. Groundwater invertebrates particularly have evolved to dark, stable and low energy conditions of groundwater environments, such that aquifers the world over contain a diversity of unique taxa not found in surface environments [3]. Bathynellid syncarids are archetypal groundwater organisms. Found only in hyporheic or groundwater environments [4], bathynellid syncarids share the common morphological traits of blindness, lack of body pigments, vermiform (wormlike) body shape, and enhanced sensory appendages that have evolved independently across multiple lineages of groundwater fauna [3]. These traits and other adaptations make groundwater fauna morphologically and physiologically different from even related surface water species, and consequently, they may be expected to respond differently to toxicants compared to surface-dwelling species [5].

As in any ecosystem, pollution threatens the health of the groundwater biota, with flow-on effects to other elements of the ecosystem. For groundwater ecosystems, disruption to one element of the ecosystem is likely to have significant effects on the microbial communities and subsequently impact the ability of the ecosystem to self-purify and provide clean groundwater [6]. The risk to groundwater ecosystems is that the fauna has evolved classic k-strategist traits of longevity, low metabolic and reproductive rates [3], meaning that recovery from disturbance is, at best, slow [7].

Contamination of groundwater by metals is a global concern, often occurring as a result of industrial uses, spills and land application of contaminated materials (Castano-Sanchez in press) [8]. Unfortunately, groundwater ecosystems may not be protected by water quality legislation to the same extent as other aquatic ecosystems (such as rivers and estuaries) because water quality guidelines for ecosystem protection are not based on toxicological data for the biota of that ecosystem [9,10]. Instead, existing water quality guidelines for groundwater ecosystems are based on toxicity data for surface waters, but there are a number of reasons why groundwater biota may respond differently to toxicants compared to surface water taxa [5]. In particular, the metabolic rates of groundwater macroinvertebrates tend to be lower than those of related epigean species [11]. As a result, the uptake of toxicants may be lower where uptake of toxicants is an active, metabolic-driven process [5]. Alternatively, the removal of toxicants accumulated by passive processes may be low, leading to greater body burdens in groundwater fauna compared to surface water fauna [5].

The aim of this study is to test the sensitivity of an obligate groundwater crustacean (Malacostraca: Syncarida) to metal and metalloid contaminants (Zn, As and Cr) in groundwater. The first step to mortality and harm from contaminants is the uptake of those contaminants by the organism, so we also examine here the uptake of these metals by the syncarids and relate uptake to exposure and mortality.

2. Materials and Methods

2.1. Test Species

In this study, we test an undescribed species of the family Bathynellidae (Malacostraca: Syncarida). Until recently, all Australian Bathynellidae were assigned to the genus *Bathynella*, but phylogenies using newly collected specimens suggest strong regional separation within the Australian Bathynellidae, with specimens from eastern Australia being a separate clade [12–14], and those tested here likely a novel taxon. Type specimens are held at Macquarie University. Unfortunately, taxonomic keys to allow further identification are currently unavailable. However, multiple specimens from each test were examined to ensure morphological consistency of the test organisms within and between tests. Syncarida do not have morphologically different life stages after hatching [4], making it difficult to determine life stage; however, the syncarids tested were approximately 1000 μm long and of similar size within and between tests.

Test animals were collected from a fractured sandstone aquifer at Somersby, NSW Australia, approximately 80 km north of Sydney (33°22′15.4″ S, 151°18′9″ E, Bore depth = 22 m). The sample collection area is semirural. Groundwater in the region is used heavily by beverage bottling industries and is considered of very high quality. There is no recorded history of groundwater contamination at the site and analysis of the water over time has revealed very low concentrations of As, Cr and Zn, as seen in Table 1. The groundwater is typically low in dissolved solids and has a slightly acidic pH as a consequence of iron in the sandstone aquifer, as seen in Table 1.

Syncarids were collected from groundwater as described in Korbel & Hose [15] as summarised below. Groundwater was pumped using a motorised inertia pump (Waterra, ON, Canada). Approximately 300 L of water was pumped and passed through a 63 μm mesh sieve to collect the invertebrates. The sieve contents were placed in a sealable, 1 L plastic container which was filled with clean groundwater and placed in a portable cooler for transportation to the laboratory.

Table 1. Physicochemical variables and concentrations of common metals in the groundwater at the Somersby (NSW, Australia) collection sites (from [16]).

Water Quality Variable	Units	Concentration
pH		4.2–5.6
Conductivity	μS/cm	131–195
Dissolved Oxygen	% Saturation	59–83
Hardness	mg/L as $CaCO_3$	25–44
Total Organic Carbon	mg/L	3–13
As	mg/L	<0.01
Cr	mg/L	0.03
Zn	mg/L	0.02

In the laboratory, containers containing the invertebrates were placed in a dark environmental cabinet at 18 °C, which approximates the temperature of the groundwater at the time of collection. Syncarids were acclimated to conditions in the environmental cabinet for at least 48 h prior to testing. Tests were conducted within seven days of collection and data from repeat tests were pooled for analysis.

Tests were conducted following the methods reported in Hose et al. [16], which are summarised below. Once acclimated, the syncarids were removed from the collection container using a pipette and placed individually into wells of 24-well plates (Greiner) along with a known volume of clean groundwater from the collection site. Treatment concentrations were randomly allocated to each well and the required volume of clean groundwater added such that the subsequent addition of toxicant solution would give a final test volume of 2.5 mL. The 24-well plates were covered and placed in a dark cabinet at 18 °C for the duration of the test. Mortality was monitored daily and dead animals were removed, rinsed in Milli-Q water and frozen individually for later analysis.

Toxicity tests were conducted over 14 days with the test end point being immobility, defined as failure to move within 15 s of being gently prodded. This test was repeated twice within 15 min to confirm immobility. Immobility was recorded at 96 h and 14 days. Dissolved oxygen, conductivity, pH and temperature were recorded every 48 h from a randomly selected replicate of each treatment using HANNA (Hanna Instruments Inc, Smithfield, RI, USA) handheld meters.

Each test was comprised of a negative control and six or seven test concentrations, each with four or three replicate wells per plate, respectively. The number of animals varied between tests depending on the number in field collections and controls in all tests had ≥13 individuals. The concentration ranges used for each toxicant were 0–10 mg As/L, 0–100 mg Zn/L, and 0–10 mg Cr/L. These concentrations span the ranges reported in groundwater in the literature (e.g., [17–20]). Toxicants were tested individually. Results from a test were not accepted if there was more than 20% mortality of organisms in control treatments [10].

Toxicant stock solutions were made up in Milli-Q water. Chromium(VI) was added as $K_2Cr_2O_7$ (Sigma-Aldrich, 99% purity), arsenic was added as $NaAsO_2$ (Sigma-Aldrich, 98% purity) and zinc was added as $ZnSO_4.7H_2O$ (Ajax Finechem, 99% purity). Test concentrations were measured using the S2 Picofox Total Reflection X-Ray Fluorescence (TRX-RF) Spectrometer (Bruker Nano GMBH, Berlin, Germany), using the method for aqueous samples as outlined in [21]. A 1 mL aliquot of toxicant solution was taken from randomly selected wells of each treatment at the completion of each test. The aliquot of toxicant solution was spiked with approximately 5–10 μL of standard gallium solution, which acts as an internal standard [21,22]. Aliquots of 20 μL of this solution were then pipetted onto clean, hydrophobic carriers and dried before analysis. Analyses of spectra was done using the Spectra software (version 7.5.3.0, Bruker Nano GMBH, Berlin, Germany). Toxicant concentrations were averaged across replicates and these average values were used in subsequent statistical analyses. Merck XVI multielement standard (Merck KGaA, Darmstadt, Germany) was analysed along with the samples, and indicated average (±SD) recoveries of 107 ± 8% for As, 95 ± 5% for Cr, and 80 ± 5% for Zn. Limits of detection were 1 μg/L for As and Zn, and 3 μg/L for Cr.

The accumulation of As and Zn in the syncarids was measured using Particle Induced X-ray Emission (PIXE). Individual syncarids were freeze dried and placed whole, intact on to sample holders (see below). Resource limitations precluded the analysis of Cr.

2.2. PIXE Elemental Analysis

The samples were analysed for trace and bulk element concentrations with μ-PIXE using the Heavy Ion Microprobe (HIMP) at the Australian Nuclear and Science Technology Organisation (ANSTO) [23]. The HIMP uses the Australian National Tandem Research (ANTARES) accelerator to accelerate ions to the MeV energy range. In this analysis, a 3 MeV proton beam with typical spot size of between 5 and 10 μm was used. At this spot size, beam currents between 0.4 and 0.5 nA can be achieved, which is sufficient for μ-PIXE analyses. A high-purity Germanium (Ge) detector was used with a 100 mm^2 active area to measure the characteristic x-rays emerging from the calcification samples. The detector was located 32 mm from the sample for optimum count rate and sensitivity. A 100 μm Mylar foil was used to reduce the low energy x-rays count as well as prevent scattered protons from entering the detector. Samples were fixed to the sample holder by means of double-sided carbon tape. This tape is very pure and does not contain trace elements higher than oxygen ($Z > 8$). The tape does not affect measurements and is effective in holding the samples providing good electric conductivity to allow target current measurement. Each sample was then scanned for 5–30 min for extensive emission data to be obtained. Elemental concentration maps were created using GeoPIXE [24,25] and used to calculate the average elemental concentrations.

2.3. Statistical Analysis

Measured toxicant concentrations were compared with nominal concentrations using least squares linear regression. Concentration response curves were estimated by fitting a two-parameter nonlinear regression function with a binomial error structure using the DRC package [26] in R version 3.5.3 [27]. Weibull, log-logistic and log-normal models were fit to each dataset, with the best fitting model selected based on Akaikes Information Criteria. The model parameters were estimated using maximum likelihood, with starter values determined by the programs self-starter function. Concentrations affecting 10% and 50% of the syncarids (EC10 and EC50 values, respectively) were extrapolated from the fitted curve. Data from both 96 h and 14-day exposures were analysed separately.

The concentrations of As and Zn measured in syncarids after exposure were compared between treatments using one-way analyses of variance. Post-hoc comparisons between treatments were made using SNK tests. Data were log transformed to approximate normality and homogeneity of variance, which were assessed visually using plots of residuals and Q-Q plots. Analyses of variance were done using the NCSS program (Version 10, NCSS LLC, Kaysville, Utah, USA). The significance level (α) for all analyses was 0.05.

3. Results

Measured test concentrations were mostly within 20% of the nominal concentrations, and have been used in all analyses. Dissolved oxygen concentrations remained at 60% saturation or higher throughout all tests. This level of oxygen saturation is in line with the general concentrations observed in the field, as seen in Table 1. Temperature was consistently at or close to the test temperature of 18 °C. The hardness of the diluent water was relatively low, as seen in Table 1, which reflects the low carbonate concentrations in the sandstone aquifer from which the fauna and diluent water was collected. pH values varied little across the tests, and were in the range 5–6 pH units, as expected given the naturally acidic groundwater used as the diluent water.

Control mortality at 96 h was below zero for Cr and As exposures, and 3% for Zn. At 14 days, control mortality was higher across all tests (As: 9%, Cr: 11%, Zn: 11%), but below acceptability criterion of 20%, as seen in Figure S1. A Weibull model was fitted to all data sets, with details provided in Figure S1 and Table S1. Arsenic and Cr were similarly toxic after 96 h, but As was the most toxic after 14-day exposure. Zinc was the least toxic to the syncarids across both exposure periods, as seen in Table 2. As expected, toxicity increased and EC values decreased with exposure period from 96 h to 14 days for all toxicants, as seen in Table 2.

Table 2. Toxicity of As(III), Cr(VI) and Zn to bathynellid syncarids. EC10 and EC50 values indicate the concentration (mg/L) effecting 10% and 50% of the test population, respectively. Values in parentheses indicate 95% fiducial limits, followed by the total number of animals tested (N).

		As(III)		Cr(VI)		Zn(II)	
Test Duration	**ECx**	**Estimate**	**95% Confidence Interval**	**Estimate**	**95% Confidence Interval**	**Estimate**	**95% Confidence Interval**
		N = 74		N = 71		N = 190	
96 h	10	0.17	(−0.04–0.38)	0.11	(−0.07–0.28)	0.78	(0.30–1.26)
	50	1.99	(0.79–3.19)	1.83	(−0.10–3.77)	3.61	(2.56–4.66)
14 days	10	0.01	(−0.02–0.04)	0.004	(−0.010–0.018)	0.21	(0.02–0.39)
	50	0.25	(0.07–0.44)	0.51	(−0.07–1.09)	1.77	(1.11–2.43)

The mean body concentration of As in the syncarids in controls was 2.8 μg As/g, as seen in Figure 1A. The body concentrations in all exposure treatments were significantly greater than in the controls ($p < 0.05$), and concentrations in syncarids from the 10,000 and 300 mg/L treatments were greater than those in the 100 mg/L treatment ($p < 0.05$), but were not significantly different to each other, as seen in Figure 1A. The PIXE maps showed that As was concentrated around the head, anterior thoracic and abdominal sections, as seen in Figure 2.

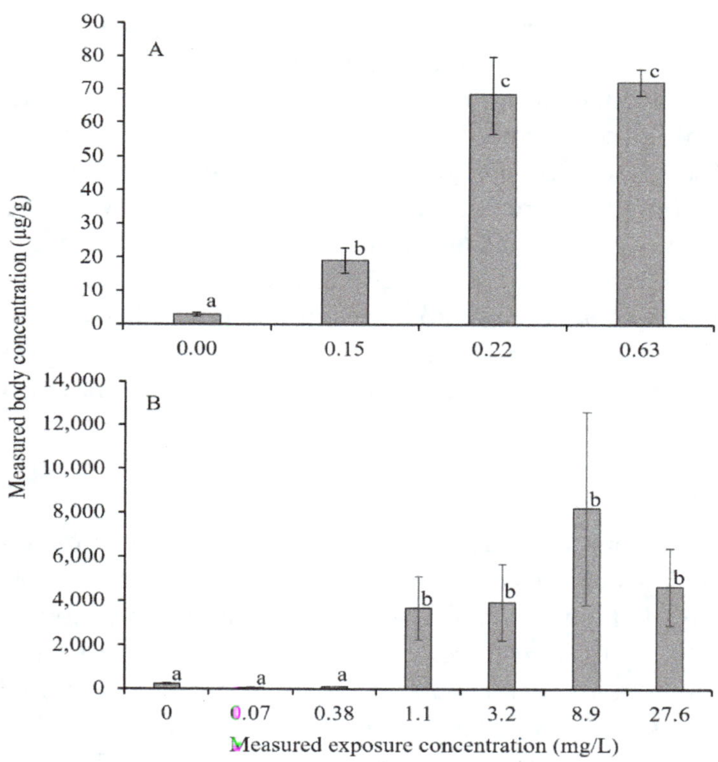

Figure 1. Mean concentration (±SE) between exposure concentration and mean body concentration in bathynellid syncarids exposed to (**A**) As and (**B**) Zn. Common lower case letters indicate no significant difference ($p > 0.05$) between those treatments.

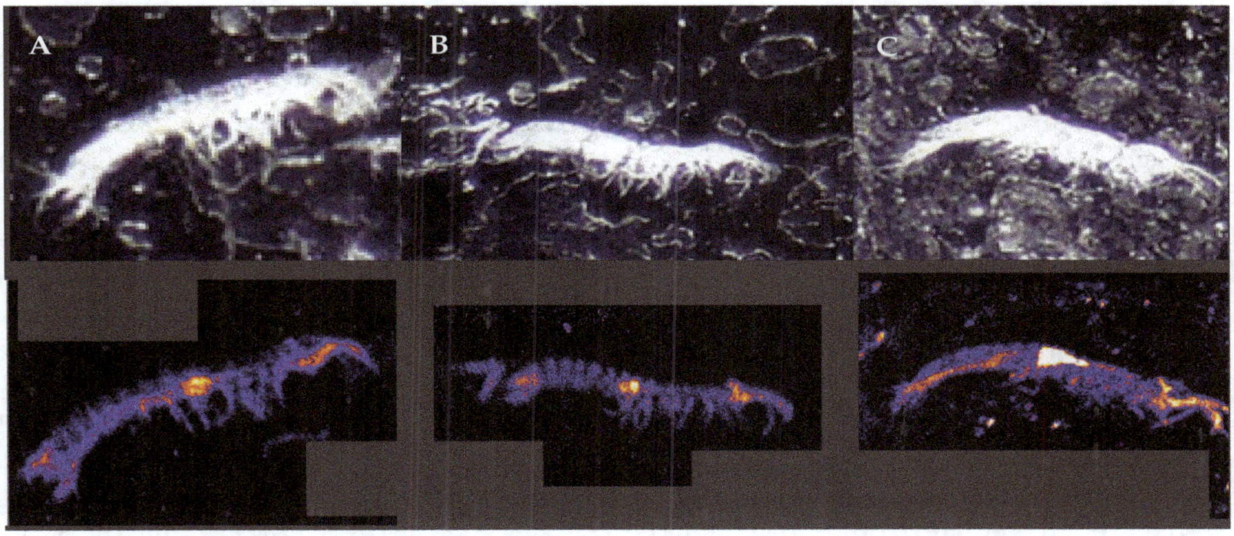

Figure 2. Images of As accumulation in bathynellid syncarids exposed to As in aqueous solution. Mean concentrations of specimens shown were (**A**) 12 ppm, (**B**) 2 ppm, and (**C**) 28 ppm.

Concentrations of Zn measured in the syncarids from the control treatments were on average 228 μg Zn/g, which increased significantly ($p < 0.001$) with exposure to all Zn treatments ≥1 mg Zn/L, as seen in Figure 1B. There was no significant difference ($p > 0.05$) in body concentrations across the Zn treatments, as seen in Figure 1B above 1 mg Zn/L. The accumulation of Zn in the syncarids appears to be concentrated in the anterior thoracic region and abdomen, as seen in Figure 3.

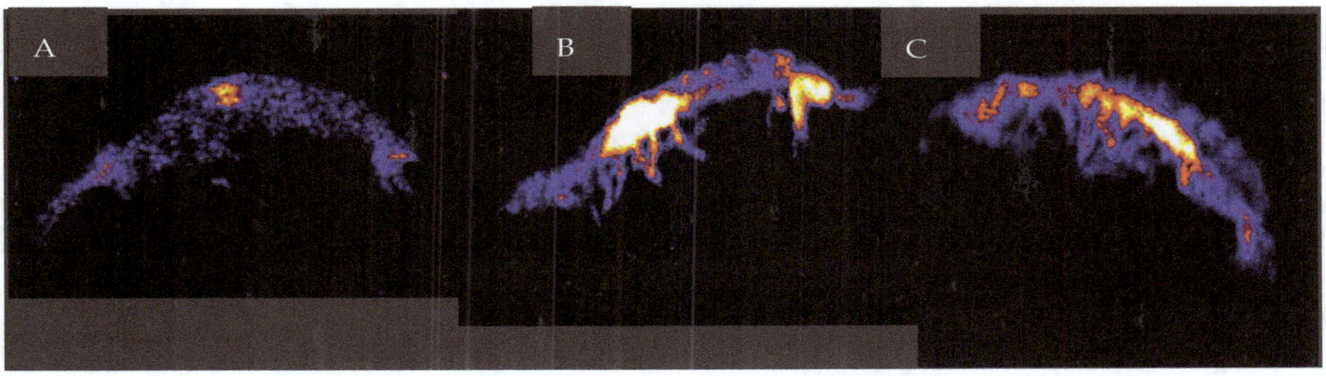

Figure 3. Images of Zn accumulation in bathynellid syncarids exposed to Zn in aqueous solution. Mean concentrations of specimens shown were (**A**) 250 ppm, (**B**) 6400 ppm, and (**C**) 8000 ppm.

4. Discussion

Our tests have provided the first toxicity and accumulation data for the Syncarida, specifically the Bathynellidae, which are an obligate subterranean taxon. We discuss below the sensitivity of the syncarids to the metals and metalloid relative to other stygobiotic taxa and surface water crustaceans. The accumulation of Zn suggests the ability of syncarids to regulate body concentrations at low exposure concentrations, but for As, we found no evidence of regulation at the exposure concentration tested. Tests were conducted in relatively low hardness, low organic content water from a fractured rock aquifer, which likely makes them more susceptible to toxicants than stygobitic taxa from hard, karstic or more organic groundwaters [28]. While this trend is not unexpected, we highlight the need to consider the aquifer types in the management and risk assessment for subterranean ecosystems.

Mortality in controls was ≤11%, below the acceptable threshold of 20%, and water quality conditions were stable during all tests, indicating we had created a suitable testing environment.

Dissolved oxygen (DO) concentrations in the tests were in line with those measured at the collection site in groundwater, as seen in Table 1, but higher than often reported in groundwaters [29]. The relatively high dissolved oxygen concentrations in the static tests is a likely consequence of the small test volume used and large surface area for gas exchange within the 24 well plates. Previous studies suggest that stygobitic crustaceans can survive well in oxygenated laboratory environments (e.g., [11,16]), and do not affect the toxic response of groundwater crustaceans to metals [30].

The toxicity of As to the syncarids after 96 h was similar to that observed for other stygobitic crustaceans. Hose et al. [16] tested the sensitivity of cyclopoid and harpacticoid copepods from the same Somersby location and showed 96-h LC50 values of similar magnitude (5.24, 3.04 mg As/L, respectively) to those estimated here for the syncarids. The syncarids were similar in sensitivity to As as the cladoceran *Daphnia magna* (96-h LC50 1.50–4.34 mg As(III)/L [31]). However, over 14 days, our syncarids (14-day LC50, 0.25 mg As/L) were slightly more sensitive than the cyclopoid and harpacticoid copepods from Somersby (14-day LC50 0.79, 1.46 mg As/L, respectively [16]). The syncarids were also more sensitive to As than the stygobitic amphipod *Niphargus rhenorhodanensis* (240-h LC50, 3.97 mg/L [32]), but similar to other freshwater crustaceans such as *Gammarus fossarum* (LC50, 0.2 mg As/L [32]), which was the most sensitive of a range of taxa in that study. Overall the syncarids were less sensitive to As than Cr, which is consistent with the findings of Canivet et al. [32]. For As and the metals tested, there are few low ECx values in the literature for comparison, nevertheless, LC10 values are reported here so that they may be available to inform the development of environmental quality criteria for groundwaters (e.g., [9]).

Chromium was the most toxic of the elements tested. The acute toxicity of chromium to the syncarids was similar to that recorded for subterranean crustacea reported previously. The ranges of 96-h LC50 values for the styogophilous *Bryocamptus echinatus* (1.26 mg Cr/L [33]), *N. rhenorhodanensis* (1.51–2.07 mg Cr/L [32]), stygobitic copepods (0.7–3.1 mg Cr/L [16]), and the stygobitic isopod *Proassellus* spp. (48-h EC50 0.396–6.35 mg Cr/L [34]) all overlap the 96-h LC50 value for the syncarids (1.8 mg Cr/L). The 96-h LC50 values for *D. magna* exposed to Cr (0.07–0.16 mg Cr/L, [35,36]) were approximately an order of magnitude lower than those for the syncarids tested here, as seen in Table 2. The 14-day LC50 value (0.5 mg Cr/L) is also within the range of values for copepods from the same study site (0.03–1.06 mg Cr/L [16]) and a 10-day LC50 for *N. rhenorhodanensis* (0.23 mg Cr/L [32]) and less than those for a range of epigean taxa [32,36]. The 14-day LC10 for Cr (0.004 mg Cr/L) was below the background concentration at the site (0.03 mg Cr/L). This anomaly is a likely consequence of the exposure-response curve fitting, as seen in Figure S1, which is often less reliable at the extremities of the curve. We do, however, believe that the overall curve fit is reliable and the inherent conservatism in this case is preferable for the application of these data to ecosystem protection.

The syncarids were less sensitive to Zn than the Harpacticoid copepod and more sensitive than the cyclopoid copepod from the same locations over both 96 h and 14 days [16]. Ninety-six hour LC50 values for the syncarids were similar to those determined for the groundwater copepod *Parastenocaris germanica* (96-h LC50 1.7 mg Zn/L) under oxic (10 mg O_2/L) conditions [30], but both *P. germanica* and the syncarid tested here were much more sensitive than several other stygobitic crustaceans such as *Niphargus aquilex* (96-h LC50 180 mg Zn/L [37]), *Asellus cavaticus* (96-h LC50 127 mg Zn/L [37]) and *Niphargus montellianus*, which had no mortality to Zn after 10-day exposure to 25 mg Zn/L [38]. The syncarids were less sensitive to zinc over 96 h than was *D. magna* (96-h LC50 0.15–0.18 mg Zn/L [36,39]).

Overall, the syncarids tested were similar, but slightly more sensitivity than other stygobitic crustaceans. The syncarids were, however, generally less sensitive over 96 h exposure than the standard cladoceran test species, *D. magna*. Current approaches for risk assessment for groundwater include the use of standard test species, such as *D. magna* [40], as a first tier in lieu of toxicity data for groundwater biota. Our results suggest that this may be a suitable interim approach until sufficient data for groundwater taxa become available.

Our syncarids were collected from a fractured rock aquifer which had low hardness, low-organic matter water, contrasting to some previous studies in which biota were collected from karstic aquifers with much greater water hardness. Hardness is a major factor affecting the toxicity of metals, particularly zinc and chromium, to organisms [41–43], and hardness correction has been used to compare organism sensitivity in ecological risk assessments. Interestingly, using the hardness correction of Warne et al. [44], the syncarids tested were the most Zn-sensitive of the stygobitic crustaceans tested to date. No hardness correction algorithm is available for Cr(VI). With likely differences in the hardness of water between aquifer types, it is critical that aquifer types are considered in risk assessments and decision making for subterranean ecosystems and when setting environmental quality standards.

The toxicity of all toxicants increased with exposure period, with decreases in EC values of up to two orders of magnitude between 96 h and 14 days. The significance of such short-term exposures in groundwater ecosystems, with the exception of perhaps karst systems (see [34]), is debatable because of the slow groundwater flow rates in most aquifers. Hose et al. [16] showed that the relative toxicities of As, Cr and Zn were consistent between 96 h, 14-day and 28-day exposures, but with up to a six-fold decrease in ECx values between 14 and 28 days. Di Lorenzo et al. [10] recommend the reporting 96-h LCx values for the purposes of comparison between studies and the existing ecotoxicological databases, but recommend that tests be extended well beyond 96 h. Unfortunately, due to the longevity and low reproductive rates of stygobitic species, continuing tests for durations that are relevant to life cycles of the organisms tested may be difficult. For this reason, incipient LC50 values (sensu [45,46]) would be useful and should be a direction of future research.

The accumulation of both As and Zn appeared greatest in the region at the anterior of the thoracic region. Previous studies using decapods have shown high concentrations of both As [47] and Zn [48,49] in the hepatopancreas, which is consistent with the location of the thoracic hotspot in the syncarids, although we recognise that it is difficult to determine the specific organs with this method, and the uncertainty of the organ location on these small animals. Cresswell et al. [49] suggest that Zn is transported from the gill to the hepatopancreas for detoxication and/or metabolism. Cresswell et al. [49] also report high concentrations of Zn in the antennal gland, which is an important excretion site in crustaceans and which are well developed in the Bathynellacea [50]. However, this was not the case for Zn in this study, although this may explain the high concentration of As in that region. It is unclear which tissues in the posterior of the abdomen are causing the high Zn, but the distribution of As throughout the length of the abdomen may occur in variety of tissues of systems that extend through the syncarid abdomen [51]. Alternatively, it may be a result of water uptake through the abdomen as done by some crustaceans (sensu [52]) but it is unclear whether syncarids do this. Both White and Rainbow [48] and Cresswell et al. [49] report high Zn concentrations in the eye of decapods (associated with zinc metalloenzymes used in vision [53]), which was clearly not the case for our blind (eyeless) stygobitic syncarid.

The body concentration of As in the syncarids was far lower than that recorded for Zn, which reflects the relative abundance of these elements in the environment and most biological systems [54]. The concentration of As in the unexposed syncarids was 2.8 μg As/g and similar to unexposed amphipods and *Daphnia* (<5 μg As/g [55]). The concentrations of As in unexposed *N. rhenorhodanensis* [32] were below detection limit; however, the mean concentration of As in amphipods exposed to 0.1 mg As/L for 10 days (22.6 μg As/g) [32] was of a similar magnitude to the concentrations in syncarids (19 μg As/g) exposed to 0.15 mg As/L for a similar period in this study. The arsenic accumulation data suggest that the syncarids have no ability to regulate As uptake, with the body concentration for syncarids exposed to 0.15 mg As/L being significantly greater than those in the syncarids from controls, as well as mortality occurring at that concentration after 14 days. Arsenic accumulation by the syncarids did not increase significantly as exposure concentration increased beyond 0.22 mg As/L. It is unlikely that this reflects an ability to regulate As uptake at these concentrations since there was no evidence at the lower concentrations, and this has not been reported elsewhere. Rather, the lack of significant increase in body concentrations beyond 0.22 mg As/L is likely a statistical artefact caused by large variability in internal concentrations within the 0.22 mg As/L treatment, as evidenced by the size of the error bars in Figure 1A.

The background concentration of Zn in the syncarids was around 228 μg Zn/g which was similar to that for *N. montellianus* (113 μg Zn/g [38]). The accumulation of Zn by syncarids exposed to 1 mg Zn/L was significantly greater than syncarids in controls and exposed to lower concentrations, as seen in Figure 1B. This trend differs from that observed for *N. montellianus*, for which significant accumulation of Zn did not occur until they were exposed to 10 mg Zn /L [38], but this difference may reflect the greater hardness of the water used in that study (150–165 mg $CaCO_3$/L), and the expected reduced bioavailability and uptake of metals by organisms with increasing hardness [56,57].

Our results suggest an ability of syncarids to regulate Zn at concentrations below 1 mg Zn/L. When considered alongside the mortality data, the increase in accumulation from the 0.38 to the 11 mg Zn/L exposures coincides approximately with a threshold change in mortality after 14 days. Rainbow and Luoma [58] detail several models of metal (particularly Zn) uptake in crustaceans, and that observed here for the syncarids most closely fits that common in the Decapoda (e.g., [59,60]), which might be expected given the phylogenetic proximity of the decapods and syncarids [61]. In this model, Zn concentration in the animals is regulated by excreting Zn until a threshold exposure concentration is reached beyond which the animal is unable to continue regulating their internal concentration [58]. Relatively little metal is stored in detoxified forms [59] and the contribution of Zn adsorbed passively to the exoskeleton to the total body concentration is small [59]. Above the threshold concentration, Zn is accumulated proportionally to the exposure concentration [58] and binds at sites where it will interfere with normal metabolic functioning, leading to toxicity [59]. While for some aquatic organisms the uptake of some metals from solution plateaus at high concentrations, this is has not been reported for invertebrates (e.g., [62]). The lack of apparent increase in mean body concentrations at exposure concentrations above 1 mg Zn/L may be a consequence of the large variability in body concentrations in those animals, which obscures an increasing trend.

Understanding the ecological significance of laboratory response data to population level effects in crustaceans in the field is difficult [63]. Our syncarids are no different, with the ecological significance of their bioaccumulation of Zn and As unclear. The accumulation of Zn by gammarid amphipods at concentrations as low as 0.3 mg Zn/L effected reproduction and energy utilisation in females [64]. At that concentration, the number of aborted broods increased, and the reduced energy absorption and reallocation of that energy away from reproduction resulted in smaller sized offspring [64]. The reproductive strategies of bathynellid syncarids are poorly known, but like other stygobiotic crustaceans they are expected to have low metabolic rates, reproductive rates and small brood sizes (i.e., k-strategy traits), a likely consequence of the low energy groundwater environment (Humphreys 2006). What is the effect of reducing energy absorption and fecundity in a population of organisms that are already at the lower extremes of such rates? Kammenga et al. [65] showed no difference in sensitivity of k-strategist and r-strategist nematodes to cadmium, but whether this holds true for other taxa is unclear.

Maltby and Naylor [64] predict that the smaller offspring resulting from zinc exposure will take longer to reach maturity and may reproduce at a smaller size, which typically means reduced fecundity, which is exacerbated by the increase in aborted broods. The net effect is expected to thus be reduced recruitment and population decline. Importantly for syncarids that feed on detrital matter, bacteria and fungi [66], changes in microbial assemblages, particularly fungal biomass, may also occur at relatively low Zn and As concentrations. Lategan et al. [67] observed 50% declines in biomass of groundwater fungi at 4.4 mg As/L and 0.61 mg Zn/L, thereby exacerbating potential sublethal effects on feeding and energy assimilation. In groundwaters where contamination is typically long-term and difficult to remediate, these impacts will be persistent, and likely lead to a decrease in population viability. The broader consequences of declining invertebrate populations in aquifers is the loss of ecosystem services [6], including changes to aquifer hydraulic properties [68].

Chromium, zinc and arsenic were acutely toxic to the stygobitic bathynellid syncarid. The accumulation of Zn was regulated at exposure concentrations below 1 mg Zn/L, above which body concentrations increased, leading to increased mortality. Arsenic was not regulated and was accumulated by the syncarids at all concentrations above the control. These are the first published toxicity data for syncarids and show them among the most sensitive of stygobitic crustaceans so far tested, likely in part due to the low hardness of the groundwater from the aquifer they inhabit and in which they were tested. The ecological significance of the toxicant accumulation and mortality may be significant given the consequent population effects and low capacity for stygobitic populations to recover.

Author Contributions: Conceptualization, G.C.H. and M.J.L.; methodology, G.C.H., M.J.L., and R.S.; formal analysis, K.S., R.S. and G.C.H.; investigation, all; resources, G.C.H. and R.S.; writing—original draft preparation, G.C.H.; writing—review and editing, all; supervision, G.C.H. and M.J.L.; project administration, G.C.H.; funding acquisition, G.C.H. and R.S.

Acknowledgments: The authors are grateful for the assistance of Ashleigh Keast with the arsenic tests and Sarah Stephenson with field collections. The insightful comments of two anonymous reviewers improved this manuscript.

References

1. Alley, W.M.; Healy, R.W.; LaBaugh, J.W.; Reilly, T.E. Flow and storage in groundwater systems. *Science* **2002**, *296*, 1985. [CrossRef] [PubMed]
2. Morris, B.L.; Lawrence, A.R.L.; Chilton, P.J.C.; Adams, B.; Calow, R.C.; Klinck, B.A. *Groundwater and Its Susceptibility to Degradation: A Global Assessment of the Problem and Options for Management*; United Nations Environment Programme: Nairobi, Kenya, 2003; p. 126.
3. Humphreys, W.F. Aquifers: The ultimate groundwater-dependent ecosystems. *Aust. J. Bot.* **2006**, *54*, 115–132. [CrossRef]
4. Camacho, A.I.; Valdecasas, A.G. Global diversity of syncarids (Syncarida; Crustacea) in freshwater. In *Freshwater Animal Diversity Assessment*; Balian, E.V., Lévêque, C., Segers, H., Martens, K., Eds.; Springer: Dordrecht, The Netherlands, 2008; pp. 257–266.
5. Hose, G.C. Response to Humphreys' (2007) Comments on Hose GC (2005) Assessing the Need for Groundwater Quality Guidelines for Pesticides Using the Species Sensitivity Distribution Approach. *Hum. Ecol. Risk Assess.* **2007**, *13*, 241–246. [CrossRef]
6. Griebler, C.; Avramov, M.; Hose, G. Groundwater ecosystems and their services—Current status and potential risks. In *Atlas of Ecosystem Services—Drivers, Risks, and Societal Responses*; Schröter, M., Bonn, A., Klotz, S., Seppelt, R., Baessler, C., Eds.; Springer: Berlin/Heidelberg, Germany, 2019; pp. 197–203.
7. Hose, G.C.; Asmyhr, M.G.; Cooper, S.J.B.; Humphreys, W.F. Down under Down Under: Austral groundwater life. In *Austral Ark: The State of Wildlife in Australia and New Zealand*; Stow, A., Maclean, N., Holwell, G.I., Eds.; Cambridge University Press: Cambridge, UK, 2015; pp. 512–536.
8. Castaño-Sánchez, A.; Hose, G.; Reboleira, A. Ecotoxicological effects of anthropogenic stressors in subterranean organisms: A review. *Chemosphere* **2019**, in press. [CrossRef]
9. Hose, G.C. Assessing the Need for Groundwater Quality Guidelines for Pesticides Using the Species Sensitivity Distribution Approach. *Hum. Ecol. Risk Assess.* **2005**, *11*, 951–966. [CrossRef]
10. Di Lorenzo, T.; Di Marzio, W.D.; Fiasca, B.; Galassi, D.M.P.; Korbel, K.; Iepure, S.; Pereira, J.L.; Reboleira, A.S.P.S.; Schmidt, S.I.; Hose, G.C. Recommendations for ecotoxicity testing with stygobiotic species in the framework of groundwater environmental risk assessment. *Sci. Total Environ.* **2019**, *681*, 292–304. [CrossRef]
11. Di Lorenzo, T.; Di Marzio, W.D.; Spigoli, D.; Baratti, M.; Messana, G.; Cannicci, S.; Galassi, D.M.P. Metabolic rates of a hypogean and an epigean species of copepod in an alluvial aquifer. *Freshw. Biol.* **2015**, *60*, 426–435. [CrossRef]
12. Little, J.; Schmidt, D.J.; Cook, B.D.; Page, T.J.; Hughes, J.M. Diversity and phylogeny of south-east Queensland Bathynellacea. *J. Aust. J. Zool.* **2016**, *64*, 36–47. [CrossRef]

13. Perina, G.; Camacho, A.I.; Huey, J.; Horwitz, P.; Koenders, A. New Bathynellidae (Crustacea) taxa and their relationships in the Fortescue catchment aquifers of the Pilbara region, Western Australia. *Syst. Biodivers.* **2019**, *17*, 148–164. [CrossRef]
14. Camacho, A.I.; Mas-Peinado, P.; Dorda, B.A.; Casado, A.; Brancelj, A.; Knight, L.R.F.D.; Hutchins, B.; Bou, C.; Perina, G.; Rey, I. Molecular tools unveil an underestimated diversity in a stygofauna family: A preliminary world phylogeny and an updated morphology of Bathynellidae (Crustacea: Bathynellacea). *Zool. J. Linn. Soc.* **2017**, *183*, 70–96. [CrossRef]
15. Korbel, K.L.; Hose, G.C. Habitat, water quality, seasonality, or site? Identifying environmental correlates of the distribution of groundwater biota. *Freshw. Sci.* **2015**, *34*, 329–343. [CrossRef]
16. Hose, G.C.; Symington, K.; Lott, M.J.; Lategan, M.J. The toxicity of arsenic(III), chromium(VI) and zinc to groundwater copepods. *Environ. Sci. Pollut. Res.* **2016**, *23*, 18704–18713. [CrossRef] [PubMed]
17. Fruchter, J. Peer Reviewed: In-Situ Treatment of Chromium-Contaminated Groundwater. *Environ. Sci. Technol.* **2002**, *36*, 464A–472A. [CrossRef] [PubMed]
18. Scheeren, P.J.H.; Koch, R.O.; Buisman, C.J.N.; Barnes, L.J.; Versteegh, J.H. New biological treatment plant for heavy metal contaminated groundwater. In *EMC '91: Non-Ferrous Metallurgy—Present and Future*; Springer: Dordrecht, The Netherlands, 1991; pp. 403–416.
19. Smedley, P.L.; Kinniburgh, D.G. A review of the source, behaviour and distribution of arsenic in natural waters. *Appl. Geochem.* **2002**, *17*, 517–568. [CrossRef]
20. Edmunds, W.M.; Shand, P. Groundwater Baseline Quality. In *Natural Groundwater Quality*; Edmunds, W.M., Shand, P., Eds.; Blackwell Publishing: Oxford, UK, 2009; pp. 1–21.
21. Klockenkämper, R.; von Bohlen, A. Elemental Analysis of Environmental Samples by Total Reflection X-Ray Fluorescence: A Review. *X-ray Spectrom.* **1996**, *25*, 156–162. [CrossRef]
22. Farías, S.S.; Casa, V.A.; Vázquez, C.; Ferpozzi, L.; Pucci, G.N.; Cohen, I.M. Natural contamination with arsenic and other trace elements in ground waters of Argentine Pampean Plain. *Sci. Total Environ.* **2003**, *309*, 187–199. [CrossRef]
23. Siegele, R.; Cohen, D.D.; Dytlewski, N. The ANSTO high energy heavy ion microprobe. *Nucl. Instrum. Methods Phys. Res. Sect. B Beam Interact. Mater. Atoms* **1999**, *158*, 31–38. [CrossRef]
24. Ryan, C.G. Developments in Dynamic Analysis for quantitative PIXE true elemental imaging. *Nucl. Instrum. Methods Phys. Res. Sect. B Beam Interact. Mater. Atoms* **2001**, *181*, 170–179. [CrossRef]
25. Ryan, C.G.; Jamieson, D.N.; Churms, C.L.; Pilcher, J.V. A new method for on-line true-elemental imaging using PIXE and the proton microprobe. *Nucl. Instrum. Methods Phys. Res. Sect. B Beam Interact. Mater. Atoms* **1995**, *104*, 157–165. [CrossRef]
26. Ritz, C.; Streibig, J.C. Bioassay analysis using R. *J. Stat. Softw.* **2005**, *12*, 1–22. [CrossRef]
27. R Core Team. *R: A Language and Environment for Statistical Computing*; R Foundation for Statistical Computing: Vienna, Austria, 2019.
28. Hyne, R.V.; Pablo, F.; Julli, M.; Markich, S.J. Influence of water chemistry on the acute toxicity of copper and zinc to the cladoceran Ceriodaphnia cf dubia. *Environ. Toxicol. Chem.* **2005**, *24*, 1667–1675. [CrossRef] [PubMed]
29. Hahn, H.J. The GW-Fauna-Index: A first approach to a quantitative ecological assessment of groundwater habitats. *Limnol. Ecol. Manag. Inland Waters* **2006**, *36*, 119–137. [CrossRef]
30. Notenboom, J.; Cruys, K.; Hoekstra, J.; Vanbeelen, P. Effect of Ambient Oxygen Concentration Upon the Acute Toxicity of Chlorophenols and Heavy-Metals to the Groundwater Copepod Parastenocaris-Germanica (Crustacea). *Ecotoxicol. Environ. Saf.* **1992**, *24*, 131–143. [CrossRef]
31. Lima, A.; Curtis, C.; Hammermeister, D.; Markee, T.; Northcott, C.E.; Brooke, L.T. Acute and chronic toxicities of arsenic(III) to fathead minnows, flagfish, daphnids, and an amphipod. *Arch. Environ. Contam. Toxicol.* **1984**, *13*, 595–601. [CrossRef]
32. Canivet, V.; Chambon, P.; Gibert, J. Toxicity and bioaccumulation of arsenic and chromium in epigean and hypogean freshwater macroinvertebrates. *Arch. Environ. Contam. Toxicol.* **2001**, *40*, 345–354. [CrossRef]
33. Di Marzio, W.D.; Castaldo, D.; Pantani, C.; Di Cioccio, A.; Di Lorenzo, T.; Saenz, M.E.; Galassi, D.M.P. Relative Sensitivity of Hyporheic Copepods to Chemicals. *Bull. Environ. Contam. Toxicol.* **2009**, *82*, 488–491. [CrossRef]

34. Reboleira, A.S.; Abrantes, N.; Oromí, P.; Gonçalves, F. Acute Toxicity of Copper Sulfate and Potassium Dichromate on Stygobiont Proasellus: General Aspects of Groundwater Ecotoxicology and Future Perspectives. *Water Air Soil Pollut.* **2013**, *224*, 1–9. [CrossRef]
35. Fargasova, A. Toxicity of Metals on Daphnia magna and Tubifex tubifex. *Ecotoxicol. Environ. Saf.* **1994**, *27*, 210–213. [CrossRef]
36. Ewell, W.S.; Gorsuch, J.W.; Kringle, R.O.; Robillard, K.A.; Spiegel, R.C. Simultaneous evaluation of the acute effects of chemicals on seven aquatic species. *Environ. Toxicol. Chem.* **1986**, *5*, 831–840. [CrossRef]
37. Meinel, W.; Krause, R. Zur Korrelation zwischen Zink und verschiedenen pH–Werten in in her toxischen Wirkung auf einige Grundwasser–Organismen. *Z. Für Angew. Zool.* **1988**, *75*, 159–182.
38. Coppellotti Krupa, O.; Toniello, V.; Guidolin, L. Niphargus and Gammarus from karst waters: First data on heavy metal (Cd, Cu, Zn) exposure in a biospeleology laboratory. *Subterr. Biol.* **2004**, *2*, 33–41.
39. Lazorchak, J.M.; Smith, M.E.; Haring, H.J. Development and validation of a Daphnia magna four-day survival and growth test method. *Environ. Toxicol. Chem.* **2009**, *28*, 1028–1034. [CrossRef] [PubMed]
40. Daam, M.A.; Silva, E.; Leitao, S.; Trindade, M.J.; Cerejeira, M.J. Does the actual standard of 0.1 mu g/L overestimate or underestimate the risk of plant protection products to groundwater ecosystems? *Ecotoxicol. Environ. Saf.* **2010**, *73*, 750–756. [CrossRef] [PubMed]
41. Paulauskis, J.D.; Winner, R.W. Effects of water hardness and humic acid on zinc toxicity to *Daphnia magna* Straus. *Aquat. Toxicol.* **1988**, *12*, 273–290. [CrossRef]
42. Pawlisz, A.V.; Kent, R.A.; Schneider, U.A.; Jefferson, C. Canadian water quality guidelines for chromium. *Environ. Toxicol. Water Qual.* **1997**, *12*, 123–183. [CrossRef]
43. Park, E.J.; Jo, H.J.; Jung, J. Combined effects of pH, hardness and dissolved organic carbon on acute metal toxicity to Daphnia magna. *J. Ind. Eng. Chem.* **2009**, *15*, 82–85. [CrossRef]
44. Warne, M.; Batley, G.; van Dam, R.; Chapman, J.; Fox, D.; Hickey, C.; Stauber, J. *Revised Method for Deriving Australian and New Zealand Water Quality Guideline Values for Toxicants—Update of 2015 Version*; Australian and New Zealand Governments and Australian State and Territory Governments: Canberra, Australia, 2018; p. 48.
45. Sprague, J.B. Measurement of pollutant toxicity to fish I. Bioassay methods for acute toxicity. *Water Res.* **1969**, *3*, 793–821. [CrossRef]
46. Avramov, M.; Schmidt, S.I.; Griebler, C. A new bioassay for the ecotoxicological testing of VOCs on groundwater invertebrates and the effects of toluene on Niphargus inopinatus. *Aquat. Toxicol.* **2013**, *130*, 1–8. [CrossRef]
47. Williams, G.; West, J.M.; Snow, E.T. Total arsenic accumulation in yabbies (Cherax destructor clark) exposed to elevated arsenic levels in Victorian gold mining areas, Australia. *Environ. Toxicol. Chem.* **2008**, *27*, 1332–1342. [CrossRef]
48. White, S.L.; Rainbow, P.S. Regulation of zinc concentration by Palaemon elegans (Crustacea: Decapoda): Zinc flux and effects of temperature, zinc concentration and moulting. *Mar. Ecol. Prog. Ser.* **1984**, *16*, 135–147. [CrossRef]
49. Cresswell, T.; Simpson, S.L.; Mazumder, D.; Callaghan, P.D.; Nguyen, A.P. Bioaccumulation Kinetics and Organ Distribution of Cadmium and Zinc in the Freshwater Decapod Crustacean Macrobrachium australiense. *Environ. Sci. Technol.* **2015**, *49*, 1182–1189. [CrossRef] [PubMed]
50. Steenken, B.; Schminke, H.K. Ultrastructure of maxillary gland of Antrobathynella stammeri (Syncarida, Malacostraca). *J. Morphol.* **1996**, *228*, 107–117. [CrossRef]
51. Smith, G. On the Anaspidacea, Living and Fossil. *J. Cell Sci.* **1909**, *s2-53*, 489–578.
52. Fox, H.M. Anal and Oral Intake of Water by Crustacea. *J. Exp. Biol.* **1952**, *29*, 583. [CrossRef]
53. White, S.L.; Rainbow, P.S. Zinc flux in Palaemon elegans (Crustacea: Decapoda): Moulting, individual variation and tissue distribution. *Mar. Ecol. Prog. Ser.* **1984**, *19*, 153–166. [CrossRef]
54. Smith, K.S.; Huyck, H. An overview of the abundance, relative mobility, bioavailabilty, and human toxicity of metals. In *The Environmental Geochemistry of Mineral Deposits: Part, A; Processes, Techniques, and Health Issues*; Plumlee, G., Logsdon, J., Eds.; Society for Economic Geologists: Littleton, CO, USA, 1999; Volume 6A, pp. 29–70.

55. Spehar, R.L.; Fiandt, J.T.; Anderson, R.L.; DeFoe, D. Comparative toxicity of arsenic compounds and their accumulation in invertebrates and fish. *Arch. Environ. Contam. Toxicol.* **1980**, *9*, 53–63. [CrossRef]
56. Winner, R.W.; Gauss, J.D. Relationship between chronic toxicity and bioaccumulation of copper, cadmium and zinc as affected by water hardness and humic acid. *Aquat. Toxicol.* **1986**, *8*, 149–161. [CrossRef]
57. Markich, S.; Brown, P.; Batley, G.; Apte, S.; Stauber, J. Incorporating metal speciation and bioavailability into water quality guidelines for protecting aquatic ecosystems. *Aust. J. Ecotoxicol.* **2001**, *7*, 109–122.
58. Rainbow, P.S.; Luoma, S.N. Metal toxicity, uptake and bioaccumulation in aquatic invertebrates—Modelling zinc in crustaceans. *Aquat. Toxicol.* **2011**, *105*, 455–465. [CrossRef]
59. Rainbow, P.S. Trace metal bioaccumulation: Models, metabolic availability and toxicity. *Environ. Int.* **2007**, *33*, 576–582. [CrossRef]
60. Vijayram, K.; Geraldine, P. Toxicology. Regulation of Essential Heavy Metals (Cu, Cr, and Zn) by the Freshwater Prawn Macrobrachium malcolmsonii (Milne Edwards). *Bull. Environ. Contam. Toxicol.* **1996**, *56*, 335–342. [CrossRef] [PubMed]
61. Koenemann, S.; Jenner, R.A.; Hoenemann, M.; Stemme, T.; von Reumont, B.M. Arthropod phylogeny revisited, with a focus on crustacean relationships. *Arthropod Struct. Dev.* **2010**, *39*, 88–110. [CrossRef] [PubMed]
62. Luoma, S.N.; Rainbow, P.S. *Metal. Contamination in Aquatic Environments: Science and Lateral Management*; Cambridge University Press: Cambridge, UK, 2008. [CrossRef]
63. Kuhn, A.; Munns, W.R., Jr.; Serbst, J.; Edwards, P.; Cantwell, M.G.; Gleason, T.; Pelletier, M.C.; Berry, W. Evaluating the ecological significance of laboratory response data to predict population-level effects for the estruarine amphipod Ampelisca abdita. *Environ. Toxicol. Chem.* **2002**, *21*, 865–874. [CrossRef] [PubMed]
64. Maltby, L.; Naylor, C. Preliminary Observations on the Ecological Relevance of the Gammarus 'Scope for Growth' Assay: Effect of Zinc on Reproduction. *Funct. Ecol.* **1990**, *4*, 393–397. [CrossRef]
65. Kammenga, J.E.; Van Gestel, C.A.M.; Bakker, J. Toxicology. Patterns of sensitivity to cadmium and pentachlorophenol among nematode species from different taxonomic and ecological groups. *Arch. Environ. Contam. Toxicol.* **1994**, *27*, 88–94. [CrossRef] [PubMed]
66. Wellborn, G.A.; Witt, J.D.S.; Cothran, R.D. Chapter 31—Class Malacostraca, Superorders Peracarida and Syncarida. In *Thorp and Covich's Freshwater Invertebrates*, 4th ed.; Thorp, J.H., Rogers, D.C., Eds.; Academic Press: Cambridge, MA, USA, 2015; pp. 781–796. [CrossRef]
67. Lategan, M.J.; Klare, W.; Kidd, S.; Hose, G.C.; Nevalainen, H. The unicellular fungal tool RhoTox for risk assessments in groundwater systems. *Ecotoxicol. Environ. Saf.* **2016**, *132*, 18–25. [CrossRef]
68. Hose, G.C.; Stumpp, C. Architects of the underworld: Bioturbation by groundwater invertebrates influences aquifer hydraulic properties. *Aquat. Sci.* **2019**, *81*, 20. [CrossRef]

14

Metallic Pollution and the use of Antioxidant Enzymes as Biomarkers in *Bellamya unicolor* (Olivier, 1804) (Gastropoda: Bellamyinae)

Mohammed Othman Aljahdali [1] and Abdullahi Bala Alhassan [1,2,*]

1 Department of Biological Sciences, Faculty of Sciences, King Abdulaziz University, Jeddah 80203, Saudi Arabia; moaljahdali@kau.edu.sa
2 Department of Biology, Faculty of Life Science, Ahmadu Bello University, Zaria 810001, Nigeria
* Correspondence: balahassan80@gmail.com or aalhassan0021@stu.kau.edu.sa

Abstract: Industrial and domestic discharges of effluent is one of the major causes of heavy metal pollution in aquatic ecosystems. Samples of benthic sediment and freshwater mollusc *Bellamya unicolor* were collected from 5 sites in the River Kaduna to determine heavy metal concentration, their ecological risk, and antioxidant enzymes activities in *Bellamya unicolor*. The results revealed the level of pollution based on heavy metal concentrations across the sites in the order S5 > S3 > S4 > S1 > S2. The ecological risk factor (ErF) revealed that Cd made the highest contribution to pollution, recording the highest ErF (2206.41). Moreover, the results of correlation base multivariate analysis showed that urban and industrial waste were the sources of Cu and Pb in the River Kaduna. The significant positive correlation between metal concentration and antioxidants catalase (CAT) and superoxide dismutase (SOD) was established, with maximum activities of antioxidants at site S5. Results from this study have revealed potential ecological risk as a result of heavy metals pollution in the River Kaduna. Hence the need for approaches and policies be put in place to prevent the discharge of untreated industrial and domestic waste into this aquatic ecosystem.

Keywords: heavy metal; pollution; antioxidant; enzyme; biomarkers; ecological risk

---

1. Introduction

Industrialization and urban development are among the major causes of metal pollution in a natural aquatic ecosystem [1]. The persistent nature and bioaccumulation ability of heavy metals in trophic levels make these metals serious pollutants of the aquatic environment [2–4]. The presence of heavy metals in an aquatic ecosystem either through adsorption or co-precipitation can threaten its biodiversity, and the health of humans depending on the resources of that ecosystem [5,6]. However, biological and chemical factors influence the mobility of heavy metals in aquatic environments by desorption from sediments into the surface water [7]. This made the surface water a major intermediate source of metal pollutants in benthic sediments, which is the definitive receptor [8,9].

Heavy metal concentrations of sediments and benthic organisms such as freshwater mollusc, speciation and several analytical techniques are often used in the evaluation of probable ecological risks in benthic sediments and their effect on the biota [10,11]. Several environmental factors and pollution indices such as geoaccumulation index (I_{geo}), ecological risk factors (ErF) and sediment quality guidelines (SQGs) have been used by several authors. It was used to measure the level or degree of pollution caused by metal and the ecological risk posed by metals in benthic sediments [12]. Although chemical speciation is ignored, it can be considered as subjective when evaluation of the level of pollution is computed based on I_{geo} and ErF. This is done because of the efficiency of these indices in utilizing both the concentration and toxic effect of metals to draw valuable conclusions on risk assessments [13,14].

The discharged industrial and domestic effluent containing toxic metals in aquatic ecosystems is a major concern to the survival of aquatic biota. Freshwater molluscs are not an exception. Although they have devised a means of bioaccumulating the metals even at high levels, due to their high tolerant ability to metal concentrations [15]. This ability makes them a good bio-monitor and bio-indicators of metal pollution in aquatic ecosystems [15,16]. The need to involve freshwater mollusc in ecological risk assessment studies does not stop at their ability to accumulate metal pollutants in high concentration. They also form an important link in the metal cycle for aquatic ecosystems and occupy a trophic level in the aquatic food chain [17,18].

Heavy metal concentrations in high levels can lead to the generation of reactive oxygen species such as H_2O_2, OH, RO_2 among others [19,20]. The manifestation of oxidative stress as a result of oxidative damage to proteins, nucleic acids, and lipids in aquatic organisms exposed to metal-polluted environments is triggered by reactive oxygen species (ROS). This happens when the process of detoxifying metals by organisms has to do with redox cycling reactions [20]. However, another important biomarker of oxidative stress known as defensive antioxidant enzymes is found in organisms [20,21], to scavenge ROS produced in organisms because of metal pollution and oxidative stress. Therefore an increase in ROS formation as a result of an increase in metal pollution triggers an increase in the production of antioxidant enzymes [22,23]. This phenomenon leads to the utilization of antioxidants as biomarkers of environmental pollution, which is a source or primary cause of oxidative stress in organisms [23].

In this study, five sites were selected to evaluate the risk assessment of heavy metals in the River Kaduna. Hence, to achieve the aim of this study, we determined concentrations of heavy metals in sediments and the freshwater mollusc *Bellamya unicolor*, pollution indices and antioxidant enzyme activities in *Bellamya unicolor* across the five sites.

2. Materials and Methods

2.1. Study Area

The study area is located in Kaduna state (Lat. 10.52° N and Lat. 12° N, Long. 7.44° E and Long. 9° E). Kaduna state is located in the northern guinea savannah zone of Nigeria as one of the most developed industrial cities (Figure 1). The textile industry, flour mill, fertilizer, plastic, agrochemical, brewery, and bottling companies are some of the major industries sited in Kaduna state. The climatic conditions like most of Nigeria are characterized by the dry and wet season. The wet season commences either in April or May and ends in October of the same year while the dry season begins towards the end of October and ends in March the following year. Annual rainfall on average is between 1450 to 2000 mm with a temperature regime average of 25 to 43 °C, and relative humidity estimated to be between 20 and 40% in January and 60 to 80% in July. The solar radiation of the Kaduna state was 25.0 Wm^{-2} day^{-1} [24].

The River Kaduna is one of the major rivers in Nigeria. It stretches southwest and south course before completing a flow of 550 KM into River Niger. During its course through Kaduna city, it stretches along the southern part of the state through Kakuri where it receives industrial and domestic wastewater. It serves as a source of domestic water supply and irrigation farming for Kaduna urban settlements and its various industries [25].

2.2. Collection of Samples

Surface sediments (0–10 cm) were collected along a stretch of River Kaduna from five sites base on the type of anthropogenic activities, stages for the stretch of the river either downstream, upper or in between, and history of sediment pollution. Coordinates of the sites selected were determined and recorded using a T10 handheld Global Positioning System (GPS) receiver. At each site, five grab hauls were sampled using an Ekman's grab. The samples were sorted in the field to separate *Bellamya unicolor* from the sediment samples. *Bellamya unicolor* were identified as described by Brown and Kristensen [26]. After sorting, the samples were placed inside zip lock bags and stored in the icebox to be transported to the laboratory for further analysis.

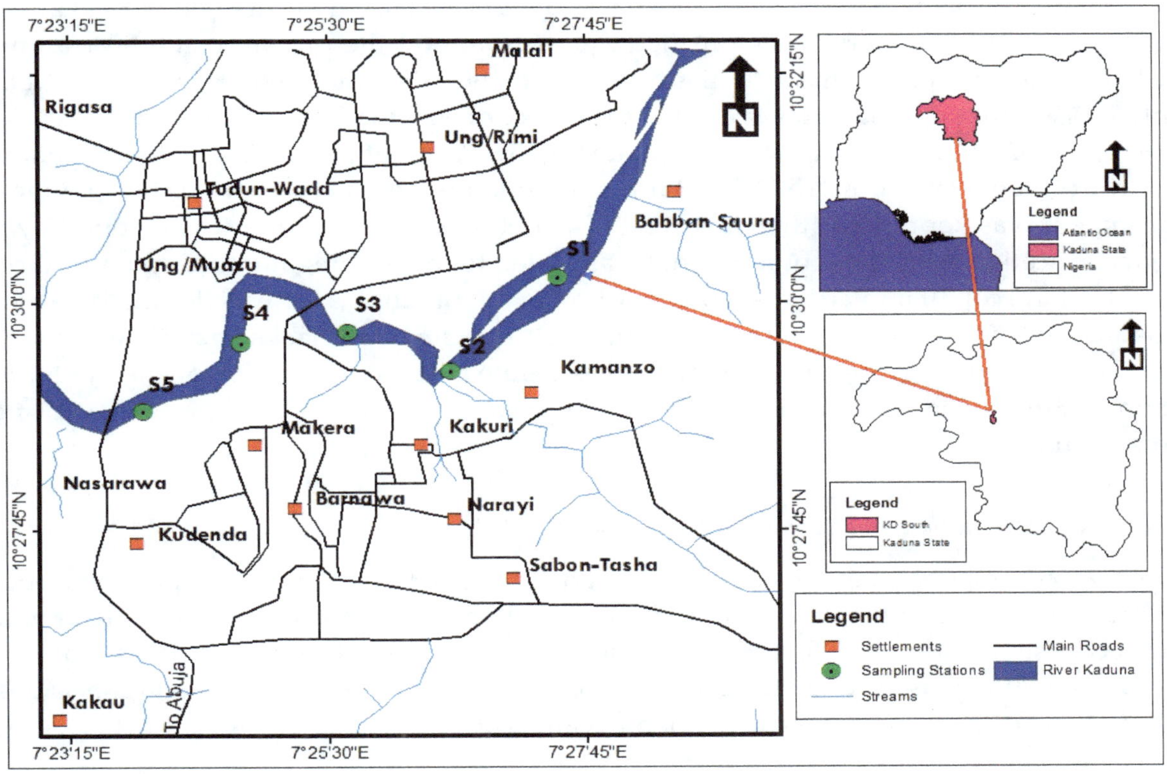

Figure 1. Map of River Kaduna showing sampling sites.

2.3. Treatment of Samples and Determination of Heavy Metals

Sediment samples were freeze-dried and grounded while samples of *Bellamya unicolor* were treated and prepared for heavy metal determination and antioxidant enzyme activities assays; 0.4 g of grounded freeze-dried samples was digested in 8 mL of 1:1 HNO_3:HCl and kept for 1 h in Anton-Paar PE Multiwave 3000 (microwave oven) set at 200 °C [1,27].

For the freshwater mollusc *Bellamya unicolor* samples, 0.3 g of the grounded samples were digested chemically in 3:1 HNO_3:H_2O_2 and maintained for 45 min at 180 °C. The volume was then made up to 50 mL using ultra water. The solutions were left to stand after the mixture until the following day. The solute from the digested solution of sediment and *Bellamya unicolor* tissue was then transferred into polypropylene vials after filtration, stored between 4–5 °C and later analyzed for heavy metals using inductively coupled plasma optical emission spectrometry (ICP-OES, Thermo, Waltham, MA, USA).

2.4. Assay of Antioxidant Enzymes Activities

Bellamya unicolor tissues were separated from the shell and washed with deionized water; 0.25 g of the tissue sample was then homogenized using 0.01 (M) chilled phosphate buffer at pH 7 with ice-cold mortar and pestle. The homogenate was later centrifuged for 25 min at 4 °C and 14,000 rpm [1,28]. The supernatant from the centrifuged samples was used for the measurement of antioxidant enzyme activities using an LT-291 Single Beam ultraviolet-visible (UV-VIS) spectrophotometer.

2.4.1. Assay of Catalase (CAT)

Catalase (CAT) activities in the supernatant of initially centrifuged samples of *Bellamya unicolor* were assayed as described by Chance and Maehly [29]. CAT activities were estimated spectrophotometrically at 230 nm. The enzyme extract prepared initially was used to measure the activity of catalase. Specific activity was expressed as Ug^{-1} protein. 1 IU = change in absorbance/min/extinction coefficient (0.021).

2.4.2. Assay of Superoxide Dismutase (SOD)

The activity of superoxide dismutase (SOD) was determined using the method of Kakkar et al. [30]. The initial supernatant prepared was purified by precipitating the protein using 90% ammonium sulphate before an assay of enzyme activities. The fraction was then dialysed against 0.0025 M tris-HCl buffer (pH 7.4) and the supernatant was used as the source of enzyme. The content of the assay solution was made up of 1.2 mL of sodium pyrophosphate buffer, 0.1 mL of phenazine methosulphate (PMS), 0.3 mL of nitro-blue tetrazolium (NBT), 1.3 mL of distilled water and 0.1 mL of the enzyme source. The supernatant was kept inside the vials at 30 °C for 90 s and the reaction was stopped by adding 1 mL of glacial acetic acid. 4.0 mL of n-butanol was added to the reacting mixture and mixed properly, then allow to stand for 10 min and the upper portion of the butanol layer was decanted. Absorbance was measured at 560 nm against a blank of n-butanol. A system that lacks enzyme was used as the control and one unit of activities of the enzyme was define as its concentration needed to inhibit 50% production of chromogen per minute. Specific activity was expressed as Ug^{-1} protein. 1 IU = change in absorbance/min/extinction coefficient (0.021).

2.5. Sediment Contamination and Ecological Risk Assessment

The geo-accumulation index (I_{geo}) provides a good explanation of the contamination or pollution status of sediment and thereby gives a better understanding of the possible threat of metal pollutants to natural ecosystem settings [31]. For that reason, I_{geo} was applied in this study as a quantitative indicator, putting into consideration the classifications that evaluate the level of pollution as described by Müller [32]. This index is expressed mathematically as:

$$I_{geo} = log_2[\frac{C_n}{1.5 \times B_n}] \quad (1)$$

where C_n and B_n are the metal concentration in the sediment and geochemical background value of the element (n).

The levels of metal contamination were classified into seven levels: uncontaminated (<0), uncontaminated to moderately contaminated (0–1), moderately contaminated (1–2), moderately to strongly contaminated (2–3), strongly contaminated (3–4), strongly to extremely contaminated (4–5) and extremely contaminated (>5) [32].

In aquatic ecosystems generally, sediment quality guidelines (SQGs) are utilized to assess the potential risk posed by pollutants to the natural ecosystem [33]. Sub variables of SQGs, which are threshold effect limit (TEL) and probable effect limit (PEL), were used for comparison with metal concentrations in sediment to conclude possible potential ecological risk. The concentrations of metals less than the threshold effect limit (TEL) denote a minimal effect, below which there is no expectation for adverse effects. However, metal concentrations at or greater than the probable effect limit (PEL) denote likely frequent occurrence of adverse biological effects (Table 1).

The ecological risk factor (ErF) was used to further assess the status of metal pollution in sediment and its possible toxicological effect with a comprehensive evaluation of ecological risk constituted by metal contaminants. The basis for the classification of ErF values for metal pollutants was as described by Hakanson [34]. For a particular metal, ErF was expressed as:

$$\text{ErF} = Tr^i \times CF^i \quad (2)$$

where Tr^i is the toxic response factor of a given metal (i) (Pb = 5, Zn = 1, Cu = 5 Cd = 30, Cr = 2, Ni = 5, Mn = 1) and CF^i is the contamination factor of metal (i).

The classification of ErF is in five classes according to Hakanson [34]: low risk (ErF < 30), moderate risk (30 ≤ ErF < 60), considerable risk (60 ≤ ErF < 120), high risk (120 ≤ ErF < 240) and very high risk (ErF ≥ 240).

2.6. Data Analysis

SPSS 22.0 and Minitab version 17.0 statistics software were used for data analysis. At 95% confidence interval ($p < 0.05$); analysis of variance (ANOVA) was used to test for significance in mean concentrations of metals across sites both in sediment and the freshwater mollusc *Bellayma unicolor*. Pearson correlation, principal component analysis (PCA) and cluster analysis were used to determine the sources of metal pollutants, the significant relationship between the metals, the relationship between the sites in terms of heavy metal concentration, and the influence of metal concentrations on antioxidant enzymes activities in *Bellamya unicolor*.

3. Results

3.1. Concentrations of Heavy Metals in Sediment and Freshwater Snail

The summary and some important statistics on heavy metal concentrations in sediment and freshwater mollusc from the River Kaduna used sediment quality guidelines (SQGs) to draw some important conclusions in our study which are presented in Table 1. The concentrations of the heavy metals were compared across the five sites studied both in sediment and freshwater mollusc. Site S5 which is the upstream recorded higher concentrations of heavy metals while site (S1) which the upstream recorded the least concentrations of the nine heavy metals studied.

The ranges for heavy metal concentrations in sediments were 34.54–165.32 mg/kg, 6.36–78.98 mg/kg, 1.63–59.01 mg/kg, 4.50–61.48 mg/kg, 0.96–58.84 mg/kg, 16.09–79.20 mg/kg, 39.43–96.08 mg/kg, 2.64–60.08 mg/kg and 11.42–62.61 mg/kg for Fe, Mn, Cu, Zn, Cd, Cr, Pb, Ni and Co respectively. Heavy metal concentrations in freshwater mollusc were lower than the concentrations in the sediment which ranged from Fe: 6.61–33.30, Mn: 1.34–22.75, Cu: 0.54–16.51, Zn: 1.74–18.63, Cd: 0.92–17.52, Cr: 4.26–17.24, Pb: 6.24–24.20, Ni: 0.63–14.17 and Co: 1.6–21.46.

A comparison was made between heavy metal concentrations in benthic sediment samples with TEL and PEL (Table 2). The result of the comparisons revealed that Cu, Cd, Cr, Pb, and Ni fell between TEL and PEL for 46.4, 20.8, 57.4, 100 and 61.2% of the samples respectively. However Cd again also had a value above PEL for 79.2% of the samples.

3.2. Ecological Risk Assessment of Heavy Metals

3.2.1. Geo-Accumulation Index (I_{geo})

In this study, the values of I_{geo} recorded for heavy metal concentrations of sediment samples from the River Kaduna are presented in Table 3. Based on the classification by Muller [32], the recorded values of I_{geo} for Cd at sites S1 (−2.39) and S2 (−2.91) fell into class zero, which means the sediments in S1 and S2 were unpolluted by Cd. In contrast, all the sites were found to be extremely polluted with the other heavy metals with $I_{geo} > 5$ falling in class 6. However, site S4 I_{geo} (0.33) for Cd fell in class 1, and S3 (3.53) and S5 (3.56) fell in class 4 indicating unpolluted to moderate pollution, and strongly polluted respectively.

Table 1. Consensus-based sediment quality guideline values (mg/kg) for heavy metals in sediment and averages for metals concentration in sediments and *Bellamya unicolor* sampled from the River Kaduna.

	Fe	Mn	Cu	Zn	Cd	Cr	Pb	Ni	Co
CSd	90.8225	42.2875	31.285	33.5475	29.4975	46.67625	67.66875	31.6575	34.6275
Range	34.54–165.32	6.36–78.98	1.63–59.01	4.50–61.48	0.96–58.84	16.09–79.20	39.43–96.08	2.64–60.08	11.42–62.61
CSn	20.8825	9.0625	7.295	7.555	7.4225	10.3725	15.0375	7.27	7.435
Range	6.61–33.30	1.34–22.75	0.54–16.51	1.74–18.63	0.92–17.52	4.26–17.24	6.24–24.20	0.63–14.17	1.6–21.46
TEL	NA	NA	35.7	123	0.596	37.3	35	18	NA
PEL	NA	NA	197	315	3.53	90	91.3	36	NA

Note: CSd = Concentration of heavy metals in sediment. CSn = Concentration of heavy metals in freshwater snail: *Bellamya unicolor*. TEL = Threshold effect level. PEL = Probable effect level, NA = Not available.

Table 2. Comparison between sediment quality guidelines (SQGs) and concentration of heavy metals in sediment of the River Kaduna.

SQGs	Fe	Mn	Cu	Zn	Cd	Cr	Pb	Ni	Co
% of samples < TEL	NA	NA	53.6	100	0	42.6	0	38.8	NA
% of samples btw TEL-PEL	NA	NA	46.4	0	20.8	57.4	100	61.2	NA
% of samples > PEL	NA	NA	0	0	79.2	0	0	0	NA

Note: TEL = Threshold effect level. PEL = Probable effect level, NA = Not available.

Table 3. Geoaccumulation (I_{geo}) values of heavy metals in sediments and their class.

	I_{geo}							
Sites	**Mn**	**Cu**	**Zn**	**Cd**	**Cr**	**Pb**	**Ni**	**Co**
S1	11.81	5.61	8.15	−2.39	9.89	9.00	6.42	5.78
S2	12.47	7.04	8.40	−2.91	10.31	9.03	7.51	7.13
S3	15.34	10.79	11.96	3.53	12.09	10.31	11.40	9.62
S4	13.68	8.18	10.07	0.33	10.96	9.72	9.24	7.84
S5	15.44	10.79	11.93	3.56	12.21	10.32	11.41	9.63
Mean	13.75	8.48	10.10	0.42	11.09	9.68	9.19	8.00
Level	6	6	6	1	6	6	6	6

Note: S = Site.

3.2.2. Ecological Risk Factor (ErF)

For ErF results of heavy metal concentrations in the River Kaduna, the level of pollution was in the order Cd > Co > Ni > Pb > Cr > Cu > Zn > Mn (Table 4). Compared to other heavy metals, Cd ErF revealed a high metal pollution level especially at site S5 (ErF = 2206.41). However, the values of ErF for Mn, Cu, Zn, Cr, Pb, Ni, and Co revealed low potential ecological risk or pollution. But in total, the multi-metal ErF as reflected for variation in sites classified the sediments into low pollution to very high pollution in the sequence S5 > S3 > S4 > S1 > S2. Sites S3, S4, and S5 were classified as having very high ecological risk with ErF ≥ 320, S2 as a low potential ecological risk with ErF <40, and S1 as a moderate potential ecological risk with ErF falling between 40–79.9.

Table 4. Enrichment factor (ErF) values of heavy metals in sediment samples from the River Kaduna.

	(Individual Metal)								(Multi-Metal) ErF
Sites	**Mn**	**Cu**	**Zn**	**Cd**	**Cr**	**Pb**	**Ni**	**Co**	**ErF**
S1	0.01	0.02	0.03	35.86	0.32	2.27	0.38	1.32	40.21
S2	0.02	0.07	0.04	25.31	0.43	2.32	0.58	3.26	32.03
S3	0.16	0.62	0.44	2160.00	1.45	5.60	8.51	17.74	2194.52
S4	0.09	0.32	0.24	1103.20	0.88	4.07	4.56	9.26	1122.61
S5	0.17	0.62	0.45	2206.41	1.58	5.65	8.58	17.89	2241.35
Mean	0.09	0.33	0.24	1106.16	0.93	3.98	4.52	9.89	1126.14

Note: S = Site.

3.3. *Identification of Pollution Sources*

Pearson correlation analysis (Table 5) revealed significant relationship between the metals except Fe–Mn (0.584), Fe–Cr (0.632), Mn–Cu (0.567), Cu–Zn (0.619), Mn–Pb (0.592), Cu–Cr (0.555), Cu–Ni (0.622), Cr–Pb (0.572) and Pb–Co (0.608). However, Cd concentration at $p < 0.05$ and $p < 0.01$ was significantly correlated with all the other metals concentrations.

Table 5. Pearson correlation analysis among heavy metals and antioxidant enzymes.

	Fe	**Mn**	**Cu**	**Zn**	**Cd**	**Cr**	**Pb**	**Ni**	**Co**	**SOD**	**CAT**
Fe	1										
Mn	0.584	1									
Cu	0.777 **	0.567	1								
Zn	0.671 *	0.984 **	0.619	1							
Cd	0.728 *	0.971 **	0.660 *	0.985 **	1						
Cr	0.632	0.988 **	0.555	0.993 **	0.981 **	1					
Pb	0.765 **	0.592	0.994 **	0.638 *	0.671 *	0.572	1				
Ni	0.672 *	0.978 **	0.622	0.993 **	0.988 **	0.992 **	0.634 *	1			
Co	0.640 *	0.982 **	0.585	0.993 **	0.983 **	0.987 **	0.608	0.986 **	1		
SOD	0.809 **	0.822 **	0.845 **	0.827 **	0.877 **	0.809 **	0.860 **	0.832 **	0.822 **	1	
CAT	0.567	0.779 **	0.758 *	0.765 **	0.769 **	0.735 *	0.801 **	0.759 *	0.777 **	0.904 **	1

Note: ** Correlation is significant at the 0.01 level (2-tailed), * Correlation is significant at the 0.05 level (2-tailed), SOD = Superoxide dismutase, CAT = Catalase.

The similarity in terms of the pattern of distribution of heavy metals was established through the utilisation of principal component analysis (PCA); PCA pulls the data together into a form that can be managed easily and takes out a small number of latent factors to analyse the relationship between the variables observed [35,36]. PCA was applied here primarily to evaluate the source origin of heavy metals in the region.

The results for PCA are presented in Table 6. The high eigenvalue is an indication of pattern and to what extent the data is spread and this leads to consideration of high eigenvalue as the principal component. The components/factors associated with the results of PCA for heavy metals in sediments and freshwater mollusc had a total variation of 92.5%. Component 1 (PC1) recorded the highest eigenvalue of 13.82, accounting for 76.80% of the total variation, and dominate the other components with more significant variation and strong positive loadings (>0.90) of Mn, Cu, Cd, Cr, and Pb in the sediment. Component 2 (PC2) accounted for 8.81% of total variation and had moderate positive loadings (>0.50) of Cu and Pb in the sediment. Component 3 (PC3) accounted for 6.89% of the total variation having moderate positive loadings (>0.50) on Co.

Table 6. Principal component analysis (PCA) loadings for heavy metals in sediment (Sd) and *Bellamya unicolor* (S).

	Component		
Heavy Metal	**PC1**	**PC2**	**PC3**
Fe_S	0.74	0.40	0.03
Mn_S	0.97	−0.20	−0.02
Cu_S	0.73	−0.23	0.01
Zn_S	0.98	−0.17	−0.02
Cd_S	0.99	−0.10	0.00
Cr_S	0.97	−0.23	−0.02
Pb_S	0.75	−0.17	0.05
Ni_S	0.98	−0.16	0.03
Co_S	0.97	−0.22	0.30
Fe_Sd	0.85	0.26	0.21
Mn_Sd	**0.97**	−0.18	0.07
Cu_Sd	**0.97**	**0.63**	−0.01
Zn_Sd	0.68	0.36	−0.05
Cd_Sd	**0.98**	−0.18	−0.05
Cr_Sd	**0.98**	−0.16	0.01
Pb_Sd	**0.97**	**0.60**	−0.09
Ni_Sd	0.67	0.33	−0.52
Co_Sd	0.24	0.07	**0.95**
Eigenvalue	13.82	1.59	1.24
Total variance (%)	76.80	8.81	6.89

Note: Method of extraction: principal component analysis. Rotation method. Bold numbers indicate a strong loading value (>0.9); moderate loading value (>0.5).

Cluster analysis reveals the relationship between sampling sites (Figure 2a) based on heavy metals concentrations and also the relationship between the metals (Figure 2b). For the relationship between the sites, three (3) clusters were formed at a similarity level of 89.09. The first cluster presented only site S1 and the second cluster had only site 2 while the third cluster grouped sites S3, S4, and S5. Four (4) clusters were formed for the relationship between the metals at a similarity level of 99.57. The clusters were Cluster 1 (Fe), Cluster 2 (Mn, Cr, Cu, Cd, and Pb), Cluster 3 (Zn and Ni) and Cluster 4 (Co).

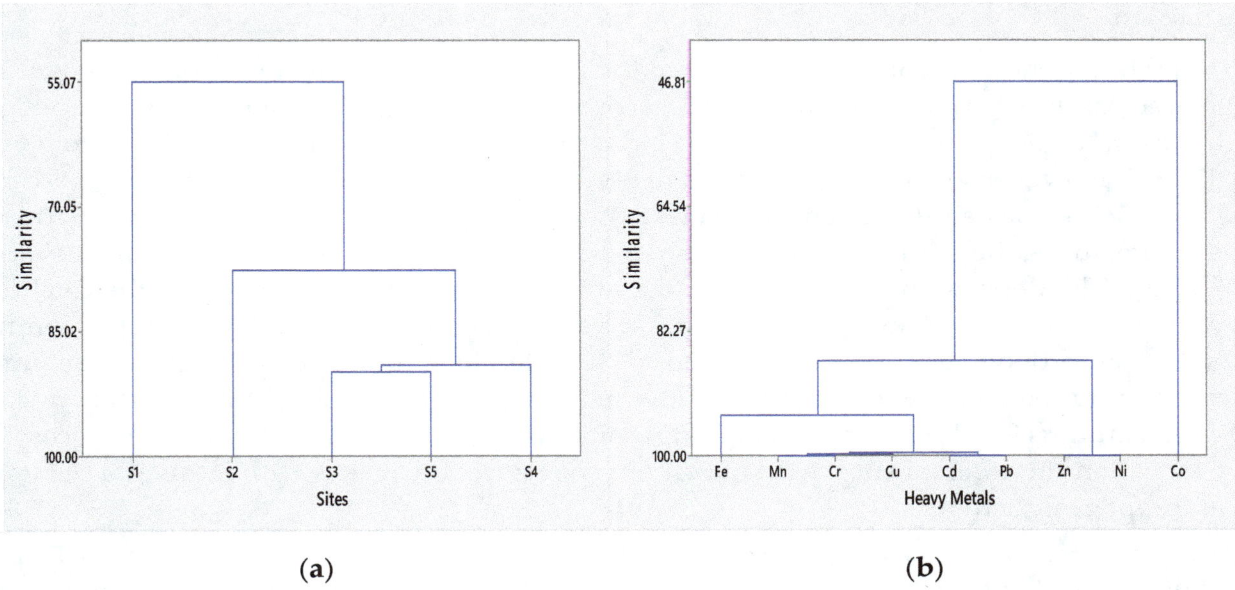

Figure 2. Cluster analysis for the relationship between (**a**) sites; (**b**) heavy metals.

3.4. *Antioxidant Enzyme Activities*

Antioxidant enzymes such as CAT and SOD can be used to monitor changes in an environment through the measurement of their activities in an organism present in that same environment. The activities of CAT and SOD in this study are shown in Figure 3a,b respectively. Significant variation existed at $p < 0.05$ across the sites for the two antioxidants with sites S5 recording the highest activities for both CAT and SOD. However, CAT activities ranged from 6.67–72.46 Ug^{-1} protein and SOD 2.70–12.49 Ug^{-1} protein. The sequence for activities of CAT and SOD across the sites; S5 > S3 > S4 > S1 > S2 corresponded with that of heavy metals concentrations, I_{geo} and ErF.

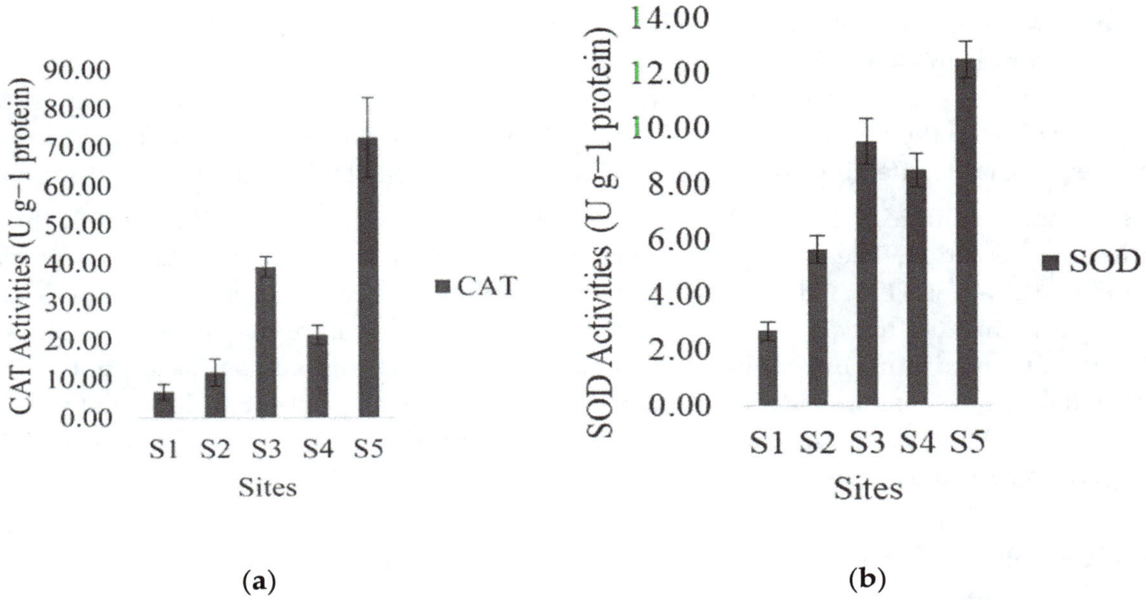

Figure 3. Antioxidant enzymes activities in *Bellamya unicolor* across the sites for (**a**) catalase (CAT); (**b**) superoxide dismutase (SOD).

A significant positive correlation ($p < 0.01$) and ($p < 0.05$) existed between CAT, SOD and all the heavy metals except Fe (0.567) which did not correlate with CAT significantly.

4. Discussion

The high concentrations of heavy metals recorded in this study downstream (S5) and the significant variations across the sites both in sediment and freshwater molluscs can be as a result of anthropogenic activities from the nearby settlements and the industries sited close to the aquatic ecosystem under study [37,38]. However, even though reduced to non-anthropogenic activities were seen or recorded at site S1 which is located upstream, concentrations of heavy metals were still recorded. This can be due to the atmospheric deposition of heavy metals and the effect of runoff that ends up in the river [17,39]. The deposition of metals from the atmosphere, industries located at the catchment, effluents of agrochemical sources and domestic sewage discharge forms part of the input for metal pollutants of anthropogenic sources. Their effect is reflected in natural aquatic ecosystem sediments with negative consequences to benthic organisms and the general wellbeing of the ecosystem [40,41].

Several authors [4,6,14] report similar findings in ecological assessment of the aquatic ecosystem having variations in different sites, with maximum concentration at sites with industrial plants and more anthropogenic activities.

The concentrations of heavy metals such as Cu, Cd, Cr, Pb and Ni that fall between TEC and PEC values for 46.4, 20.8, 57.4, 100 and 61.2% respectively from sediment quality guidelines of the freshwater ecosystem may be an indication of occasionally serious negative effects of this metals upon the ecosystem under study [4]. However, 79.2% of samples with a concentration of Cd greater than PEL reflected the frequent occurrence of biological effect [42]. Comparison is made between metal concentrations with TEC and PEC values to identify if heavy metals present in the sediment has the possibility of threatening the aquatic life [14,42].

This site has industries sited close to it and receives effluents from the industries. Hence, serious attention might be required at this site due to high concentrations of metal pollutants in comparison with the other sites. Pollution as a result of industrial effluent and domestic waste discharge contributes to the rise in concentrations of metal pollutants leading to contamination of the ecosystem [4,43].

The I_{geo} of heavy metals in sediments were generally higher in site S5 and lower in sites S1 and S2. I_{geo} scoring of S1 and S2 into class zero [32] for Cd indicate no contamination with Cd. Cd is well known as one of the major pollutants with industrial effluents and domestic sewage as its source [43]. However, extreme pollution of the sites with heavy metals especially S5 with I_{geo} >5 having a score of 6 reflects on the source variations of the metal pollutants and the nature of catchments. Site S5 is downstream where the other stretch of the river ends up, receiving some of its contents apart from the effluents received from the nearby industrial plants. Downstream has been reported in several studies to have more pollutants affecting the natural wellbeing of that particular ecosystem setting [44–46].

Considerable ecological risk exhibited by the metals across the sites is linked with the level of geoaccumulation recorded. The ecological risk was consistent with I_{geo} for both metals and site variation as revealed by multi-metal ErF. Cd posed a high ecological risk in River Kaduna with an expectation of an adverse effect expected to occur suggesting that they are present in high concentrations, especially at site S5 with the maximum ErF value. This major ecological risk in surface sediment of the River Kaduna by Cd must have been seriously influenced by anthropogenic activities, leading to a high toxic response factor [6]. The low and moderate potential ecological risk in sites S1 and S2 may be a result of metals present in residual forms and minimal concentrations [31,47,48].

Source and migration of metals may be reflected by correlation analysis and PCA [49,50]. Non-significant correlation between Fe–Mn, Fe–Cr, Mn–Cu, Cu–Zn, Mn–Pb, Cu–Cr, Cu–Ni, Cr–Pb, and Pb–Co indicates different factors controlling the availability and concentrations of these metals [4,9]. This implies the metals might have originated from different sources. This was also revealed in the PCA; the groupings of metals, and component 1 and 2 accounting for 76.80% and 8.81%, respectively, with strong positive loading of Mn, Cu, Cd, Cr and Pb in the sediment and moderate positive loading of Cu and Pb in the sediment. This implies urban and industrial waste to be the source of Mn, Cu, Cd, Cr and Pb [51,52].

Similarity and dissimilarity in groups are effectively represented using cluster analysis. Three cluster formations for sites under study revealed a close relationship between these sites. Grouping of S3, S4, and S5 together at a similarity level of 89.09 indicates possible similarities in human activities, site morphology and the pristine background types of sources [53]. The significant relationship revealed by cluster analysis at 99.57 similarity level for the relationship between the metals, grouping Mn, Cr, Cu, Cd, and Pb together makes it possible that they have the same natural and anthropogenic sources [54,55]. This is also a link to the multiple effects of these metals on benthic organisms. For example, *Bellamya unicolor* collected during our sampling for this study. Our result agrees with the finding of Peijnenburg et al. [56]; Li et al. [55]; Chung et al. [54]. These authors assess and monitor heavy metal sources and their risk in different aquatic environments.

CAT and SOD activities were used in this study as biomarkers to measure the extent of stress in the River Kaduna as reflected in *Bellamya unicolor* sampled from the same environment at different sites. CAT and SOD are a type of antioxidant enzyme that scavenges reactive oxygen species (ROS) (H_2O_2, OH, O_2, etc.) produced by organisms as a result of oxidative stress caused as a result of unfavorable environmental conditions such as metal pollution [1]. Significant variation in CAT and SOD activities across the sites with high activities in site S5 may be a result of different concentrations of ROS production at the sites with high concentrations at site S5 due to high metal pollution [57]. The scavenging ability of antioxidants increases through their activities in response to increased ROS to reduce or prevent membrane lipid peroxidation by ROS and to improve membrane stability of the cell [58,59].

The significant positive correlation between the antioxidants and the metals showed that the organism *Bellamya unicolor* is under stress as a result of an increase in metal contamination. Bakshi et al. [1] report similar findings for the biological response of an aquatic organism to metal contamination through antioxidant activities. They revealed an increase in antioxidants with a spatial and temporal increase in heavy metal concentrations.

Our results in this study tend to bridge the gap in knowledge for metal pollution risk assessment, and the use of CAT and SOD as primary biomarkers in benthic organisms of a tropical ecosystem. There has so far been no report, to the best of our knowledge, assessing risk assessment using ecological indices, antioxidant enzymes and *Bellamya unicolor* as a biomonitor. Our study will provide data to be used as a baseline for studies in the River Kaduna and other related tropical rivers in the same region.

5. Conclusions

The results from this study provide information on the contamination of heavy metals Fe, Mn, Cu, Zn, Cd, Cr, Pb, Ni, and Co in surface sediment of the River Kaduna and their bioavailability in the freshwater mollusc *Bellamya unicolor* sampled from the same ecosystem.

At a 95% confidence interval, significant variation existed in the concentration of metals in surface sediment and freshwater molluscs across the sampling sites with the site S5 (downstream) having the maximum metal concentrations.

Cd concentration in sediment was greater than PEL, which implies likely frequent occurrence of biological effect by Cd in the River Kaduna with the possibility of threatening the aquatic life. This was reflected in Cd ErF values that are higher than the values of ErF for other metals. I_{geo} and ErF values follow the same sequence with maximum values at site S5 revealing very high ecological risk with $I_{geo} > 5$ and ErF > 320. Mn, Cu, Cd, Cr, and Pb originate mainly from urban and industrial waste as revealed by PC1 and PC2 analysis while S3, S2, and S3 have similar anthropogenic activities and natural metal inputs as shown in cluster analysis. Significant variations existed in CAT and SOD activities in *Bellamya unicolor* across the sites with maximum activities recorded at sites S5 for both antioxidants. This also coincides with the high values of I_{geo} and ErF at site S5. However, there is a significant positive correlation between the antioxidants and metal concentrations.

Important approaches and policies should be put in place to prevent the discharge of untreated industrial and domestic waste into the River Kaduna. The approach should involve the prevention of irrigation farming close to the river and, in that way, non-point sources of pollution can be abated and there can be a decrease in ecological risk associated with metal pollutants.

Author Contributions: M.O.A., and A.B.A. conceived the idea and performed the laboratory analysis. All of the authors contributed to give the manuscript its present shape. All authors have read and agreed to the published version of the manuscript.

Acknowledgments: The authors acknowledge the Deanship of Scientific Research (DSR), King Abdulaziz University, Jeddah for their technical and financial support.

References

1. Bakshi, M.; Ghosh, S.; Chakraborty, D.; Hazra, S.; Chaudhuri, P. Assessment of potentially toxic metal (PTM) pollution in mangrove habitats using biochemical markers: A case study on Avicennia officinalis L. in and around Sundarban, India. *Mar. Pollut. Bull.* **2018**, *133*, 157–172. [CrossRef]
2. Bastami, K.D.; Bagheri, H.; Kheirabadi, V.; Zaferani, G.G.; Teymori, M.B.; Hamzehpoor, A.; Soltani, F.; Haghparast, S.; Harami, S.R.M.; Ghorghani, N.F. Distribution and ecological risk assessment of heavy metals in surface sediments along southeast coast of the Caspian Sea. *Mar. Pollut. Bull.* **2014**, *81*, 262–267. [CrossRef] [PubMed]
3. Xiao, R.; Bai, J.; Lu, Q.; Zhao, Q.; Gao, Z.; Wen, X.; Liu, X. Fractionation, transfer, and ecological risks of heavy metals in riparian and ditch wetlands across a 100-year chronosequence of reclamation in an estuary of China. *Sci. Total Environ.* **2015**, *517*, 66–75. [CrossRef] [PubMed]
4. Ke, X.; Gui, S.; Huang, H.; Zhang, H.; Wang, C.; Guo, W. Ecological risk assessment and source identification for heavy metals in surface sediment from the Liaohe River protected area, China. *Chemosphere* **2017**, *175'* 473–481. [CrossRef]
5. Suresh, G.; Ramasamy, V.; Sundarrajan, M.; Paramasivam, K. Spatial and vertical distributions of heavy metals and their potential toxicity levels in various beach sediments from high-background-radiation area, Kerala, India. *Mar. Pollut. Bull.* **2015**, *91*, 389–400. [CrossRef] [PubMed]
6. Xu, M.; Sun, W.; Wang, R. Spatial Distribution and Ecological Risk Assessment of Potentially Harmful Trace Elements in Surface Sediments from Lake Dali, North China. *Water* **2019**, *11*, 2544. [CrossRef]
7. Hill, N.A.; Simpson, S.L.; Johnston, E.L. Beyond the bed: Effects of metal contamination on recruitment to bedded sediments and overlying substrata. *Environ. Pollut.* **2013**, *173*, 182–191. [CrossRef] [PubMed]
8. Bermejo, J.S.; Beltrán, R.; Ariza, J.G. Spatial variations of heavy metals contamination in sediments from Odiel river (Southwest Spain). *Environ. Int.* **2003**, *29*, 69–77. [CrossRef]
9. Kükrer, S.; Şeker, S.; Abacı, Z.T.; Kutlu, B. Ecological risk assessment of heavy metals in surface sediments of northern littoral zone of Lake Çıldır, Ardahan, Turkey. *Environ. Monit. Assess.* **2014**, *186*, 3847–3857. [CrossRef]
10. Yang, Z.; Wang, Y.; Shen, Z.; Niu, J.; Tang, Z. Distribution and speciation of heavy metals in sediments from the mainstream, tributaries, and lakes of the Yangtze River catchment of Wuhan, China. *J. Hazard. Mater.* **2009**, *166*, 1186–1194. [CrossRef]
11. Yu, G.; Liu, Y.; Yu, S.; Wu, S.; Leung, A.; Luo, X.; Xu, B.; Li, H.; Wong, M.H. Inconsistency and comprehensiveness of risk assessments for heavy metals in urban surface sediments. *Chemosphere* **2011**, *85'* 1080–1087. [CrossRef] [PubMed]
12. Zahra, A.; Hashmi, M.Z.; Malik, R.N.; Ahmed, Z. Enrichment and geo-accumulation of heavy metals and risk assessment of sediments of the Kurang Nallah—Feeding tributary of the Rawal Lake Reservoir, Pakistan. *Sci. Total Environ.* **2014**, *470*, 925–933. [CrossRef] [PubMed]
13. Zhu, H.N.; Yuan, X.Z.; Zeng, G.M.; Jiang, M.; Liang, J.; Zhang, C.; Juan, Y.; Huang, H.J.; Liu, Z.F.; Jiang, H.W. Ecological risk assessment of heavy metals in sediments of Xiawan Port based on modified potential ecological risk index. *Trans. Nonferrous Met. Soc. China* **2012**, *22*, 1470–1477. [CrossRef]
14. Rahman, M.S.; Hossain, M.B.; Babu, S.O.F.; Rahman, M.; Ahmed, A.S.; Jolly, Y.; Choudhury, T.; Begum, B.; Kabir, J.; Akter, S. Source of metal contamination in sediment, their ecological risk, and phytoremediation ability of the studied mangrove plants in ship breaking area, Bangladesh. *Mar. Pollut. Bull.* **2019**, *141'* 137–146. [CrossRef]

15. Waykar, B.; Petare, R. Studies on monitoring the heavy metal contents in water, sediment and snail species in Latipada reservoir. *J. Environ. Biol.* **2016**, *37*, 585.
16. Zhou, Q.; Zhang, J.; Fu, J.; Shi, J.; Jiang, G. Biomonitoring: An appealing tool for assessment of metal pollution in the aquatic ecosystem. *Anal. Chim. Acta* **2008**, *606*, 135–150. [CrossRef]
17. Chen, M.; Boyle, E.A.; Switzer, A.D.; Gouramanis, C. A century long sedimentary record of anthropogenic lead (Pb), Pb isotopes and other trace metals in Singapore. *Environ. Pollut.* **2016**, *213*, 446–459. [CrossRef]
18. Shalaby, B.; Samy, Y.M.; Mashaly, A.O.; El Hefnawy, M.A.A. Comparative Geochemical Assessment of Heavy Metal Pollutants among the Mediterranean Deltaic Lakes Sediments (Edku, Burullus and Manzala), Egypt. *Egypt. J. Chem.* **2017**, *60*, 361–378. [CrossRef]
19. Halliwell, B.; Gutteridge, J.M. *Free Radicals in Biology and Medicine*; Oxford University Press: Oxford, UK, 2015.
20. Gawad, S.S.A. Concentrations of heavy metals in water, sediment and mollusc gastropod, Lanistes carinatus from Lake Manzala, Egypt. *Egypt. J. Aquat. Res.* **2018**, *44*, 77–82. [CrossRef]
21. Buhler, D.R.; Williams, D.E. The role of biotransformation in the toxicity of chemicals. *Aquat. Toxicol.* **1988**, *11*, 19–28. [CrossRef]
22. Regoli, F.; Principato, G.; Bertoli, E.; Nigro, M.; Orlando, E. Biochemical characterization of the antioxidant system in the scallop Adamussium colbecki, a sentinel organism for monitoring the Antarctic environment. *Polar Biol.* **1997**, *17*, 251–258. [CrossRef]
23. Siwela, A.H.; Nyathi, C.; Naik, Y.S. A comparison of metal levels and antioxidant enzymes in freshwater snails, Lymnaea natalensis, exposed to sediment and water collected from Wright Dam and Lower Mguza Dam, Bulawayo, Zimbabwe. *Ecotoxicol. Environ. Saf.* **2010**, *73*, 1728–1732. [CrossRef] [PubMed]
24. NIMET. *Nigeria Meteorological Agency*; CRC Press: Kaduna, Nigeria, 2010.
25. KEPA. Kaduna State Environmental Protection Authority, Revised in 1998:1998. Available online: http://www.kepa.org.ng/ (accessed on 31 December 2019).
26. Brown, D.; Kristensen, T. A Field Guide to African Freshwater Snails I. West African species. *Dan. Bilharz. Lab. Charlottenlund* **1993**, *32*.
27. United States Environmental Protection Agency (USEPA). *Method 3051 A. Microwave Assisted Acid Digestion of Sediments, Sludge's, Soils and Oils*; USEPA, U.S. Government Printing Office: Washington, DC, USA, 1997. Available online: http://www.epa.gov/SW-846/pdfs/3051a (accessed on 31 December 2019).
28. Banaee, M.; Sureda, A.; Taheri, S.; Hedayatzadeh, F. Sub-lethal effects of dimethoate alone and in combination with cadmium on biochemical parameters in freshwater snail, Galba truncatula. *Comp. Biochem. Physiol. Part C Toxicol. Pharm.* **2019**, *220*, 62–70. [CrossRef] [PubMed]
29. Chance, B.; Maehly, A. [136] Assay of catalases and peroxidases. *Methods Enzymol.* **1955**, *2*, 764–775.
30. Kakkar, P.; Das, B.; Viswanathan, P. A modified spectrophotometric assay of superoxide dismutase. *Indian J. Biochem. Biophys.* **1984**, *21*, 130–132.
31. Cheng, H.; Li, M.; Zhao, C.; Yang, K.; Li, K.; Peng, M.; Yang, Z.; Liu, F.; Liu, Y.; Bai, R. Concentrations of toxic metals and ecological risk assessment for sediments of major freshwater lakes in China. *J. Geochem. Explor.* **2015**, *157*, 15–26. [CrossRef]
32. Muller, G. Schwermetalle in den sedimenten des Rheins-Veranderungen seit. *Umschav* **1979**, *79*, 133–149.
33. Wang, Y.; Yang, L.; Kong, L.; Liu, E.; Wang, L.; Zhu, J. Spatial distribution, ecological risk assessment and source identification for heavy metals in surface sediments from Dongping Lake, Shandong, East China. *Catena* **2015**, *125*, 200–205. [CrossRef]
34. Hakanson, L. An ecological risk index for aquatic pollution control. A sedimentological approach. *Water Res.* **1980**, *14*, 975–1001. [CrossRef]
35. Loska, K.; Wiechuła, D. Application of principal component analysis for the estimation of source of heavy metal contamination in surface sediments from the Rybnik Reservoir. *Chemosphere* **2003**, *51*, 723–733. [CrossRef]
36. Ma, X.; Zuo, H.; Tian, M.; Zhang, L.; Meng, J.; Zhou, X.; Min, N.; Chang, X.; Liu, Y. Assessment of heavy metals contamination in sediments from three adjacent regions of the Yellow River using metal chemical fractions and multivariate analysis techniques. *Chemosphere* **2016**, *144*, 264–272. [CrossRef] [PubMed]
37. Chai, M.; Shi, F.; Li, R.; Shen, X. Heavy metal contamination and ecological risk in Spartina alterniflora marsh in intertidal sediments of Bohai Bay, China. *Mar. Pollut. Bull.* **2014**, *84*, 115–124. [CrossRef] [PubMed]
38. Cui, J.; Zang, S.; Zhai, D.; Wu, B. Potential ecological risk of heavy metals and metalloid in the sediments of Wuyuer River basin, Heilongjiang Province, China. *Ecotoxicology* **2014**, *23*, 589–600. [CrossRef]

39. Lin, Q.; Liu, E.; Zhang, E.; Nath, B.; Shen, J.; Yuan, H.; Wang, R. Reconstruction of atmospheric trace metals pollution in Southwest China using sediments from a large and deep alpine lake: Historical trends, sources and sediment focusing. *Sci. Total Environ.* **2018**, *613*, 331–341. [CrossRef]
40. Sutherland, R. Bed sediment-associated trace metals in an urban stream, Oahu, Hawaii. *Environ. Geol.* **2000**, *39*, 611–627. [CrossRef]
41. Liu, E.; Birch, G.F.; Shen, J.; Yuan, H.; Zhang, E.; Cao, Y. Comprehensive evaluation of heavy metal contamination in surface and core sediments of Taihu Lake, the third largest freshwater lake in China. *Environ. Earth Sci.* **2012**, *67*, 39–51. [CrossRef]
42. Zhang, R.; Zhou, L.; Zhang, F.; Ding, Y.; Gao, J.; Chen, J.; Yan, H.; Shao, W. Heavy metal pollution and assessment in the tidal flat sediments of Haizhou Bay, China. *Mar. Pollut. Bull.* **2013**, *74*, 403–412. [CrossRef]
43. Zhang, J.; Wang, S.; Xie, Y.; Wang, X.; Sheng, X.; Chen, J. Distribution and pollution character of heavy metals in the surface sediments of Liao River. *Huan Jing Ke Xue Huanjing Kexue* **2008**, *29*, 2413–2418.
44. Mummullage, N.; Wasanthi, S. Source Characterisation of Urban Road Surface Pollutants for Enhanced Water Quality Predictions. Ph.D. Thesis, Queensland University of Technology, Brisbane, Australia, 2015.
45. Brady, J.P.; Kinaev, I.; Goonetilleke, A.; Ayoko, G.A. Comparison of partial extraction reagents for assessing potential bioavailability of heavy metals in sediments. *Mar. Pollut. Bull.* **2016**, *106*, 329–334. [CrossRef]
46. Duodu, G.O.; Goonetilleke, A.; Ayoko, G.A. Potential bioavailability assessment, source apportionment and ecological risk of heavy metals in the sediment of Brisbane River estuary, Australia. *Mar. Pollut. Bull.* **2017**, *117*, 523–531. [CrossRef] [PubMed]
47. Brady, J.P. Heavy Metals in the Sediments of Northern Moreton Bay, Queensland, Australia. Ph.D. Thesis, Queensland University of Technology, Brisbane, Australia, 2015.
48. Liu, R.; Bao, K.; Yao, S.; Yang, F.; Wang, X. Ecological risk assessment and distribution of potentially harmful trace elements in lake sediments of Songnen Plain, NE China. *Ecotoxicol. Environ. Saf.* **2018**, *163*, 117–124. [CrossRef] [PubMed]
49. Wang, Y.; Hu, J.; Xiong, K.; Huang, X.; Duan, S. Distribution of heavy metals in core sediments from Baihua Lake. *Procedia Environ. Sci.* **2012**, *16*, 51–58. [CrossRef]
50. Suresh, G.; Ramasamy, V.; Meenakshisundaram, V.; Venkatachalapathy, R.; Ponnusamy, V. Influence of mineralogical and heavy metal composition on natural radionuclide concentrations in the river sediments. *Appl. Radiat. Isot.* **2011**, *69*, 1466–1474. [CrossRef]
51. Mohiuddin, K.; Otomo, K.; Ogawa, Y.; Shikazono, N. Seasonal and spatial distribution of trace elements in the water and sediments of the Tsurumi River in Japan. *Environ. Monit. Assess.* **2012**, *184*, 265–279. [CrossRef]
52. Chandra, R.; Yadav, S.; Yadav, S. Phytoextraction potential of heavy metals by native wetland plants growing on chlorolignin containing sludge of pulp and paper industry. *Ecol. Eng.* **2017**, *98*, 134–145. [CrossRef]
53. Varol, M.; Şen, B. Assessment of nutrient and heavy metal contamination in surface water and sediments of the upper Tigris River, Turkey. *Catena* **2012**, *92*, 1–10. [CrossRef]
54. Chung, C.Y.; Chen, J.J.; Lee, C.G.; Chiu, C.Y.; Lai, W.L.; Liao, S.W. Integrated estuary management for diffused sediment pollution in Dapeng Bay and neighboring rivers (Taiwan). *Environ. Monit. Assess.* **2011**, *173*, 499–517. [CrossRef]
55. Li, J.; He, M.; Han, W.; Gu, Y. Analysis and assessment on heavy metal sources in the coastal soils developed from alluvial deposits using multivariate statistical methods. *J. Hazard. Mater.* **2009**, *164*, 976–981. [CrossRef]
56. Peijnenburg, W.J.; Zablotskaja, M.; Vijver, M.G. Monitoring metals in terrestrial environments within a bioavailability framework and a focus on soil extraction. *Ecotoxicol. Environ. Saf.* **2007**, *67*, 163–179. [CrossRef]
57. Harish, S.; Murugan, K. Oxidative stress indices in natural populations of Avicennia alba Blume. as biomarker of environmental pollution. *Environ. Res.* **2011**, *111*, 1070–1073. [CrossRef] [PubMed]
58. Shahid, M.; Pourrut, B.; Dumat, C.; Nadeem, M.; Aslam, M.; Pinelli, E. Heavy-metal-induced reactive oxygen species: Phytotoxicity and physicochemical changes in plants. In *Reviews of Environmental Contamination and Toxicology*; Springer: New York, NY, USA, 2014; Volume 232, pp. 1–44.
59. Asaeda, T.; Barnuevo, A. Oxidative stress as an indicator of niche-width preference of mangrove Rhizophora stylosa. *Forest Ecol. Manag.* **2019**, *432*, 73–82. [CrossRef]

15

Potential of a Trait-Based Approach in the Characterization of an N-Contaminated Alluvial Aquifer

Tiziana Di Lorenzo [1,*], Alessandro Murolo [2], Barbara Fiasca [2], Agostina Tabilio Di Camillo [2], Mattia Di Cicco [2] and Diana Maria Paola Galassi [2]

1 Research Institute on Terrestrial Ecosystems (IRET-CNR), Via Madonna del Piano 10, 50019 Sesto Fiorentino, Firenze, Italy

2 Department of Life, Health and Environmental Sciences, University of L'Aquila, Via Vetoio 1, Coppito, 67100 L'Aquila, Italy; alessandro.murolo92@gmail.com (A.M.); barbara.fiasca@univaq.it (B.F.); ago.tabilio@gmail.com (A.T.D.C.); diciccomattiauni@yahoo.it (M.D.C.); dianamariapaola.galassi@univaq.it (D.M.P.G.)

* Correspondence: tiziana.dilorenzo@cnr.it

Abstract: Groundwater communities residing in contaminated aquifers have been investigated mainly through taxonomy-based approaches (i.e., analyzing taxonomic richness and abundances) while ecological traits have been rarely considered. The aim of this study was to assess whether a trait analysis adds value to the traditional taxonomy-based biomonitoring in N-contaminated aquifers. To this end, we monitored 40 bores in the Vomano alluvial aquifer (VO_GWB, Italy) for two years. The aquifer is a nitrate vulnerable zone according to the Water Framework Directive. The traditional taxonomy-based approach revealed an unexpectedly high biodiversity (38 taxa and 5725 individuals), dominated by crustaceans, comparable to that of other unpolluted alluvial aquifers worldwide. This result is in contrast with previous studies and calls into question the sensitivity of stygobiotic species to N-compounds. The trait analysis provided an added value to the study, unveiling signs of impairments of the groundwater community such as low juveniles-to-adults and males-to-females ratios and a crossover of biomasses and abundances curves suggestive of an intermediate alteration of the copepod assemblages.

Keywords: crustaceans; copepods; stygobiotic; traits; groundwater; nitrate; ammonium; nitrite; nitrogen; contamination

---

1. Introduction

The taxonomy-based approach, which is traditionally focused on the analysis of taxonomic richness and abundances, has long been used to characterize and assess the impairment of freshwater ecosystems [1]. Following this approach, biological samples are collected from impaired sites and the taxa are identified, enumerated, and then compared to those of pristine reference sites (e.g., [2]). Trait-based approaches have been developed over the years alternatively to the taxonomy-based one or, more often, in combination with it. Traits are measurable physiological, morphological, and ecological characteristics of organisms, populations, or communities, such as the body size, the trophic role, the duration of life cycle, the reproductive rate, and so forth [3]. The trait-based approach allows understanding ecosystems in ways often unattainable by studies that are exclusively focused on taxonomic diversity. In the first place, traits i) reflect the species (population or community) adaptation level to their environment [4] and ii) can be used as measures of functional diversity [5]. Secondly, trait-based approaches work well across geographies since many traits are shared by widely distributed taxa [3]. From this perspective, some traits can be even applied across ecoregions [6]. Other strengths are that they add diagnostic knowledge, do not require new sampling methodology, and can supplement taxonomic analyses, while weaknesses are redundancy, autocorrelation, and difficulty in the measurements [3]. One of the most useful and fascinating aspects of trait-based approaches is that they can reveal a substantive trait variation (e.g., metabolic changes) before the taxonomic composition of the community is impaired by the loss of some species [1]. Similarly, some traits may have an earlier diagnostic potential regarding ecosystem recovery [1]. However, the decision to prefer a trait-based approach over a taxonomy-based one must be taken with caution; rather, the best option is the combination of the two approaches, at least for freshwater ecosystems [1]. In fact, trait-based approaches may fail to diagnose impacts on imperiled species which are rare in terms of abundance, because of functional redundancy [1]. On the other hand, the diagnostic power of a taxonomy-based approach decreases if the community is not characterized at the species level, but rather described at the genus or even family resolution, as required in several biomonitoring programs [7,8].

Although trait-based approaches have an old tradition in freshwater ecology (reviewed in [1]), they have not been widely applied to groundwater ecosystems. Groundwater habitats, and alluvial aquifers in particular, are characterized by lack of light and by environmental conditions that are more stable compared to surface water environments [9]. The absence of light precludes primary production by photosynthesis, so groundwater ecosystems are generally reliant on allochthonous organic carbon from surface environments [10,11]. However, particulate organic matter is not always available to groundwater food webs because it is intercepted in the soil and the unsaturated zone [10]. Consequently, the basal food source for stygofauna (obligate groundwater-dweller fauna) consists mainly of bacteria and the organic matter that is assimilated by microorganisms [12–14]. Unlike surface waters, groundwater ecosystems support only a few vertebrate species and the invertebrate assemblages are generally dominated by crustaceans [15,16]. Groundwater invertebrate assemblages also may include mites, oligochaetes and, rarely, insects and mollusks [9,17]. Copepods, amphipods, isopods, and syncarids are crustacean taxa commonly collected in healthy groundwater ecosystems [18,19]. The response of stygobiotic communities to impairments due to anthropic pressures has often been examined through taxonomy-based approaches [20–23]. Recently, some stygofaunal traits have proven to be useful in assessing the response of groundwater communities to groundwater flooding [24], carbon input [25,26], habitat structure [27,28] and chemical contamination [20,29–31]. However, the selection of a set of traits that best characterize groundwater communities is still to come.

Alluvial aquifers are particularly prone to organic pollution as their outcrops are frequently cultivated or occupied by urban settlements [21,32]. As a result, an increase in the concentrations of nitrate, nitrite, and ammonium is often observed in groundwater, with long-lasting exceedances of the legal limits [22]. Organic contamination is known to cause a change in the structure and composition of stygobiotic assemblages. Elevated concentrations of N-compounds, due to contamination through sewage and septic tank leaks, or due to the overuse of natural and synthetic fertilizers, alter the structure of groundwater communities through favoring the proliferation of surface water organisms [22,33,34]. However, the ecological studies conducted on alluvial aquifers contaminated by N-compounds are still rather scarce and reliant on the taxonomy-based approach [21,35].

The aim of this study was to evaluate the performance of a trait-based approach in the ecological characterization of an alluvial aquifer with a ten-year history of N-compound contamination. To this end, we analyzed the biological assemblages of the Vomano alluvial aquifer (central Italy) that, in 2010, was designated as a Nitrate Vulnerable Zone (NVZ) under the Water Framework Directive (EC, 2000) due to a nitrate contamination affecting over 50% of the groundwater body volume [36]. We measured five ecological traits and compared the results with those deriving from the traditional approach based on taxonomic richness and abundance distribution.

2. Materials and Methods

2.1. Study Area

The study was performed on an alluvial aquifer (VO_GWB; Figure 1) located in the hydrographic basin of the River Vomano (Italy). The outcrop of the aquifer is 30 km^2. VO_GWB is mainly fed by the River Vomano waters and rainfalls which reach the base of the aquifer in a few weeks [36,37]. The alluvial deposits of VO_GWB consist mainly of gravel and sand interbedded by silty-clay lenses of limited extension [37]. The basement of the aquifer (at a depth of about 35 m) is made up of silty-clay deposits of very low permeability [37]. The groundwater flow is NW–SE directed, and the aquifer mean discharge is 1.3×10^{-2} m^3/s [36]. VO_GWB is used for drinking and industrial and agricultural purposes. The chemical and quantitative status of VO_GWB is routinely monitored by the local environmental authority according to the requirements of the Water Framework Directive [38] and Groundwater Daughter Directive [39]. In 2010, VO_GWB was designated as a Nitrate Vulnerable Zone (NVZ) under the Water Framework Directive [38] due to a persisting (>10 years) nitrate contamination [36].

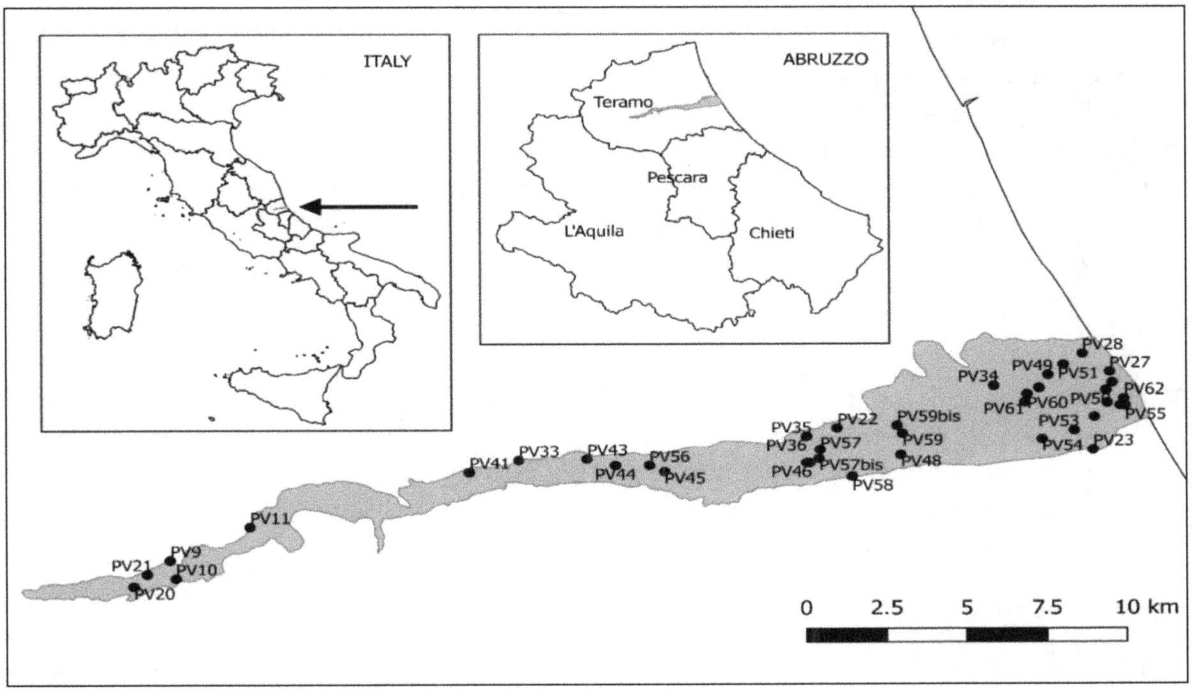

Figure 1. Clockwise: the aquifer VO_GWB in Italy and in the Abruzzo region; sampling sites (black dots) location.

Groundwater sampling occurred in 40 bores used for agricultural purposes (Figure 1). Groundwater organisms and water samples for chemical analyses were collected from shallow bores (up to 30 m deep) in autumn 2014 and spring and winter 2015 for a total of 97 samples. The altitude of the bores varied from 0 to 240 m above sea level and the depth from 2 to 30 m below the soil level. The piezometric level, measured using a freatimeter (PASI, Turin), was in the range 0.5–25.9 m below ground level. The bores' distance from the River Vomano varied from a minimum of 167 m to a maximum of 1733 m.

Groundwater flow of VO_GWB is considered basal, meaning that the groundwater undergoes remixing at all aquifer depths [37]. Preliminary investigations in which the aquifer volume was arbitrarily subdivided into three (0–10 m, 11–20 m, and 21–30 m deep) or two (0–15 m, 16–30 m deep) sectors did not reveal any significant environmental or biological pattern (Supplementary Materials: PRELIMINARY). For this reason, the aquifer was investigated as a unique groundwater unit.

2.2. Biological Survey

Bores were sampled with either an immersion pump (IP), a manual pump (MP), or a modified Cvetkov net [40], depending on the bore type (drilled or hand-dug). Five hundred liters of groundwater were withdrawn using IP and 50 L using MP. The volumes were subsequently filtered by a 60 μm mesh net to collect groundwater fauna. The modified Cvetkov net, equipped with a weight anchored to the tail, was lowered into the hand-dug bores to the bottom and subsequently hauled to filter the entire column of bore water. Ten hauls were performed in each bore [41]. The collected samples were transported in 1 L plastic vials to the laboratory within 6 h and then fixed in 70% ethanol solution. In the laboratory, specimens were sorted under a stereomicroscope at 16× magnification and identified to the lowest taxonomic level possible. Copepods were identified to species level after dissection and subsequent observation with a Leica DM 2500 optical microscope, following dichotomous keys [42,43] and more recent taxonomic papers.

Depending on the degree of adaptation to groundwater, each taxon was classified either as stygobiont (SB: species strictly associated to groundwater habitats, where it completes the whole life cycle; [44]) or non-stygobiont (nSB: epigean species accidentally or occasionally occurring in groundwater). Each specimen of the dominant taxon (Copepoda; see par. 4.) was individually photographed under a Leica M205C stereomicroscope. Length and width were measured with the software LAS (Leica Application Suite, Version 4.7.1 of Leica Microsystems, Wetzlar, Germany). Copepod lengths (in mm) were measured from the tip of the cephalic shield to the end of caudal rami, while width (in mm) was measured at the larger somite bearing legs. Sex was identified in adult individuals. Afterwards, copepod body dimensions (length and width) were converted to species–specific biovolume (in nL) using Equation (1) [45]:

$$\mathrm{BV} = \mathrm{a} \times \mathrm{b}^2 \times \mathrm{CF} \tag{1}$$

where BV = biovolume, a = length (mm), b = width (mm), and CF is a correction factor that accounts for the body shape. CF varied from 230 to 750 according to the body shape outlines of each specimen [46]. The biovolumes were then converted into individual fresh weights assuming a specific gravity of 1.1 [24]. The dry mass (mg) was estimated assuming a dry/wet weight ratio of 0.25 and the dry weight carbon content (biomass) was assumed to be 40% of the dry mass [47].

2.3. Environmental Survey

Physicochemical parameters, such as electrical conductivity (μS/cm), pH, dissolved oxygen (mg/L), and temperature (°C), were measured in situ with a multiparametric probe (WTW 3430 SET G). Two liters of groundwater were collected and stored in high-density polyethylene bottles for further chemical analyses aimed at measuring the concentrations of the 110 chemical compounds indicated in the Supplementary Materials (ENV). Particulate organic matter (POM) was obtained as the difference between the sediment weight deprived of the faunal component and dehydrated at 105 °C and that obtained after combustion of the organic fraction at 540 °C.

2.4. Data Analysis: Environmental Variables

As subsequent samples from the same bore must be considered temporal pseudoreplicates producing bias in the statistics [48], the mean values of each abiotic variable were used in the statistical analyses in the place of the raw data. Only 22 variables out of 110 were analyzed, that is, only those showing a standard deviation (SD) ≠ 0 (Table 1) so as to exclude irrelevant variables in the statistical models. Potential environmental patterns were investigated through a principal component analysis (PCA) on the basis of the Euclidean distance matrix. Prior to PCA, data were normalized and a Draftsman's plot was used to examine the correlation among the environmental variables [49,50].

Table 1. Mean and standard deviations (SD) of the 22 environmental variables (Var.) retained in the statistical analyses. Alt: altitude (m above sea level); W.t.: water table (m below ground level.); T: temperature (°C); EC: electrical conductivity (μS/cm); O_2: dissolved oxygen (mg/L); POM: particulate organic matter (mg/L); TOC: total organic carbon (mg/L); DOC: dissolved organic carbon (mg/L); NO_2^-, NO_3^-; NH_4^+, SO_4^{2-}; Cl^-, PO_4^{3-}, Ca^{2+}, K^+, Na^+ (mg/L); DIC: cis-1,2-dichloroethylene (μg/L); TCE: 1,1,2,2-tetrachloroethylene (μg/L); CHL: trichloromethane (μg/L); THC: total hydrocarbons (expressed as n-hexane in μg/L). TV: legal threshold value. Mean values exceeding legal limits are indicated in bold.

Var.	Mean	SD	TV
Alt	72.7	72.3	
W.t.	8.3	5.3	
T	16.9	1.7	
EC	1166	231	
pH	7.3	0.2	
O_2	4.7	1.6	
POM	0.73	2.32	
TOC	3.02	2.39	
DOC	2.18	1.70	
NO_2^-	**10.25**	11.19	0.5
NO_3^-	**61.60**	30.07	50
SO_4^{2-}	103.75	51.97	250
Cl^-	55.76	37.49	
PO_4^{3-}	0.27	0.42	
NH_4^+	0.23	0.50	0.5
Ca^{2+}	144.99	30.13	
K^+	5.22	2.312	
Na^+	60.17	33.31	
DIC	0.03	0.043	0.05
TCE	0.07	0.132	1.1
CHL	0.04	0.046	1.5
THC	25.7	4.480	350

2.5. *Taxonomy-Based Approach*

The exhaustiveness of the biological sampling effort was assessed through taxonomic richness estimations, using nonparametric estimators, namely, Chao1, Chao2, Jackknife1 and Jackknife2, Bootstrap, Michaelis-Menton (MM), and Ugland-Gray-Ellingsen (UGE) [51]. Values were estimated by means of 999 randomizations without replacement.

Abundances were expressed as ind/L dividing the number of individuals of each sample by its volume (liters). Raw data are available in the Supplementary Materials (BIO_VOGWB). Abundances of temporal replicates were summed per each sampling site prior to being used in the statistical analyses. Potential biological patterns were examined using a non-metric multidimensional scaling (nMDS) analysis, incorporating Bray-Curtis similarities after adding a dummy variable equal to 1 [49]. Prior to nMDS, data were log(x + 1)-transformed to approximately conform the data to normality. BIO-ENV analyses [52], based on Spearman's rank correlations, were performed to find the 'best' match between the biological and the environmental data. The statistical significance of the results of the BIO-ENV routine was tested by the global BIO-ENV match permutation test whereby each set of samples was randomly permuted 99 times relative to the others [52]. In addition to multivariate analyses, univariate statistics was performed to assess correlation between individual environmental variables and cumulative abundances per sampling site. The significance was assessed through a permutation test performed by randomly shuffling 999 times the independent variable (i.e., the environmental variables). The permutation test was selected because it is more powerful than tests based on normal theory assumptions when data distribution departs substantially from normality, as it was the case for the data of this study. Regressions were performed with the *ape* library in R 3.0.2 [53].

Finally, the structure of the biological assemblages of N-compound-contaminated alluvial aquifers other than VO_GWB was obtained from literature by using i) ISI Web of Knowledge SM (© 2010 Thomson Reuters, Toronto, ON, CA) with the following strings applied to the topic field: "aquifer" and "stygo*" and "nitrate" or "nitrite" or "ammon*"; ii) Google with the same setting as for i).

2.6. Trait-Based Approach

Potential biological patterns were examined using the following traits and the statistical settings as for the taxonomy-based approach (par. 2.5): Trait#1 (crustaceans vs. non-crustaceans); Trait#2 (SB vs. nSB); Trait#3 (adults vs. juveniles); Trait#4 (males vs. females); Trait#5 (biomass). Since copepod crustaceans accounted for 98% of the aquifer biodiversity, Traits#2–5 were analyzed with respect to copepods only. We analyzed these traits since previous studies have demonstrated that they are reliable indicators of impairment of groundwater communities, with particular references to organic contamination. In detail: i) groundwater assemblages dominated by non-crustacean taxa and a high nSB/SB ratio should be considered impaired [34]; ii) the sex ratio of some populations of copepods may be skewed in favor of female individuals in impaired water bodies [54–56]; iii) the development of juvenile copepods is slowed down in populations exposed to ammonium with respect to control populations [57]; and the survival of copepodids under N-compound exposure is significantly lower than that of adult individuals of the same species [58]. Finally, we analyzed the biomass because its relationship with abundances provides key information about community impairment [49,50]. Distributions of copepod species abundances and biomasses were compared on the same terms with *k*-dominance curves, namely applying the Abundance/Biomass Comparison method (ABC) with the following assumptions [52]: i) in an undisturbed community, the *k*-dominance curve for biomass lies above the curve of abundances for its entire length; ii) as impairment occurs and becomes more severe, the abundance curve lies above the biomass curve, in part, or completely. The W statistic was used to algebraically interpret the position of the curves, the metric values being in the range −1,1 (though neither limit is likely to be attained), with W close to 1 indicating undisturbed condition, W close to 0 indicating intermediate impairment, and W close to −1 indicating severe impairment [52].

Eventual significant patterns of each trait were explored by nMDS plots performed on the Bray–Curtis dissimilarity matrices as in the taxonomy-based approach (par. 2.5). Similarly, BIO-ENV analyses [52] and permutational regressions were used to explore the correlations between the environmental variables and the data of the individual traits.

3. Results

3.1. Environmental Variables

Mean and SD values of the 22 variables with SD $\neq 0$ are shown in Table 1.

Mean values of nitrites and nitrates were indicative of contamination according to the legal threshold values (respectively: 0.05 and 50 mg/L) set by Italian regulation [59].

Nitrite contamination was detected in 24 out of 40 bores, 20 of which also showed a nitrate contamination. Overall, bores contaminated by nitrates were 30 out of 40. Fifteen out of 40 bores were contaminated by ionized ammonium. Seven of these bores were also contaminated by nitrates while nine were contaminated by nitrites. Overall, contamination by N-compounds (nitrites, nitrates, and ionized ammonium) involved 36 bores out of 40. The PCA (matrix: 22 variables × 40 sampling sites; plot not shown) explained only 34% of the cumulative variance of the environmental data on the first two axes, failing to highlight a clear pattern of the environmental variables. This result implies that the chemical and physicochemical conditions of VO_GWB were uniform through the aquifer and there were no areas with environmental peculiarities.

3.2. Taxonomy-Based Approach

The stygofauna collected in VO_GWB accounted for 5724 individuals belonging to 38 taxa (35 crustaceans, 2 insect instars, and 1 water mite). Crustaceans were dominant in terms of abundances (99.70%), followed by insects (0.23%) and water mites (0.07%). Among crustaceans, the taxon Copepoda accounted for 97.93% of the total abundances, Amphipoda for 1.52%, Isopoda for 0.02%, Syncarida for 0.16%, and Ostracoda for 0.37%. Two out of seven non-parametric estimators reached the asymptotes, indicating the exhaustiveness of the sampling effort (Supplementary Materials: SRC). The remaining estimators indicated that the sampling effort was such to cover from 67% (Chao1) to 90% (Bootstrap) of the expected biodiversity of the aquifer.

Literature data related to N-compound concentrations and groundwater taxonomic richness in 11 alluvial aquifers worldwide are reported in Table 2.

Table 2. Taxonomic richness (T: number of taxa; SB: number of stygobiotic taxa; nSB: number of non-stygobiotic taxa; C: number of crustacean taxa; Co: number of copepod taxa) and nitrate, nitrite, and ionized ammonium concentrations (mean values in mg/L) relating to 11 alluvial aquifers worldwide obtained from this study and from literature. Ref: references. N-compound exceedances are in bold. (ts): this study.

Country	Alluvial Aquifer	Area	T	SB	nSB	C	Co	NO_3^-	NO_2^-	NH_4^+	REF
Australia	Murray	97,000	11			8	2	2.4		0.04	[60]
Australia	Gwydir	26,600	21		>5	8	2	8.0			[61]
France	Ariège	250	10	10	0	10	0	**90.0**		**31**	[62]
France	Ariège and Hers	540	36	14	22		4	**>50.0**			[35]
Germany	Rhine	5000	35	16	22	35	23	4.2		0.33	[63]
Italy	Vibrata	48	18	9	9	14	12	**72.5**			[21]
Italy	Adige	45	15	9	6	15	15	7.8	0.01	0.16	[22]
Italy	Lessinia	428		21		20	10	29.7			[64]
Italy	Vomano	30	38	21	17	35	28	**61.6**	**10.25**	0.23	(ts)
Slovenia	Brest	6	38	22	16	23	15	2.5			[65]
Spain	Tajo	201	11	3	8	11	6	11.4	0.02	0.19	[66]

The stygobiotic cyclopoid *Diacyclops cosanus* (Dco) occurred in 15 bores out 40 and was the species with the widest distribution in VO_GWB. The remaining 35 taxa occurred in less than 12 bores out of 40 and 24 taxa (12 of which were stygobiotic) occurred in less than 4 bores. Bore PV28 (13 m deep) showed the highest biodiversity, with 11 taxa, 7 of which were stygobiotic, followed by PV48 (20 m deep) with 10 taxa, 9 of which were stygobiotic. More than 50% of the bores (namely, 23 out of 40) were characterized by a maximum of three taxa and no taxa were collected from three bores, namely, from P26 (18 m deep), PV53 (10 m deep), and PV62 (8 m deep).

The nMDS plot (Figure 2; 2D stress: 0.03; matrix: 38 variables × 40 sampling sites) showed a group of 27 bores of different depths (from 4 to 30 m deep), almost completely overlapping at the center of the plot and surrounded by 11 bores (from 2 to 25 m deep) located at different distances. Thirteen out of 38 taxa were exclusive of the bores overlapping in the center of the plot, namely, the copepods *Paracyclops fimbriatus* (Pfi), *Bryocamptus pygmaeus* (Bpy), *Eucyclops serrulatus* (Ese), *Diacyclops maggii* (Dma), *Graeteriella unisetigera* (Gun), *Nitokra hibernica* (Nhi), *Nitocrella fedelitae* (Nfi), *Nitocrella achaiae* (Nac), *Bryocamptus dentatus* (Bde), *Elaphoidella plutonis* (Epl), *Pseudectinosoma reductum* (Pre), the amphipod *Niphargus* sp. 2 (Nsp2), and a species belonging to the Coleoptera Elmidae (Elm). Three species, namely, the copepods *Diacyclops bicuspidatus* (Dbic) and *Nitocrella kunzi* (Nku), and the amphipod *Niphargus sodalis* (Nso), occurred exclusively in the not-overlapping bores (Figure 2 and Supplementary Materials: BIO_VOGWB).

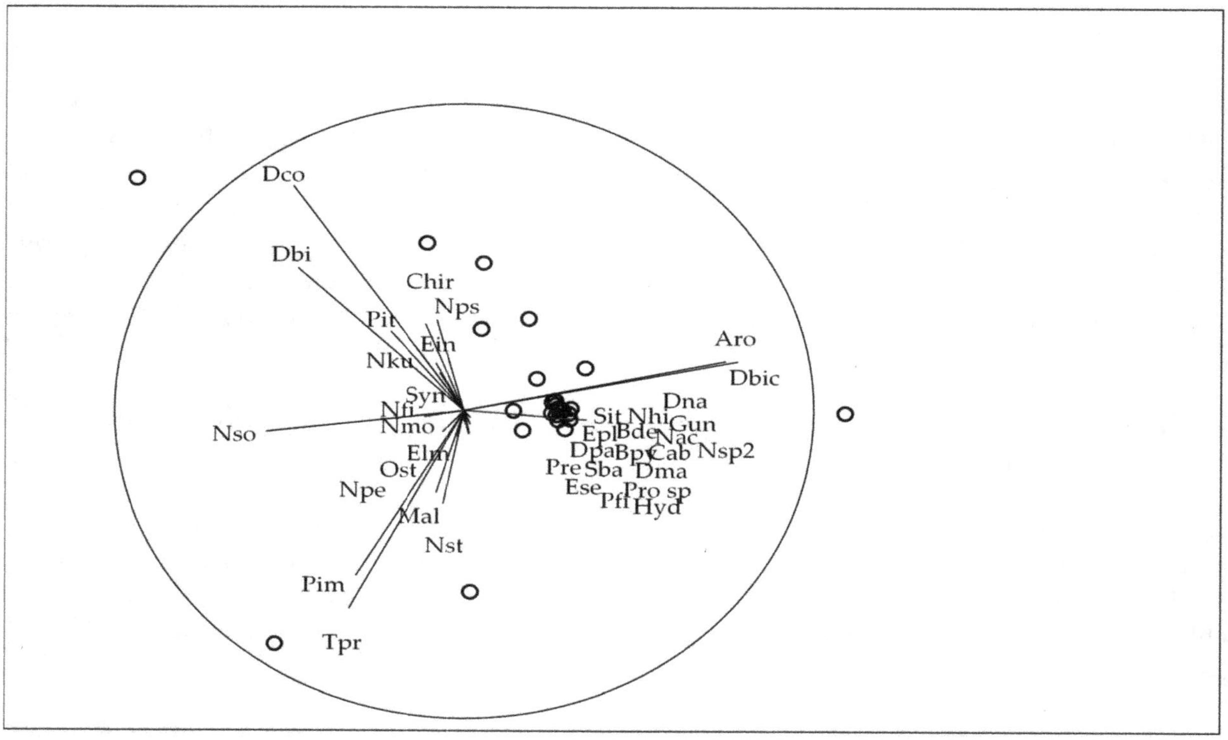

Figure 2. nMDS (2D stress: 0.03) ordination plot following the taxonomy-based approach in VO_GWB. Acronyms are as in the Supplementary Materials: BIO_VOGWB. Vector overlays are reported as exploratory tools to visualize potential linear monotonic relationships, based on Pearson correlation, between the taxa abundances and the nMDS ordination axes [52].

The BIO-ENV analysis showed that there was no significant explanatory relationship between the environmental variables and the matrix of the 38 taxa collected in VO_GWB. In fact, the sole environmental variable which best correlated to the taxa abundances was POM ($\rho = 0.436$) and the best n-dimensional combination consisted of five variables, namely, electrical conductivity, pH, nitrites, nitrates, and ionized ammonium. However, the real value of ρ fell within the 95% confidence limit of the normal distribution of randomly permuted values of ρ (p-value = 14%), indicating that the null hypothesis of no relationship between the best subset of environmental variables and the community structure should not be rejected. Univariate analyses gave significant results. In fact, cumulative densities were negatively correlated, although weakly, to electrical conductivity, dissolved oxygen, nitrates, sulphates, calcium, and sodium (Supplementary Materials: REGR). However, R^2 values were <20% for all parameters except for electrical conductivity ($R^2 = 57\%$), meaning that the regression models had a low predictive potential.

3.3. Trait-Based Approach

Trait#1 (crustacean vs. non-crustacean taxa). VO_GWB was dominated by crustaceans, occurring with 35 out of 38 taxa. The nMDS returned an arched, or horseshoe, pattern. This effect is very common in datasets in which there is a progressive turnover of a dominant variable [67]. In fact, the nMDS bubble plot showed that the arch effect was due to the turnover of crustacean taxa (Supplementary Materials: nMDS_T1). The BIO-ENV analysis indicated no significant relationship between the best subsets of environmental variables (POM: $\rho = 0.435$; p-value = 18%) and the abundances of crustacean and non-crustacean taxa. Univariate analyses did not highlight significant correlation between the cumulative abundances of crustaceans (or non-crustaceans) and the individual environmental variables.

Trait#2 (SB vs. nSB taxa). Seventeen out of 38 taxa were nSB and 21 were SB (Supplementary Materials: BIO_VOGWB). The nMDS (2D Stress: 0.03) showed that SB and nSB taxa occurred almost in all bores (Figure 3a), however, with different abundances.

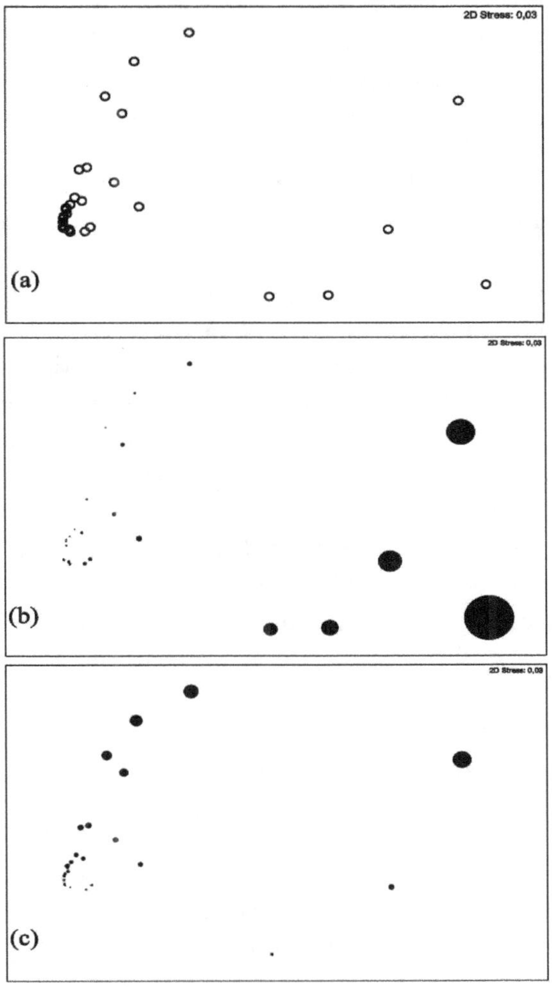

Figure 3. (**a**) nMDS plots (stress: 0.03) performed on copepod abundances data of both nSB and SB species; (**b**) nMDS bubble plot of the nSB species (ind/L); (**c**) nMDS bubble plot of the SB species (ind/L).

SB abundances (Figure 3b) were higher in the eight bores located in the upper sector of the plot while nSB abundances (Figure 3c) were higher in the lower sector of the plot, and in particular, in five bores. The group of eight bores with the highest abundances of SB taxa differed from the group of the five bores in the mean values of some environmental variables. Specifically, the average values of POM (0.44 mg/L), nitrite (9 mg/L), and ionized ammonium (0.031 mg/L) were lower in the bores with high densities of SB than in those with high densities of nSB (POM: 1.33 mg/L, nitrite: 13 mg/L, ionized ammonium: 0.644 mg/L). On the other hand, the mean values of electrical conductivity (1237 μS/cm), nitrates (79 mg/L), sulphates (109 mg/L), chloride (51 mg/L), calcium (156 mg/L), and sodium (53 mg/L) were higher in the sites with high densities of SB than in those with high densities of nSB (electrical conductivity: 835 μS/cm, nitrates: 43 mg/L, sulphate: 45 mg/L, chloride: 30 mg/L, calcium: 105 mg/L, and sodium: 31 mg/L). However, the BIO-ENV analysis indicated no significant relationship between the best subsets of environmental variables (POM: $\rho = 0.443$; p-value = 16%) and the densities of the two ecological categories. The univariate analyses highlighted a significant negative correlation between SB copepod abundances and bore depth (Supplementary Materials: REGR), however, with an R^2 equal to 11%. The abundances of nSB copepods were negatively correlated to electrical conductivity, dissolved oxygen, nitrate, sulphates, calcium, and sodium (Supplementary Materials: REGR). R^2 values were <20% for all parameters except for electrical conductivity ($R^2 = 57\%$).

Trait#3 (adults vs. juveniles). All copepod species were collected with adults and juveniles with a few exceptions, such as the nSB species *Nitokra hibernica* and the SB species *Nitocrella fedelitae*, *Nitocrella morettii*, *Nitocrella kunzi*, *Bryocamptus dentatus*, and *Pseudectinosoma reductum*, which were collected with adults only. The nMDS plot (Supplementary Materials: nMDS_T3) showed no clear separation of the bores, however, the bubble plot highlighted that adult copepods were slightly more numerous than juveniles almost in every bore. The univariate analyses highlighted a significant negative correlation between adult copepod abundances and electrical conductivity, dissolved oxygen, nitrate, sulphates, calcium, and sodium (Supplementary Materials: REGR). A significant negative correlation was assessed between juvenile abundances and electrical conductivity, sulphates, calcium, and sodium as well (Supplementary Materials: REGR). However, R^2 values were all <20%, except for electrical conductivity (R^2 > 50%).

Trait#4 (males vs. females). Twelve copepod species occurred with more females than males, nine species with more males than females, and seven species with the same number of males and females. Overall, the average abundance ± SD of males was 0.26 ± 0.70 ind/L and that of females was 0.67 ± 2.13 ind/L, with a sex ratio equal to 0.38. The nMDS did not show a clear pattern (Supplementary Materials: nMDS_T4). The BIO-ENV analysis indicated no relationship between the best subsets of environmental variables (POM: $\rho = 0.437$; *p*-value = 18%) and the abundances of males and females. The univariate analyses highlighted a significant negative correlation between male abundances and electrical conductivity, dissolved oxygen, nitrate, sulphates, and calcium (Supplementary Materials: REGR). A significant correlation was assessed between female abundances and electrical conductivity, dissolved oxygen, nitrates, sulphates, calcium, and sodium as well (Supplementary Materials: REGR). However, R^2 values were <20% for all parameters and for both males and females, except for electrical conductivity (R^2 > 43%).

Trait#5 (biomass). The total dry weight carbon (mean ± SD) was 0.012 ± 0.061 mg/L C. Multivariate analyses did not reveal significant patterns. The univariate analyses highlighted a significant negative correlation between biomass and electrical conductivity, dissolved oxygen, and calcium (Supplementary Materials: REGR). However, R^2 values were all <20%, except for electrical conductivity (R^2 = 42%). The biomass and abundance curves partially crossed at about the 20th species but since both curves were close to 100% at this point, the crossover was unclear (Figure 4). However, the W metric equal to 0.096 was indicative of a moderate impairment of the copepod assemblages.

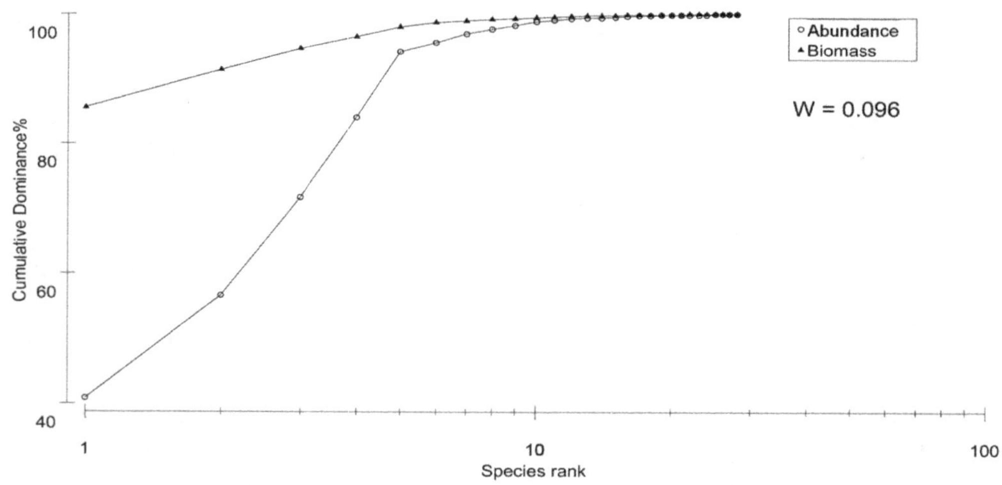

Figure 4. Abundances-biomasses curves of copepod species in GWB_VO aquifer. The two curves' crossover occurring at the 20th species and the metric W = 0.096 were indicative of moderate impairment of the copepod assemblages [52].

4. Discussion

The taxonomic richness of VO_GWB was high and consistent with previous studies in other European and non-European alluvial aquifers [60,64,66,68,69], both at the local and the regional scales [15,70]. However, the number of species collected in VO_GWB is surprisingly high if considering the persisting N-compound contamination which occurred in 90% of the aquifer. This study highlighted the high taxonomic richness and abundances of the assemblages of VO_GWB with respect to those of other alluvial aquifers worldwide contaminated by N-compounds. The taxonomic richness (38 taxa) and the abundances (5725 individuals) of VO_GWB were comparable to those of pristine alluvial aquifers. Such a result is in contrast with previous data showing that freshwater invertebrates are highly sensitive to N-compounds [71]. From this point of view, the results of this study represent a novelty and require further ecotoxicological investigations. In fact, to date, ecotoxicological studies with N-compounds have been focused mainly on non-stygobiotic invertebrates, with the exception of the study carried out by [58], who analyzed the sensitivity of the stygobiotic cyclopoid *Diacyclops belgicus* (not found in VO_GWB) to ionized ammonium. The low number of ecotoxicological data concerning stygobiotic species, and the reasons behind the difficulty in carrying out such studies, have recently been reviewed by [72], who have also claimed as unrealistic the probability of gathering new data in the near future. Thus, the high taxonomic richness and abundances observed in VO_GWB will likely remain unsupported by ecotoxicological considerations for some more time.

The traditional taxonomy-based approach showed that most taxa were collected only in a few bores and showed, therefore, restricted distributions as already observed in other studies (e.g., [64,66]). The most widespread species in VO_GWB, *D. cosanus*, is known to be highly tolerant to elevated values of salinity and electrical conductivity, being collected in other N-contaminated alluvial aquifers, in brackish coastal groundwater in southern Italy, and in sulfidic saturated karst [21]. The occurrence of epigean species in the deeper sectors of VO_GWB is recurrent in alluvial aquifers, especially in those polluted by contaminants of organic origin (e.g., [22,64]). The weak correlations of the environmental and biological variables are a common result in groundwater studies, in particular, in aquifers that are highly homogeneous from a physicochemical point of view [22,66,73]. In this study, the variance explained by the univariate models was low and this indicated that further factors, in addition to those already investigated, must be taken into account in the future to better explain the observed biological patterns. For instance, copepod species interactions could play a role in determining the biological assemblages in VO_GWB as already observed in other aquifers [74]. Similarly, previous studies showed that the grain size composition of sediments [75–77], and the lowering of the groundwater table [21,75], could influence the biological assemblage's patterns in alluvial aquifers.

Trait#1 (crustaceans vs. non-crustaceans) highlighted the predominance of crustaceans in VO_GWB. This result is in contrast with previous studies [21,33,62] which showed that crustacean abundances are lower than those of non-crustaceans in aquifers polluted by N-enriched infiltration waters.

Trait#2 provided additional information to this study, highlighting that stygobiotic species seemed to thrive in the bores where non-stygobiotic species were less abundant. The environmental variables did not entirely explain such a pattern. Hence, additional factors should be investigated such as the isolation of the bores from the surface. In fact, groundwater habitats characterized by intense water exchanges with the surface environment often exhibit a predominance of generalist non-stygobiotic fauna, while isolated groundwater sites show a predominance, or even an exclusive occurrence, of stygobiotic species [19,61,78,79].

Trait#3 showed that juveniles were less abundant than adult copepods and that some species of stygobiotic copepods were collected exclusively with adults. This is a peculiar result because stygobiotic copepod species are known to reproduce in groundwater, albeit showing a low fertility [80]. In unaltered ecosystems, juveniles are usually more abundant than adults, while the opposite was observed in this study, at least for a few species. This result could be related to the N-compound contamination of the aquifer. In fact, juvenile stages are generally more sensitive to pollution than adults [81,82] as it was observed for the stygobiotic cyclopoid *Diacyclops belgicus* when exposed to ammonium nitrate [58].

Trait#4 highlighted that the sex ratio departed from 1 and was equal to 0.38 in favor of females. In diploid populations, sex ratio is usually equal to 1 [83,84], however, surface water populations' sex ratio tends to be skewed as result of seasonality, environmental conditions, and contamination level [54–56].

In general, seasonality does not affect stygobiotic populations [44]. However, N-compounds could have a role in determining a sex ratio <1 in VO_GWB. For example, the sex ratios of the populations of *Cyclops vicinus* and *Thermocyclops crassus* were skewed in favor of female individuals in some altered surface water bodies [54]. Although no correlations were assessed between nitrites, nitrates, or ionized ammonium and abundances of adult male or female copepods in this study, these pollutants may nevertheless have had an effect on the populations in the past, reducing the number of male copepods during the ten-year-long contamination (still persisting in 2015) and inducing the skewed sex ratio <1 observed in 2014–2015.

Trait#5 highlighted that VO_GWB had an average carbon content equal to 0.012 ± 0.061 mg/L C, accounting for 0.049 g C. With respect to the carbon content (0.004 g) measured in an unpolluted chalk aquifer in southern England [24], the amount measured in VO_GWB is high. However, a fair part of this value, namely 0.047 g C, was contributed by the individuals of *Acanthocyclops robustus*, a large non-stygobiotic copepod species collected in bore PV36. Excluding the carbon contributed by this species, the average dry carbon found in VO_GWB was equal to 0.006 g C which is comparable to the amount measured in southern England [24]. However, the partial crossover of the biomasses and abundances curves and the W metric indicated impairment of copepod assemblages although not as severe as expected in a long-term contaminated alluvial aquifer as VO_GWB is.

Based on these results, we can conclude that the trait-based approach provided an added value to this study, allowing a better characterization of the biological assemblages of VO_GWB. The main difficulties that we encountered were i) the high taxonomic expertise required and ii) the long time needed for measuring individuals. In fact, the attribution of the two categories "stygobiotic" and "non-stygobiotic", at least for copepods, necessitated identifications at species level and a solid knowledge of the autecology of each species.

5. Conclusions

In this study, we pursued two objectives, namely, i) to provide new data concerning the biodiversity of an alluvial aquifer affected by a persistent N-compound contamination, and ii) to assess whether an approach based on some selected ecological traits may provide an added value to the traditional taxonomy-based methodology. The alluvial aquifer investigated in this study showed a surprisingly high biodiversity, dominated by crustaceans, comparable to that of alluvial aquifers in good chemical state. This result is in contrast with the observations of previous studies. The results call into question the real sensitivity of stygobiotic species (copepods, in particular) to N-compounds, which turned out to be lower than expected. The trait-based approach provided new details, unveiling signs of impairments of the groundwater community such as low juveniles-to-adults and males-to-females ratios and a crossover of biomasses and abundances curves suggestive of an intermediate alteration of the copepod assemblages. The signs of community alteration were not clearly imputable to the N-compound contamination affecting the aquifer for 10 years so that further investigations are needed, including additional factors such as sediment size and species interactions, in the analyses. The application of a trait-based approach to copepod-dominated groundwater assemblages requires a long time for data acquisition and high taxonomic skills. A trait database could ease the application of such approaches in studies concerning groundwater ecosystems and its setup should be strongly encouraged among researchers.

Author Contributions: Conceptualization, T.D.L. and D.M.P.G.; methodology, T.D.L.; validation, T.D.L., B.F. and D.M.P.G.; formal analysis, A.M., A.T.D.C. and M.D.C.; investigation, B.F., A.M. and A.T.D.C.; resources, D.M.P.G.; data curation, B.F., A.M., A.T.D.C. and M.D.C.; writing-original draft preparation, T.D.L. and A.M.; writing-review and editing, T.D.L.; project administration, B.F.; funding acquisition, D.M.P.G.

Acknowledgments: We thank the students of the MESVA Department of the University of L'Aquila (IT) for the support in sampling activities.

References

1. Culp, J.M.; Armanini, D.G.; Dunbar, M.J.; Orlofske, J.; Poff, L.N.; Pollard, A.I.; Yates, A.G.; Hose, G.C. Incorporating traits in aquatic biomonitoring to enhance causal diagnosis and prediction. *Integr. Environ. Assess. Manag.* **2010**, *7*, 187–197. [CrossRef] [PubMed]
2. Larras, F.; Coulaud, R.; Gautreau, E.; Billoir, E.; Rosebrery, J.; Usseglio-Polatera, P. Assessing anthropogenic pressures on streams: A random forest approach based on benthic diatom communities. *Sci. Total Environ.* **2017**, *15*, 1101–1112. [CrossRef] [PubMed]
3. Van den Brink, P.J.; Alexander, A.; Dessrosiers, M.; Goedkoop, W.; Goethals, P.; Liess, M.; Dyer, S. Traits-based approaches in bioassessment and ecological risk assessment: Strengths, weaknesses, opportunities and threats. *Integr. Environ. Assess. Manag.* **2011**, *7*, 198–208. [CrossRef] [PubMed]
4. McGill, B.J.; Enquist, B.J.; Weiher, E.; Westboy, M. Rebuilding community ecology from functional traits. *Trends Ecol. Evol.* **2006**, *21*, 178–185. [CrossRef]
5. Petchey, O.L.; Gaston, K.J. Functional diversity: Back to basics and looking forward. *Ecol. Lett.* **2006**, *9*, 741–758. [CrossRef]
6. Usseglio-Polatera, P.; Bournaud, M.; Richoux, P.; Tachet, H. Biological and ecological traits of benthic freshwater macroinvertebrates: Relationships and definition of groups with similar traits. *Freshw. Biol.* **2000**, *43*, 175–205. [CrossRef]
7. Carter, J.L.; Resh, V.H. After site selection and before data analysis: Sampling, sorting, and laboratory procedures used in stream benthic macroinvertebrate monitoring programs by USA state agencies. *J. N. Am. Benthol. Soc.* **2001**, *20*, 658–682. [CrossRef]
8. Jones, C.F. Taxonomic sufficiency: The influence of taxonomic resolution on freshwater bioassessments using benthic macroinvertebrates. *Environ. Rev.* **2008**, *16*, 45–69. [CrossRef]
9. Gibert, J.; Culver, M.J.; Dole-Oliver, F.; Malard, F.; Christman, M.; Deharveng, L. Assessing and conserving groundwater biodiversity: Synthesis and perspectives. *Freshw. Biol.* **2009**, *54*, 930–941. [CrossRef]
10. Pabich, W.J.; Valiela, I.; Hemond, H.F. Relationship between DOC concentration and vadose zone thickness and depth below water table in groundwater of Cape Cod, U.S.A. *Biogeochemistry* **2001**, *55*, 247–268. [CrossRef]
11. Shen, Y.; Chapelle, F.H.; Benner, R. Origins and bioavailability of dissolved organic matter in groundwater. *Biogeochemistry* **2015**, *122*, 61–78. [CrossRef]
12. Boulton, A.J. River ecosystem health down under: Assessing ecological condition in riverine groundwater zones in Australia. *Ecosyst. Health* **2001**, *6*, 108–118. [CrossRef]
13. Francois, C.M.; Mamillod Blondin, F.; Malard, F.; Fourel, F.; Lécuyer, C.; Douady, C.J.; Simon, L. Trophic ecology of groundwater species reveals specialization in a low-productivity environment. *Funct. Ecol.* **2016**, *30*, 262–273. [CrossRef]
14. Hofmann, R.; Griebler, C. DOM and bacterial growth efficiency in oligotrophic groundwater: Absence of priming and co-limitation by organic carbon and phosphorus. *Aquat. Microb. Ecol.* **2018**, *81*, 55–71. [CrossRef]
15. Stoch, F.; Galassi, D.M.P. Stygobiotic crustacean species richness: A question of numbers, a matter of scale. *Hydrobiologia* **2010**, *653*, 217–234. [CrossRef]
16. Strona, G.; Fattorini, S.; Fiasca, B.; Di Lorenzo, T.; Di Cicco, M.; Lorenzetti, W.; Galassi, D.M.P. AQUALIFE software: A new tool for a standardized ecological assessment of groundwater dependent ecosystems. *Water* **2019**. (accepted).
17. Humphreys, W. Aquifers: The ultimate groundwater dependent ecosystem. *Aust. J. Bot.* **2006**, *54*, 115–132. [CrossRef]
18. Galassi, D.M.P.; Huys, R.; Reid, J. Diversity ecology and evolution of groundwater copepods. *Freshw. Biol.* **2009**, *54*, 691–708. [CrossRef]
19. Di Lorenzo, T.; Cipriani, D.; Fiasca, B.; Rusi, S.; Galassi, D.M.P. Groundwater drift monitoring as a tool to assess the spatial distribution of groundwater species into karst aquifers. *Hydrobiologia* **2018**, *813*, 137–156. [CrossRef]

20. Malard, F.; Mathieu, J.; Reygrobellet, J.L.; Lafont, M. Biomonitoring groundwater contamination: Application to a karst area in Southern France. *Aquat. Sci.* **1996**, *58*, 158–187. [CrossRef]
21. Di Lorenzo, T.; Galassi, D.M.P. Agricultural impact in Mediterranean alluvial aquifers: Do groundwater communities respond? *Fundam. Appl. Limnol.* **2013**, *182*, 271–282. [CrossRef]
22. Di Lorenzo, T.; Cifoni, M.; Lombardo, P.; Fiasca, B.; Galassi, D.M.P. Ammonium threshold value for groundwater quality in the EU may not protect groundwater fauna: Evidence from an alluvial aquifer in Italy. *Hydrobiologia* **2015**, *743*, 139–150. [CrossRef]
23. Caschetto, M.; Barbieri, M.; Galassi, D.M.P.; Mastrorillo, L.; Rusi, S.; Stoch, F.; Di Cioccio, A.; Petitta, M. Human alteration of groundwater-surface water interactions (Sagittario River, Central Italy): Implication for flow regime, contaminant fate and invertebrate response. *Environ. Earth Sci.* **2014**, *7*, 1791–1807. [CrossRef]
24. Reiss, J.; Perkins, D.M.; Fussmann, K.E.; Krause, S.; Canhoto, C.; Romeijn, P.; Roberton, A.L. Groundwater flooding: Ecosystem structure following an extreme recharge event. *Sci. Total Environ.* **2019**, *625*, 1252–1260. [CrossRef] [PubMed]
25. Wilhelm, F.M.; Taylor, S.J.; Adams, G.L. Comparison of routine metabolic rates of the stygobite, *Gammarus acherondytes* (Amphipoda: Gammaridae) and the stygophile *Gammarus troglophilus*. *Freshw. Biol.* **2006**, *51*, 1162–1174. [CrossRef]
26. Mezek, T.; Simčič, T.; Arts, M.T.; Bracelj, A. Effect of fasting on hypogean (*Niphargus stygius*) and epigean (*Gammarus fossarum*) amphipods: A laboratory study. *Aquat. Ecol.* **2010**, *44*, 397–408. [CrossRef]
27. Descloux, S.; Datry, T.; Usseglio-Polatera, P. Trait-based structure of invertebrates along a gradient of sediment colmation: Benthos versus hyporheos responses. *Sci. Total Environ.* **2014**, *466–467*, 265–276. [CrossRef]
28. Mathers, K.L.; Hill, M.J.; Wood, C.D.; Wood, P.J. The role of fine sediment characteristics and body size on the vertical movement of a freshwater amphipod. *Freshw. Biol.* **2018**, *64*, 152–163. [CrossRef]
29. Sarkka, J.; Levonen, L.; Makela, J. Meiofauna of springs in Finland in relation to environmental factors. *Hydrobiologia* **1997**, *347*, 139–150. [CrossRef]
30. Mauclaire, L.; Gibert, J.; Claret, C. Do bacteria and nutrients control faunal assemblages in alluvial aquifers? *Arch. Hydrobiol.* **2000**, *148*, 85–98. [CrossRef]
31. Pacioglu, O.; Moldovan, O.T. Response of invertebrates from the hyporheic zone of chalk rivers to eutrophication and land use. *Environ. Sci. Pollut. Res.* **2016**, *23*, 4729–4740. [CrossRef] [PubMed]
32. Boy-Roura, M.; Nolan, B.T.; Menció, A.; Mas-Pla, J. Regression model for aquifer vulnerability assessment of nitrate pollution in the Osona region (NE Spain). *J. Hydrol.* **2013**, *505*, 150–162. [CrossRef]
33. Malard, F. Groundwater contamination and ecological monitoring in a Mediterranean karst ecosystem in Southern France. In *Groundwater Ecology: A Tool for Management of Water Resources*; Griebler, C., Danielopol, D., Gibert, J., Nachtnebel, H.P., Notenboom, J., Eds.; Official Publication of the European Communities: Luxembourg, 2001; pp. 183–194.
34. Korbel, K.L.; Hose, G.C. A tiered framework for assessing groundwater ecosystem health. *Hydrobiologia* **2011**, *661*, 329–349. [CrossRef]
35. Marmonier, P.; Maazouzi, C.; Baran, N.; Blanchet, S.; Ritter, A.; Saplairoles, M.; Dole-Olivier, M.J.; Galassi, D.M.; Eme, D.; Dolédec, S.; et al. Ecology-based evaluation of groundwater ecosystems under intensive agriculture: A combination of community analysis and sentinel exposure. *Sci. Total Environ.* **2018**, *613–614*, 1353–1366. [CrossRef] [PubMed]
36. Regione Abruzzo. Piano di Tutela delle Acque. Relazione Generale e Allegati. 2010. Available online: http://www.regione.abruzzo.it/pianoTutelaacque/index.asp?modello=elaboratiPiano&servizio=lista&stileDiv=elaboratiPiano (accessed on 10 October 2019).
37. Desiderio, G.; Nanni, T.; Rusi, S. La pianura del fiume Vomano (Abruzzo): Idrogeologia, antropizzazione e suoi effetti sul depauperamento della falda. *Boll. Soc. Geol. Ital.* **2003**, *122*, 421–434.
38. EC (European Community). Directive 2000/60/EC of the European Parliament and of the Council of 23 October 2000 establishing a framework for Community action in the field of water policy. *Off. J.* **2000**, *L 327*, 1–73.
39. EC (European Community). Directive 2006/118/EC of the European Parliament and of the Council of 12 December 2006 on the protection of groundwater against pollution and deterioration. *Off. J.* **2006**, *L 327/19*, 1–31.
40. Cvetkov, L. Un fi let phréatobiologique. *Bulletin de l'Institut de Zoologie et Musée Sofia* **1968**, *22*, 215–219.

41. Hancock, P.J.; Boulton, A.J. Sampling groundwater fauna: Efficiency of rapid assessment methods tested in bores in eastern Australia. *Freshw. Biol.* **2009**, *54*, 902–917. [CrossRef]
42. Dussart, B.; Defaye, D. *World Directory of Crustacea Copepoda of Inland Waters. II—Cyclopiformes*; Backhuys Publishers: Leiden, The Netherlands, 2009; pp. 1–276.
43. Boxshall, G.A.; Halsey, S.H. *An Introduction to Copepod Diversity*; The Ray Society: London, UK, 2004.
44. Gibert, J.; Stanford, J.A.; Dole-Oliver, M.J.; Ward, J. Basic attributes of groundwater ecosystems and prospects for research. In *Groundwater Ecology*; Gibert, J., Danielopol, D., Stanford, J., Eds.; Academic Press: California, CA, USA, 1994; pp. 7–40.
45. Reiss, J.; Schmid-Araya, J.M. Existing in plenty: Abundance, biomass and diversity of ciliates and meiofauna in small streams. *Freshw. Biol.* **2008**, *53*, 652–668. [CrossRef]
46. Warwick, R.M.; Gee, J.M. Community structure of estuarine meiobenthos. *Mar. Ecol. Prog. Ser.* **1984**, *18*, 97–111. [CrossRef]
47. Feller, R.J.; Warwick, R.M. Introduction to the study of meiofauna. In *Energetics*; Higgins, R.P., Thiel, H., Eds.; Smithsonian Institution Press: Washington, DC, USA, 1988; pp. 181–196.
48. Hurlbert, S.H. Pseudoreplication and the design of ecological field experiments. *Ecol. Monog.* **1984**, *54*, 187–211. [CrossRef]
49. Clarke, K.R.; Gorley, R.N. *PRIMER v6: User Manual/Tutorial*; PRIMER-E: Playmouth, UK, 2006.
50. Clarke, K.R.; Green, R.H. Statistical design and analysis for a 'biological effects' study. *Mar. Ecol.* **1988**, *46*, 213–226. [CrossRef]
51. Magurran, A.E.; McGill, B.J. *Biological Diversity: Frontiers in Measurement and Assessment*; University Press: Oxford, UK, 2011.
52. Clarke, K.R.; Warwick, R.M. *Change in Marine Communities: An Approach to Statistical Analysis and Interpretation*, 2nd ed.; Playmouth Marine Laboratory: Playmouth, UK, 2001.
53. R Core Team. *R: A Language and Environment for Statistical Computing*; R Foundation for Statistical Computing: Vienna, Austria, 2013; Available online: http://www.R-project.org/ (accessed on 2 December 2019).
54. Krupa, E.G. Population densities, sex ratios of adults, and occurrence of malformations in three species of cyclopoid copepods in waterbodies with different degrees of eutrophy and toxic pollution. *J. Mar. Sci. Technol.* **2005**, *13*, 226–237.
55. Peschke, K.; Jonas, G.; Kohler, H.-R.; Karl, W.; Rita, T. Invertebrates as indicators for chemical stress in sewage-influenced stream systems: Toxic and endocrine effects in gammarids and reactions at the community level in two tributaries of Lake Constance, Schussen and Argen. *Ecotoxicol. Environ. Saf.* **2014**, *106*, 115–125. [CrossRef]
56. Ozga, V.A.; da Silva de Castro, V.; da Silva Castiglioni, D. Population structure of two freshwater amphipods (Crustacea: Peracarida: Hyalellidae) from southern Brazil. *Nauplius* **2018**, *26*, e2018025. [CrossRef]
57. Di Marzio, W.D.; Castaldo, D.; Di Lorenzo, T.; Di Cioccio, A.; Sáenz, M.E.; Galassi, D.M.P. Developmental endpoints of chronic exposure to suspected endocrine-disrupting chemicals on benthic and hyporheic freshwater copepods. *Ecotoxicol. Environ. Saf.* **2013**, *96*, 86–92. [CrossRef]
58. Di Lorenzo, T.; Di Marzio, W.D.; Sáenz, M.E.; Baratti, M.; Dedonno, A.A. Iannucci, A.; Cannicci, S.; Messana, G.; Galassi, D.M. Sensitivity of hypogean and epigean freshwater copepods to agricultural pollutants. *Environ. Sci. Pollut. Res.* **2014**, *21*, 4643–4655. [CrossRef]
59. Repubblica Italiana. Decreto Legislativo 16 Marzo 2009, n. 30: Attuazione della direttiva 2006/118/CE, relativa alla protezione delle acque sotterranee dall'inquinamento e dal deterioramento. *Gazzetta Ufficiale* **2009**, *49*, 1–45.
60. Korbel, K.; Chartion, A.; Stephenson, S.; Greenfield, P.; Hose, G.C. Wells provide a distorted view of life in the aquifer: Implications for sampling, monitoring and assessment of groundwater ecosystems. *Sci. Rep.* **2017**, *7*, 40702. [CrossRef] [PubMed]
61. Menció, A.; Korbel, K.L.; Hose, G.C. River–aquifer interactions and their relationship to stygofauna assemblages: A case study of the Gwydir River alluvial aquifer (New South Wales, Australia). *Sci. Total Environ.* **2014**, *479–480*, 292–305. [CrossRef] [PubMed]
62. Dumas, P.; Bou, C.; Gibert, J. Groundwater macrocrustaceans as natural indicators of the Ariège alluvial aquifer. *Int. Rev. Hydrobiol.* **2001**, *86*, 619–633. [CrossRef]
63. Hahn, H.J. The GW-Fauna-Index: A first approach to a quantitative ecological assessment of groundwater habitats. *Limnologica* **2006**, *36*, 119–137. [CrossRef]

64. Galassi, D.M.P.; Stoch, F.; Fiasca, B.; Di Lorenzo, T.; Gattone, E. Groundwater biodiversity patterns in the Lessinian Massif of northern Italy. *Freshw. Biol.* *54*, 830–847. [CrossRef]
65. Brancelj, A.; Zibrat, U.; Jamnik, B. Differences between groundwater fauna in shallow and in deep intergranular aquifers as an indication of different characteristics of habitats and hydraulic connections. *J. Limnol.* **2016**, *75*, 248–261. [CrossRef]
66. Iepure, S.; Rasines-Ladero, R.; Meffe, R.; Carreno, F.; Mostaza, D.; Sundberg, A.; Di Lorenzo, T.; Barroso, J.L. The role of groundwater crustaceans in disentangling aquifer type features—A case study of the Upper Tagus Basin, central Spain. *Ecohydrology* **2017**, *10*, e1876. [CrossRef]
67. Shaw, P.J.A. *Multivariate Statistics for the Environmental Sciences*; Oxford University Press Inc.: Oxford, UK, 2003; pp. 1–231.
68. Deharveng, L.; Stoch, F.; Gibert, J.; Bedos, A.; Galassi, D.M.P.; Zagmajster, M.; Brancelj, A.; Camacho, A.; Fiers, F.; Martin, P.; et al. Groundwater biodiversity in Europe. *Freshw. Biol.* **2009**, *54*, 709–726. [CrossRef]
69. Mahi, A.; Di Lorenzo, T.; Haicha, B.; Belaidi, N.; Taleb, A. Environmental factors determining regional biodiversity patterns of groundwater fauna in semi-arid aquifers of northwest Algeria. *Limnology* **2019**, *20*, 309–320. [CrossRef]
70. Stoch, F.; Artheau, M.; Brancelj, A.; Galassi, D.M.P.; Malard, F. Biodiversity indicators in European ground waters: Towards a predictive model of stygobiotic species richness. *Freshw. Biol.* **2009**, *54*, 745–755. [CrossRef]
71. Camargo, J.A.; Alonso, A. Ecological and toxicological effects of inorganic nitrogen pollution in aquatic ecosystems: A global assessment. *Environ. Int.* **2006**, *32*, 831–849. [CrossRef]
72. Di Lorenzo, T.; Di Marzio, W.D.; Fiasca, B.; Galassi, D.M.P.; Korbel, K.; Iepure, S.; Pereira, J.L.; Reboleira, A.S.; Schmidt, S.I.; Hose, G.C. Recommendations for ecotoxicity testing with stygobiotic species in the framework of groundwater environmental risk assessment. *Sci. Total Environ.* **2019**, *681*, 292–304. [CrossRef] [PubMed]
73. Korbel, K.L.; Hose, G.C. Habitat, water quality, seasonality, or site? Identifying environmental correlates of the distribution of groundwater biota. *Freshw. Sci.* **2015**, *34*, 329–343. [CrossRef]
74. Fattorini, S.; Lombardo, P.; Fiasca, B.; Di Cioccio, A.; Di Lorenzo, T.; Galassi, D.M.P. Earthquake-related changes in species spatial niche overlaps in spring communities. *Sci. Rep.* **2017**, *7*, 443. [CrossRef] [PubMed]
75. Stumpp, C.; Hose, G.C. Groundwater amphipods alter aquifer sediment structure. *Hydrol. Process.* **2017**, *31*, 3452–3454. [CrossRef]
76. Hose, G.C.; Stumpp, C. Architects of the underworld: Bioturbation by groundwater invertebrates influences aquifer hydraulic properties. *Aquat. Sci.* **2019**, *81*, 20. [CrossRef]
77. Korbel, K.L.; Stephenson, S.; Hose, G.C. Sediment size influences habitat selection and use by groundwater macrofauna and meiofauna. *Aquat. Sci.* **2019**, *81*, 39. [CrossRef]
78. Bork, J.; Berkhoff, S.; Bork, S.; Hahn, J.H. Using subsurface metazoan fauna to indicate groundwater–surface water interactions in the Nakdong River floodplain, South Korea. *Hydrogeol. J.* **2009**, *17*, 61–75. [CrossRef]
79. Hahn, H.J.; Fuchs, A. Distribution patterns of groundwater communities across aquifer types in south-western Germany. *Freshw. Biol.* **2008**, *54*, 848–860. [CrossRef]
80. Galassi, D.M.P. Groundwater copepods: Diversity patterns over ecological and evolutionary scales. *Hydrobiologia* **2001**, *454–453*, 227–253. [CrossRef]
81. Migliore, L.; de Nicola Giudici, M. Toxicity of heavy metals to *Asellus aquaticus* (L.) (Crustacea, Isopoda). *Hydrobiologia* **1990**, *203*, 155–164. [CrossRef]
82. Kadiene, U.E.; Bialais, C.; Ouddane, B.; Hwang, J.-S.; Souissi, S. Differences in lethal response between male and female calanoid copepods and life-cycle traits to cadmium toxicity. *Ecotoxicology* **2017**, *26*, 1227–1239. [CrossRef]
83. Fisher, R.A. *The Genetical Theory of Natural Selection*; Clarendon Press: Oxford, UK, 1930.
84. Leigh, E.G. Natural Selection and Mutability. *Am. Nat.* **1970**, *104*, 301–305. [CrossRef]

PERMISSIONS

All chapters in this book were first published in MDPI; hereby published with permission under the Creative Commons Attribution License or equivalent. Every chapter published in this book has been scrutinized by our experts. Their significance has been extensively debated. The topics covered herein carry significant findings which will fuel the growth of the discipline. They may even be implemented as practical applications or may be referred to as a beginning point for another development.

The contributors of this book come from diverse backgrounds, making this book a truly international effort. This book will bring forth new frontiers with its revolutionizing research information and detailed analysis of the nascent developments around the world.

We would like to thank all the contributing authors for lending their expertise to make the book truly unique. They have played a crucial role in the development of this book. Without their invaluable contributions this book wouldn't have been possible. They have made vital efforts to compile up to date information on the varied aspects of this subject to make this book a valuable addition to the collection of many professionals and students.

This book was conceptualized with the vision of imparting up-to-date information and advanced data in this field. To ensure the same, a matchless editorial board was set up. Every individual on the board went through rigorous rounds of assessment to prove their worth. After which they invested a large part of their time researching and compiling the most relevant data for our readers.

The editorial board has been involved in producing this book since its inception. They have spent rigorous hours researching and exploring the diverse topics which have resulted in the successful publishing of this book. They have passed on their knowledge of decades through this book. To expedite this challenging task, the publisher supported the team at every step. A small team of assistant editors was also appointed to further simplify the editing procedure and attain best results for the readers.

Apart from the editorial board, the designing team has also invested a significant amount of their time in understanding the subject and creating the most relevant covers. They scrutinized every image to scout for the most suitable representation of the subject and create an appropriate cover for the book.

The publishing team has been an ardent support to the editorial, designing and production team. Their endless efforts to recruit the best for this project, has resulted in the accomplishment of this book. They are a veteran in the field of academics and their pool of knowledge is as vast as their experience in printing. Their expertise and guidance has proved useful at every step. Their uncompromising quality standards have made this book an exceptional effort. Their encouragement from time to time has been an inspiration for everyone.

The publisher and the editorial board hope that this book will prove to be a valuable piece of knowledge for researchers, students, practitioners and scholars across the globe.

LIST OF CONTRIBUTORS

Enzo Goretti, Matteo Pallottini, Sarah Pagliarini, Marianna Catasti, Gianandrea La Porta, Roberta Selvaggi and Elda Gaino
Dipartimento di Chimica, Biologia e Biotecnologie, Università degli Studi di Perugia, 06123 Perugia, Italy

Alessandro Maria Di Giulio
Servizio Disinfestazione, USL Umbria1, 06127 Perugia, Italy

Arshad Ali
Entomology and Nematology Department, University of Florida, Gainesville, FL 32611, USA

Meng-Yuan Ou, Yu Ting, Boon-Lek Ch'ng, Chi Chen, Yung-Hua Cheng and Hsing-Cheng Hsi
Graduate Institute of Environmental Engineering, National Taiwan University, No. 1, Sec. 4, Roosevelt Rd, Taipei 10671, Taiwan

Tien-Chin Chang
Institute of Environmental Engineering and Management, National Taipei University of Technology, No. 1, Sec. 3, Zhongxiao E. Rd., Taipei 10608, Taiwan

Giovanni Strona
Research Centre for Ecological Change, University of Helsinki, P.O. Box 4 FI-00014 Helsinki, Finland

Simone Fattorini, Barbara Fiasca Mattia Di Cicco and Diana M. P. Galassi
Department of Life, Health and Environmental Sciences, University of L'Aquila, 67100 L'Aquila, Italy

Tiziana Di Lorenzo
Italian National Research Council, Research Institute on Terrestrial Ecosystems, Via Madonna del Piano 10, Sesto Fiorentino, 50019 Florence, Italy

Walter Lorenzetti and Francesco Boccacci
Gis3W S.r.l, Viale Verdi 24, Montecatini Terme, 51016 Pistoia, Italy

Yu-Lin Kuo
Department of Civil Engineering, National Pingtung University of Science and Technology, Pingtung 91201, Taiwan

Chia-Hisng Lee
Center for Sustainability Science, Academia Sinica, Taibei 11529, Taiwan

Shih-Hao Jien
Department of Soil and Water Conservation, National Pingtung University of Science and Technology, Pingtung 91201, Taiwan

Marelize Labuschagne, Victor Wepener and Nico J. Smit
Water Research Group, Unit for Environmental Sciences and Management, North-West University, 2520 Potchefstroom, South Africa

Sonja Zimmermann
Water Research Group, Unit for Environmental Sciences and Management, North-West University, 2520 Potchefstroom, South Africa
Department of Aquatic Ecology and Centre forWater and Environmental Research, University of Duisburg-Essen, 45141 Essen, Germany

Milen Nachev and Bernd Sures
Department of Aquatic Ecology and Centre for Water and Environmental Research, University of Duisburg-Essen, 45141 Essen, Germany

Chu-Wen Yang, Chien Liu and Bea-Ven Chang
Department of Microbiology, Soochow University, Taipei 11101, Taiwan

Marco Melita, Stefano Amalfitano, Elisabetta Preziosi, Stefano Ghergo, Eleonora Frollini, Daniele Parrone and Annamaria Zoppini
Water Research Institute, National Research Council (IRSA-CNR), 00015 Monterotondo, Rome, Italy

Boon-Lek Ch'ng, Che-Jung Hsu, Yu Ting, Ying-LinWang, Chi Chen and Hsing-Cheng Hsi
Graduate Institute of Environmental Engineering, National Taiwan University, No. 1, Sec. 4, Roosevelt Rd, Taipei 106, Taiwan

Tien-Chin Chang
Institute of Environmental Engineering and Management, National Taipei University of Technology, No. 1, Sec. 3, Chung-Hsiao E. Rd., Taipei 106, Taiwan

Mattia Iannella, Giulia Console, Paola D'Alessandro, Francesco Cerasoli, Cristina Manton and Maurizio Biondi
Department of Life, Health and Environmental Sciences, Environmental Sciences Sect, University of L'Aquila, Via Vetoio, Coppito, 67100 L'Aquila, Italy

Fabrizio Ruggieri and Francesca Di Donato
Department of Physical and Chemical Sciences (DSFC), University of L'Aquila, Via Vetoio, Coppito, 67100 L'Aquila, Italy

Kuei-San Chen
Department of Soil and Environmental Sciences, National Chung Hsing University, Taichung 40227, Taiwan

Chun-Yu Pai
Department of Agronomy, National Taiwan University, Taipei 10617, Taiwan

Hung-Yu Lai
Department of Soil and Environmental Sciences, National Chung Hsing University, Taichung 40227, Taiwan
Innovation and Development Center of Sustainable Agriculture, National Chung Hsing University, Taichung 40227, Taiwan

Leonardo Piccini and Pilario Costagliola
Department of Earth Science, University of Florence, 50121 Florence, Italy

Tiziana Di Lorenzo
Research Institute on Terrestrial Ecosystems of the CNR, National Research Council of Italy, 50019 Florence, Italy

Diana Maria Paola Galassi
Department of Life, Health and Environmental Sciences, University of L'Aquila, 67100 L'Aquila, Italy

Claudia Campos, Juliana Arias, Jennifer López, Lorena Santos and Camilo Venegas
Department of Microbiology, Laboratorio de Indicadores de Calidad de Agua y Lodos (LIAL), Pontificia Universidad Javeriana, Carrera 7 No. 43-82, Bogotá 110231, Colombia

Crispín Celis Zambrano
Department of Chemistry, Pontificia Universidad Javeriana, Carrera 7 No. 43-82, Bogotá 110231, Colombia

Alison Martín
Department of Microbiology, Laboratorio de Indicadores de Calidad de Agua y Lodos (LIAL), Pontificia Universidad Javeriana, Carrera 7 No. 43-82, Bogotá 110231, Colombia
Department of Chemistry, Pontificia Universidad Javeriana, Carrera 7 No. 43-82, Bogotá 110231, Colombia

Marcela Duarte and Nubia de Parra
Independient Investigator, Bogotá 110111, Colombia

Andrés Ortíz-Ardila
Hydraulic and Environmental Engineering Department, Pontificia Universidad Católica de Chile, Av. Vicuña Mackenna 4860, Macul, Santiago 7820436, Chile

Grant C. Hose and Katelyn Symington
Department of Biological Sciences, Macquarie University, Sydney, NSW 2109, Australia

Maria J. Lategan
Department of Molecular Sciences, Macquarie University, Sydney, NSW 2109, Australia

Rainer Siegele
Australian Nuclear Science and Technology Organisation, Locked Bag 2001, Kirrawee DC, NSW 2232, Australia

Mohammed Othman Aljahdali
Department of Biological Sciences, Faculty of Sciences, King Abdulaziz University, Jeddah 80203, Saudi Arabia

Abdullahi Bala Alhassan
Department of Biological Sciences, Faculty of Sciences, King Abdulaziz University, Jeddah 80203, Saudi Arabia
Department of Biology, Faculty of Life Science, Ahmadu Bello University, Zaria 810001, Nigeria

Tiziana Di Lorenzo
Research Institute on Terrestrial Ecosystems (IRET-CNR), Via Madonna del Piano 10, 50019 Sesto Fiorentino, Firenze, Italy

Alessandro Murolo, Barbara Fiasca, Agostina Tabilio Di Camillo, Mattia Di Cicco and Diana Maria Paola Galassi
Department of Life, Health and Environmental Sciences, University of L'Aquila, Via Vetoio 1, Coppito, 67100 L'Aquila, Italy

Index

A

Active Capping, 18, 29, 118, 131

Adsorption, 19-21, 23-24, 27-31, 60, 117-120, 122-123, 125, 130, 132-133, 185, 207

Alkalinity, 58, 151, 191

Ammonium, 40, 42, 48, 50, 52-53, 55, 60, 62, 66, 119, 134-135, 137, 140, 143, 170, 210, 221-222, 226-229, 231-232, 234

Antibiotics, 84-86, 88, 90, 94, 96, 98-100

Antioxidant, 158, 207-209, 211, 213, 215, 217, 219

Aquatic Environment, 64, 79-81, 84, 111-112, 133, 147, 207

Aquatic Life, 46, 216-217

Aqueous Batch, 18-20, 29, 118-119, 122, 126, 130

Aquifers, 32, 34, 37, 42-46, 102, 113-116, 161-163, 167, 170-174, 194, 201-204, 221-222, 226-227, 230-234, 236

B

Benthic Organisms, 117-118, 207, 216-217

Bioaccumulation, 1, 13, 64-65, 67, 74-76, 78-82, 117, 157-158, 179, 192, 194, 202, 204-207

Bioassays, 2, 177, 179-182, 184, 186-187, 190-191

Biochar, 30-31, 49-55, 58-63, 149-150, 152-155, 157-159

Biodegradation, 84-85, 87-88, 96-101, 112-113, 115

Biodiversity, 32-33, 38-42, 44-47, 113, 115, 143-147, 161-163, 172-175, 178, 207, 221, 226-227, 232-233, 236

Biolog Assay, 111, 113

Biomarkers, 81, 191, 207-208, 217

C

Cadmium, 14-15, 71, 132, 146, 149-150, 157-160, 180-181, 184-185, 188, 191-192, 202, 205-206, 219, 236

Chemical Contaminants, 13, 177, 179

Chemical Forms, 149-151, 155-156, 158, 160

Chironomidae, 1-2, 4-5, 7, 13-17, 138

Co-precipitation, 19, 60, 207

Contamination, 1-3, 11-13, 15, 33, 43, 50, 69, 76, 82, 99, 102, 107, 112, 114-115, 118, 143-144, 151, 170, 177-179, 181, 183, 185-187, 189, 194-195, 202, 204, 206, 210, 216-223, 226, 230-232, 234

Control Dam, 66-67, 70, 73-75, 77-78, 80

Copepods, 46, 170-172, 200, 204, 221-222, 224, 226-227, 229-233, 235-236

Corbicula Fluminalis Africana, 64-65, 67

Crustaceans, 34, 162, 170-171, 173, 194, 199-203, 206, 221-222, 226-228, 231-232, 236

Cumulative Quantity, 57-58, 61

D

Domestic Wastewater, 178, 180, 208

E

Ecological Risk, 32-33, 35, 41-42, 46, 118, 191, 201, 207-208, 210-211, 213, 216-220, 233

Ecosystem Services, 44, 113, 162, 171-172, 174, 202-203

Environmental Concentrations, 34, 64-65, 80, 82

Environmental Contamination, 1-2, 13, 99, 220

F

Flow Cytometry, 102, 104

Flow Rates, 37, 104, 201

G

Gold Extraction, 177-178, 186

Gold Mining, 78-79, 177-179, 184, 186-187, 189, 191-192, 205

Groundwater Quality, 48, 102-103, 113, 161, 166, 203-204, 234

H

Heavy Metals, 1-2, 13, 15, 30-31, 81, 154, 157, 159-160, 177, 179, 181, 184-186, 189, 191, 206-209, 211-220, 236

Husk Biochar, 149, 152-155

Hyporheic Zones, 32, 35, 41, 43-44, 170

I

Impoundments, 65, 67, 70, 73-80, 113

Incubation, 20, 24-25, 28, 51, 55-58, 60, 88, 90, 98, 104-105, 108-112

Inhibitory Effect, 57, 60-61

K

Karst Aquifer, 161, 166-167

Karst Areas, 161, 163

Karst Habitats, 161-162

L

Lake Trasimeno, 1-3, 7-16

Leaching, 31, 49-51, 57, 59-62, 103, 130, 132

M

Mentum Deformities, 1-2, 10-13, 15

Metal Concentrations, 23, 26-27, 65-66, 69, 77, 80, 82, 207-208, 210-211, 213, 216-217

Methylmercury, 18-19, 31, 117, 132-133

Microbial Assemblages, 194, 202

Microbial Community, 87, 98, 102, 104-105, 107-108, 110, 112-116

Microbiological Indicators, 177, 179, 184, 193

Microplastics, 134, 138, 140-142, 144, 148

Mining Areas, 76-77, 177, 192, 205
Morphology, 16, 35, 47, 88-89, 204, 217
Mutagenicity, 177, 179-180, 185, 187, 190-191

O
Organic Carbon, 22, 27, 43, 49-50, 52-53, 55-56, 59, 62-63, 103-104, 106, 115-116, 119, 126, 133, 150, 159, 176, 196, 205, 222, 225, 233

P
Phosphorus, 49, 52-53, 58, 63, 115, 150-151, 158-160, 233
Platinum, 64-66, 73, 76, 78, 81-83
Pollution, 3, 14-16, 29-30, 40, 42, 45, 49, 59-61, 64, 66-67, 70, 73-78, 80-83, 113, 115, 134-135, 143-144, 146, 149, 157, 161-163, 170, 173-174, 176, 178, 189, 192, 195, 207-208, 210-211, 213, 216-220, 222, 231, 234-236
Pollution Control, 66-67, 70, 73-75, 77-78, 80, 219
Porosity, 51-53, 162, 171, 173
Potassium, 49-50, 52-53, 55, 58, 85, 150-151, 159, 205

R
Remediation, 18-19, 29, 31, 100, 117-118, 127, 130-133, 149-150, 157, 159
Risk Assessments, 201, 206, 208, 218

S
Salinity, 64-65, 99, 117-120, 124-127, 130-131, 133, 159, 231
Sediment, 2, 6, 12-15, 18-22, 24-25, 28-31, 37, 47, 49, 67-68, 111, 115-122, 126-133, 135, 165-166, 168-171, 175, 190, 207-214, 216-220, 224, 232, 234, 236
Sludge, 77, 84-87, 90-100, 115, 185, 219-220
Soil Acidification, 49, 58, 61
Soil Column, 49, 51, 55, 59-60
Sorbents, 117-120, 122-125, 127-131, 133
Stygofauna, 43, 161-162, 166, 170-172, 194, 204, 222, 227, 235
Syncarida, 194-195, 199, 203, 205-206, 227

T
Toxic Metal, 18-19, 29, 218
Toxicity Data, 34, 194-195, 200, 203
Transplanted Organisms, 64-65, 78-80
Triturus Carnifex, 134-135, 142, 147

www.ingramcontent.com/pod-product-compliance
Lightning Source LLC
LaVergne TN
LVHW081025210926
848485LV00004B/949